AF180710

To Ruth & J —

Two great people
that I have
known quite
a few years -
All my best -

Lev

The Genes of Ambition

A Multi-Generational Saga by

F. Levering Neely, M.D.

Exposition Press of Florida, Inc. **Pompano Beach, Florida**

FIRST EDITION

© 1986 by F. Levering Neely

All rights reserved. No part of this book may be reproduced, in whole or in part, in any form or by any means, electronic or mechanical, including photocopying, recording, or by any information storage and retrieval system, without permission in writing from the publisher. Address inquiries to Exposition Press of Florida, Inc., 1701 Blount Road, Pompano Beach, Florida 33069.

Library of Congress Catalog Card Number: 86-91114
ISBN 0-682-40292-3

Printed in the United States of America

The Genes of Ambition

Contents

To Margot Neely (1917–1983)
and
Little Lev Neely (1971–1985)

This painting of the author's wife, Margot Neely, was done by Lu Steed.

Preface

Everyone has curiosity as to what kind of persons his forebears were. Most of us have known at least one grandparent but it is unusual to have known all four of them. I knew only one of mine and was a young man of 16 when she died. My other three grandparents, who died before I was born, were said to be unusual people. One of them, Ben, had such an eventful life that I have made him the main focus of this book. There are a considerable number of true facts about him. I have taken these facts and embroidered them to make a likely story and have woven in his adventures throughout the book.

The book, however, is by no means about just Ben. I have gone forward in time and described, as truthfully as I could, many of his descendants. I have reached down to my children and my own grandchildren to the youngest, age 2, Margot Jackson, my daughter Virginia's latest child.

My son, Fred, never really knew either of his grandfathers. They were both very interesting men. One was a great lawyer and a great personality. The other was a famous scientist and also an admirable person. Fred has heard a little about them but has only a vague idea of what kind of people they were. I knew them both well and have done my best to describe them accurately. My grandchildren may never get to know my generation well and I have tried to recapture unusual events and simple family life and write it down.

Ben, my grandfather, had nine children—four boys and five girls. Every one of them was a remarkable person. Almost all of them had a large family, so I had twenty-one cousins. Many of these I had not seen for many years. Several had died. I looked up and visited some cousins from each of the families, getting information on all of my aunts and uncles, and have tried to record it accurately. One of my uncles, Frank Henry Neely, is recognized as one of the outstanding men of Atlanta, Georgia. He had a long and unusual career with a multifaceted record of community service, but my interest lies in preserving his sparkling personality. He had to have ambition to have arrived at his position in life.

One of the purposes of *The Genes of Ambition* is to try to add some facts to the controversy about the effects of heredity versus environment. I have not made any startling conclusions but hopefully I have added some information. Ben, my grandfather, had to have ambition to have begun the first high school in Rome, Georgia. All of his children had varying degrees of a drive to do something useful in this world. I have tried to trace this trait and see if it might be inherited.

I have taken the liberty of inserting my philosophy of life in odd places throughout the book. Some might consider this inappropriate, but I have insisted that it be left in. I want my great-grandchildren to know what I thought about life.

Perhaps posterity will be helped by computers and video tapes which could be stored away for a few hundred years and thus the past could stay alive. We shall see.

Nota bene: I am deeply grateful to Dr. Roy Hand without whose genealogical research this book could not have been written.

1
Benjamin Neely, Age Thirteen

"You leave her alone—you are not her mother!"

"You hold your tongue, Benjamin. She's been a bad girl, and deserves this switching."

"Euphemia has never been a bad girl, and you better not switch her one more time or I'll, I'll, I'll . . . ."

Loud, agonizing cries from Euphemia bounce off the old yellow kitchen walls.

"You'll, You'll, You'll do nothing, and when your father comes home you'll learn what a real licking is. Go to your room."

Benjamin grabs the switch and holds on to it, and then Mrs. Neely slaps his face with her free hand. In a white rage, Benjamin shoves her with both hands, uncertainly at first and then with one big push. She is right at the top of the basement steps, becomes unbalanced, and disappears down the stairs with a cry. A loud clatter follows as she bounces down the flight—quiet for a moment—and then a low moaning.

Euphemia stops crying, but now screams out, "Benjamin, Benjamin, look what you have done!"

Benjamin, pale-faced, goes to the top of the stairs, looks down. His rage turns to fear: "You see about her, Feemy, I'm leaving."

Upstairs he runs, two steps at a time. Grabbing his jacket and piggy bank, he heads out of the house and trots down the road towards the city, sobbing all the way.

* * *

Baltimore in 1847 was a flourishing seaport located in the Chesapeake Bay on the Patapsco River. It had a huge waterfront with large warehouses.

Steamships were just beginning to appear, but sailboats were the mainstay of shipping. The famous Baltimore clippers were making the trip to Europe in record times. Tobacco, sugar, and cotton were exported while manufactured goods from England were brought back. Railroads connected seaports to the inland cities. The Baltimore and Ohio Railroad, started in 1828 with horse drawn cars, was now beginning to use steam locomotives. It was a bustling waterfront twenty-four hours a day, with loading and unloading going on constantly, and crowds of people swarming about saloons, restaurants, and the red-light districts. Sailors were everywhere, some in and some out of uniform. There was plenty of work for all.

Benjamin was fascinated by ships. He watched the big boats come into the bay and saw some of the first paddlewheel steamships. He had done a little sailing on the river and had even been to the waterfront before with his father. The comings and goings of the big boats charmed him. Now he headed to the waterfront. After a mile of running and crying, he slowed to a resigned walk, but his resolve was getting stronger.

"I hate her. I hate her. I can't live there any longer—switching my sister and slapping me—who does she think she is? I'll show her. But she's right. Father won't forgive me, so I must leave."

Just thirteen years old, Benjamin had grown fast in the last year, his voice beginning to crack. He had always been an emotional child, subject to fits of anger, depression, or joy, true to the Irish tradition of his forefathers. In the last years, all of his ups and downs seemed more pronounced. School came easily for him, and he read everything he could get his hands on. His father, John, attended the University of Dublin as well as Oxford, learning the art of teaching. He had come to America in 1831 while still a young man, and taught at the private school in Bel-Air, a suburb of Baltimore. His first wife, Susan Henry, followed him to America and bore eight children. She died in childbirth in 1845. John married Isabella McAlister in 1847, and she became the stepmother of the seven surviving children. It was the new stepmother of only a few months whom Benjamin pushed down the stairs.

Benjamin arrived on the waterfront about 5:00 p.m., October 5, 1847. Amidst the crowds of people, no one noticed as he walked briskly along. He came to the dock where several Baltimore clipper ships were loading and unloading. Restrained by their tow ropes, these huge, beautiful boats, with their sleek lines and tall masts, were gently moving. Looking about, he read a sign on the *Nerrisa:* "Cabin Boy needed: must be fifteen years old and have permission of family."

Ben walked up the gang plank. At the top a sailor asked Ben about his business. After replying that he wanted to apply for the job of cabin boy, the sailor sent Ben to the Danish boatswain.

Speaking in poor English the boatswain asked, "How old?"

"Fifteen."

"Do you have a letter from parent?"

"No," said Ben, "they are both dead."

"Well, I don't know, I send you to the first mate, Mr. Rumans."

He took Ben up to the bridge. Mr. Rumans, tall, well-groomed in uniform and cap, said in a sneering English accent, "How do we know you are fifteen? You look about twelve."

"I've done the ninth grade in school," said Ben. (He had really done only the seventh.)

"Well," said Rumans, "how do we know that your parents are really dead, and that you are not just running away from home?"

"Well," reasoned Ben, "if I had a home, I would not be running away. I have been in the church orphan asylum for three years, and they said I could go find a job."

Hc sounded very self-assured, even plausible.

"You know," said Rumans, "we may be gone several months; first stop is Cuba, then Haiti, and then Liverpool. And we're sailing tonight if the tide and wind are right."

"That's fine with me," said Ben.

"Boswain—find this boy a berth and take him to the cook. By the way, the pay is four dollars a month."

"Thank you very much, sir," Ben rejoined, with his heart dancing.

* * *

Ah genes, ah chromosomes! You work in wondrous ways. Forty-eight chromosomes in every cell. How many genes in each chromosome? Hundreds? Thousands? Who knows? But we do know that we pass down characteristics to our offspring. He looks just like his father, or his mother. He has the Neely nose, or ears, or lips. But the egg and sperm each have twenty-four chromosomes and the fertilized egg has forty-eight, twenty-four from each side of the family. So the child may have his mother's hands, his father's feet. What a wondrous and amazing fact this is—from joining of two single cells, all of these characteristics are passed along. It's the genes that are the directors or determinants of what the child will look like. But what about some of those other characteristics? Will two bright parents produce a child brighter than either? The geneticists say no—that there is a leveling. People with high intelligence also carry recessive genes which may become dominant in the child. There apparently is considerable chance involved as the genes are settled into a new offspring. Generally, intelligent parents have intelligent children, but that is not necessarily true.

What of ambition? Is this a characteristic passed along from generation to generation? Great men or women who have accomplished much in this world often have children who have little desire to do great things. What about environment? Is this as important as genes? In an oppressed society (and the world seems always to have some group oppressed), ambition may be thwarted, held back, or destroyed. This was often true in Europe in the eighteenth and nineteenth centuries. Some group was always being persecuted, religious differences the most common, but racial and ethnic groups in the minority were forced to leave their native lands and seek new homes in America where they could be free. Once here, if they had the right genes, there was at least an opportunity for heredity and environment to work together—ambition could flourish—sometimes on a small scale, sometimes to produce greatness.

Let us consider a large family. The children are all different. Their physical characteristics are different unless they are identical twins, but they do resemble each other. Some are more intelligent than others, but generally their levels of intelligence are somewhat similar. What about ambition, the desire or motivation to do something? Is this passed out equally to all the children of one set of parents? The answer is no. Ambition is usually more pronounced in children of ambitious parents, but often only one of a large group of children will have a burning ambition.

What is this ambition or drive or motivation? Sometimes the goal is small. Just to be happy? But what makes one man or woman happy may not apply to another. Security, recognition, and response are said to be what we all seek. But how much of each? This drive to do something in this world is in us all. Some pursue it, but to some it comes naturally. Thus intelligence and drive work well in the proper environment, but other human characteristics may be equally important. The ability to get along with people, the ability to inspire other people to have drive, may be important. And creativity—is that a gene that is passed along? Is the ability to birth new ideas important? Definitely. A very intelligent person may have a great grasp of what others have done and yet his mind will be sterile. Finally, a sense of humor may be the added quality that will make a really great man or woman.

For intelligent people with drive, creativity, and a good sense of humor working in the proper environment, there is no limit to the heights they may attain. Genes could cause a great mix of all these characteristics, but if one had them all it would be most unusual. Even those with the best genes might find that the environment was not fertile soil to nurture the genes.

What about in different generations? If we find one ambitious man, can we trace back on each side of the family one, two, three, or four generations? That's two, four, eight, sixteen people in four generations at whom we would have to look. Who knows their great-grandfather or great-great-grandmother and what kind of person they were—how they lived, how they looked at life, what their thoughts were, and what their intelligence quotient was.

* * *

And what of Benjamin? What about his parents? Let's look at Ben, then his siblings, his offspring, and finally their offspring.

What manner of man were you, oh father John—born in Northern Ireland in 1804, graduated from Trinity College, Dublin, and married to Susan Henry in May of 1831? And then, in August of the same year, you sailed to Baltimore. It took you almost two months on the ocean to get here. Susan, your wife, sailed soon thereafter, landed in Quebec, and came over land—probably by stagecoach—to join you in Baltimore. You possibly had already obtained a position in the Bel-Air Academy, and you taught the classics as well as science to the young men of this private, preparatory school. Directed by the Reverend Reuben H. Davis, the school must have stressed religion, and you, John, later became a minister yourself. You were admitted to the Deacon's Orders in Christ Church, St. Peters Parish, Easton, Talbot County, Maryland in 1848. This was seventeen years after your arrival in America.

Meanwhile, you and Susan had a large family of eight children, the first born in 1832 and the last in 1845. There were six girls and two boys, and Benjamin was your eldest son. Three of your daughters died young, either shortly after birth or before they were grown. Finally, your wife Susan died in 1845 while giving birth to her ninth child. She was still a young woman, and you were only forty-one years old at the time. You waited two full years before remarrying. Isabella McAlister, a widow of Irish descent, became the stepmother of your large family of children. By now you were the principal of the academy in Easton, Maryland.

What sort of person were you? Were you intelligent? Were you full of fun? Were you ambitious? And if so, to what end? You don't give us many clues. However, we are certain the day you came home and found your new wife of only a few months with a broken arm, a cut on her forehead, your girls all crying, and your oldest son gone, you probably did show a little surprise, not to say consternation. What other factors influenced you?

The fact that you were Irish is important. The Irish are emotional, and you probably were no exception. You had the drive and intelligence to go to Trinity College, Dublin, graduating in 1825. Some say you also attended Oxford, and perhaps you met your fiancée, Susan Henry, there. You saw the turmoil in Ireland. You may have taught a few years in your native country, but you must have heard the stories about America. Something in those stories stirred you up so much that you resolved to go there. Before your marriage you must have told your fiancee that you planned to go. There is a report that Susan Henry was an English woman of "no small means." We are not sure whether this had a good or bad effect on your ambition. She might have had a good effect on the genes of your offspring. New bloodlines seem to be better than the inbred lines of a local area. You settled right down close to Baltimore, teaching in the academy and beginning your large family. Strange that you did not become a minister until 1848, three years after your first wife had died and one year after your son had run away. You must have taken some time off from teaching to study for the ministry. Apparently you continued to teach and moved to Augusta, Georgia, in 1852 to become the director or principal of Richmond Academy, the best school in that city. You also were an ordained minister. You remained there until your death in 1872 at the age of sixty-eight.

You must have had intelligence, considerable ambition, the ability to get along with people, an Irish emotional nature, and possibly a good sense of humor. Now let us see what happened to your genes.

Susan, your first wife, and mother of most of your children, is an unknown quantity, but surely she passed down good qualities of character to your children. John, you were the one who made the decision to leave your native Ireland. That decision, accomplished in 1831, brought you and Susan to America to begin the family which in direct descent, five generations later, numbers in the hundreds, scattered all over the United States, many within the fourth and fifth generations unknown to one another. Of primary interest is what happened to your genes, especially the genes of ambition.

2

John Neely; Ben to Sea

John Neely, age forty-three, walked home from the academy in the late afternoon of October 3, 1847. He had had a difficult day as director of the academy, which prepared male students both for the world and for the colleges available in the east. He was tired both physically and emotionally; his patience had been tried to the utmost. He believed in excellence in his teachers and in his students. Having given the word to the teachers to be tough, they had followed his instructions and failed several children of very prominent families in the area. John had been forced to cope with the enraged parents. Reason and logic did no good—they were furious. Why should they pay money to the academy for the education of their children only to find that their children were being expelled? No amount of soothing statements helped calm them. John stood firm; the parents had been told at the beginning of the year that unless the young men measured up, they could no longer stay. John was pleasant, polite, but firm. The young men simply had not worked at their studies. They had been warned, the parents had been warned, and now the boys could stay no longer. Four groups of parents had come in, each a little more upset than the last, and John received his share of vituperation.

His blue eyes were steely as he repeated, "You had been warned several times."

Finally they were gone, and John was exhausted.

He relaxed somewhat while walking home and began looking forward to a pleasant supper and chatting with his new wife and his children. It was getting dark, and the large, white, two-story house was ablaze with lanterns and candles. Euphemia came rushing out. She was fifteen, but quite small for her age; she looked about eleven.

"Papa, Papa, Miss Isabella is bad hurt—and Ben, Ben has run away!"

"What do you mean? What has happened?" her father asked.

"Ben pushed her down the stairs, then he ran away, and we haven't seen him since."

John had mixed emotions. Isabella, the new wife, was the center of his concern. But Ben, his oldest son and only thirteen years old, already showed great promise in school. He earned all A's, had a good curiosity, and was a great joy to his father in spite of a strong temper. He exuded a mercurial personality with many emotional highs and, at times, depths of depression. His father had said many times, "He will settle down or explode." This time he had truly exploded. Isabella had a cut on her forehead, which had stopped bleeding, and a silver-fork deformity of the left wrist with swelling. She wept softly, and the children began a great chorus of crying as he came in.

"Stop this crying right now!" shouted John, and they all changed to suppressed sobs.

John looked at Isabella's head and arm, saying, "We need to see Dr. McRae right away."

Dr. McRae lived very close. The good doctor dressed her wound, set the arm, and splinted the wrist. When she stopped crying, John asked her what happened.

She told him, truthfully, and he said, "Ben is a very tempestuous boy, but that is no excuse. I will give adequate punishment when he returns."

Isabella, a gentle but emotional soul, said, "Forgive him, John. I was in the wrong also."

When they arrived home, John asked Euphemia where she thought Ben had gone.

Euphemia said, "He has run away and will not come back."

"How do you know?" her father asked.

"Because he took his piggy bank with him."

* * *

What traits are passed down from generation to generation? Is kindness dominant or recessive? What of meekness? Can there be five generations of aggressiveness? Does a trait get diluted as the generations move along, or does it hide and skip two generations, then reappear? Who knows? Mendel doesn't tell us if selfishness has a special place on the chromosome. Honesty surely is dependent on environment, but who has tried to trace it for four or more generations? Then, there is initiative, inventiveness, and the ability to get along with your fellow man. Red hair can go back ten generations. What about the turbulent swinging emotions of the Irish? It is thought to be ingrained for many generations; or is this a characteristic that is acquired because everyone around a child is emotional? And, finally, this drive, this steadiness of purpose that burns within a few individuals—sometimes the light flares briefly and goes out; sometimes it slowly increases in intensity. Like a fire that is stoked, it begins to blaze and burn at a steady heat until the fuel is gone. If the important character genes are properly proportioned, the person with ambition or inner drive may be destined to accomplish greatness and make history. Another person with the same genes may have different goals and his or her achievements will appear minor to the world. Of course, environment is very important. A man may have great aspirations and be on his way up, but if he marries a rich wife, he may become a dilettante and simply enjoy himself. Yet to another man with burning ambition, a rich wife may be a great help in his career.

Some suggest that one can predict the future by examining the traits of children in the first grade. We all know children who excelled in early school and did great things later, but we also know those who did poorly but were successful later. What is actually meant by "accomplishing things"? Achieving goals? Is that so important? Is not the little man who is a good carpenter as great as a very successful businessman? Who are we to judge? It would be interesting to see how some generations may resemble others. We have all seen big families, and a parent of such a group will say, "Oh, they are each and every one entirely different, completely." Someone not in the family will say, "All fourteen children have many of the same traits, not just in looks but in their characters." They will all have a fairly similar level of intelligence, and a similar outlook. We can examine fourteen children in one family, but suppose we look back fourteen generations. Would we find a progenitor who resembles the last offspring? That would be roughly 300. Ulysses clearly was a man of tremendous ambition and drive, as we can see from Tennyson's poem. What manner of man was he who was the ancestor?

Push off and sitting well in order
Smite the sounding furrows.
For my purpose holds
To sail beyond the sunset
And all the westerns stars
Until I die.

It may be that the gulfs
Will wash us down.
It may be that
We shall see the
Great Achilles
Whom we knew.

One equal temper
Of heroic hearts
To strive, to seek, to find
But not to yield.

What happened to his children? History does not tell us. Ulysses had great genes, but did he pass them down with DNA? Desoxyribo-nucleic acid, how do you work? They say you make the pattern, you make the mold. What are your secrets? One tiny cell carries forty-eight chromosomes and each chromosome has a thousand or more genes, each a separate characteristic or trait. There is a wondrous quality to it all, a quality difficult to comprehend. All living things, both plants and animals, are so amazing in their complexity, in the way they function, that it's difficult to see how they could be shaped without some thinking person behind it all. They say no, it was simply the process of evolution, and the fittest survived to reproduce. The eye, with its camera-like structure, has the ability to record visual experience in the mind. A rose, a rose; how delicate its beauty when coupled with its un-reproducible scent. Did these things just happen? Perhaps. Roses haven't changed much in three thousand years. Have people changed in three thousand years? Were the Greeks of 500 B.C. like the people of today? There were individuals then who had a purpose in life, a directing force in life surely stemming from genes or environment. What kind of men were Aristotle's father and grandfather? Did Plato have children and grandchildren? What were they like? Did they have the same genes as these great men?

What about Alexander the Great? History presents few facts on his father. Young Alexander almost conquered the known Western world at the time before dying at the age of twenty-six. His inner fire was white hot, and remained that way until death. How about Moses? What happened to his grandchildren and great grandchildren? He had strong genes, but we don't know what his grandchildren grew up to do. Then, there is Jesus; how well he organized his life. A few simple concepts were his banners. Their clarity and simplicity were their charm and appeal. Jesus must have had strong genes of purpose from his father or mother. He was not just a simple soul. Deep within him burned the fire that drove him until his death. Then Paul picked up the torch which spread and multiplied and finally filled the world with its light.

What kind of man was Charlemagne? Kings came and went, and a few were outstanding. Jump forward to Napoleon, whose combination of genes worked well up to a point. What was Napoleon's *hubris?* Genes of selfishness and personal ambition, what Sophocles years before dramatized, pulled him down. Personal ambition is different from the inner drive to do something for the world. And Lincoln—there was a man—he kept the faith. He followed some purpose that changed a part of America. Think of Einstein—what a mind! His purpose: to think. And think he did, hour after hour, and he was able to write it down. Who was his father? Did he have a daughter? If he did, she probably was not famous. Freud's daughter, on the other hand, wrote about her father's work and herself became a well-known psychiatrist.

Very few famous father-son combinations exist. The Bible has a long list of begats, but mostly it is a list. The history of royal families often gives excellent insight into the character and personality of a long string of generations. Perhaps here we can get some insight into how traits of character may be inherited and how environment may have an effect. Royal families seem to follow a very irregular pattern through the generations. A great king often has a weak son whose offspring may revert to the grandfather's qualities or simply peter out. There is no pattern, so we have to look elsewhere for our answers.

We do not intend to neglect the distaff side—we know that half of the genes come from there. It is simply that a family tree follows the male line. It is equally important to try to trace the female lines back if such information is available.

* * *

The crew battened down the hatches on the *Nerrisa*. Time to sail. Thick wooden planks were fitted across the hatch, then a strong canvas cover was spread. Iron bars held the edges, while wedges were pounded between two bars to secure the canvas; it must be held tightly to stay in place in spite of buffeting by strong winds and seas. The boatswain, Mr. Swenson, supervised. The crew was somewhat motley. Many were Americans, but there was a large group of Newfoundlanders—tough, but good sailors. Then there were scatterings of Irish, Italians, and Scotch. Although discipline on cargo ships varied, Mr. Rumans, the first mate, was a very firm man. He insisted uniforms be clean, that work be done properly. The boatswain saw that orders were carried out, but Rumans often came down among the crew, admonishing the men himself. Now he was up on the bridge, conferring with the captain, Mr. Phillips.

The crew saw very little of the captain. A man of sixty, he had been at sea all of his life. He had been beset with tragedies, had lost several wives, and his children had disappeared. When one goes to sea and stays months at a time, there isn't much time for family life. The captain was bright, worldly, but liked to stay in his cabin and read. Mr. Rumans, "the First" as he was known to the crew, really ran the ship.

The two men examined the maps. First to Cuba to pick up sugar. If there was not enough there, they would proceed to Haiti for more. From there it would be on to Liverpool; they had made the trip before. it was

October, and the weather was good in Baltimore. As they travelled south, they expected better weather. However, with the hurricane season in the Caribbean, they would have to be careful to watch for any signs of a storm.

"Well, Rumans," said the captain, "how far from shore do you propose to sail as we go down the coast?"

"The winds should be very good close to shore, so I propose this route, sir," replied the first mate. "If we should have a hurricane, we can run for shore."

"I think you are right, but the hurricanes often come up that coast, and we could be driven onto the beach. If we catch the wind at sea, we could founder, but it's a gamble either way. By the way, where did you get that cabin boy? I saw him earlier with the cook—are you sure he is fifteen?"

"He's a husky lad, seems bright; talked as if he were fifteen."

"Did he say why he took the job?"

"No, but he seemed very anxious to get the job, and I think he is honest."

"We shall see before the trip is over. What's his name?"

"Benjamin. Benjamin Neely.

"Irish I would guess, but perhaps enough Scotch to make him possible. We'll soon see how honest he is. The last one stole from everyone, even me."

3

Ben at Sea

Bjorn was from Sweden. A ship's cook for many years, he had been on the *Nerrisa* for over two years. He could cook in spite of the ship's lurching on the high seas.

"Vell, young man, you tink you gotta easy chob? Vell, you are very motch mistaken. Dis chob is vork, vork, vork, and you might as well get off de ship now if you don't like vork. Not only you help me, but you clean da crew quarters, de bosun's quarters, and de captain's quarters. An' you clean it every day. If Mr. Rumans don't like de vay you do it, you do it again. He's tough guy; don't take no nonsense. Vell, vot you tink?"

"I want the job, and I don't mind working," Ben said firmly.

"Okay Ben, dot's vot I vant to hear. Ve are making the soup now for supper."

The ship was ready to sail. All hands were on deck. Each man had his job and knew what to do. The tide was going out; the sails were ready to be unfurled. The tide itself would take the ship out from the harbor, and the gentle breeze would then take it down the Chesapeake and out to sea. The ropes were loosened, the anchor pulled up, and the tide slowly pulled the ship from the dock. Mr. Rumans was directing it from the bridge with a megaphone, and the boatswain was seeing that the orders were carried out.

Winches were used to raise the heavy sails, and the winches were turned by manpower. Ropes were everywhere, and the correct one had to be pulled at the right time to give the boat direction. It was an orchestrated effort. Rumans was the conductor, although Captain Phillips gave suggestions from time to time. The second mate, Mr. Crocket, was at the helm. The ship was now directed toward the mouth of the bay. Half the sails were up, holding the breeze and pushing the boat. The sea was calm in the bay.

Bjorn and Ben were working in the kitchen as the ship slowly plied the Chesapeake. They worked steadily at cutting vegetables, firing the stove, and making the bread for supper. But Ben was close to a porthole, and he frequently looked out to watch the land go by. Soon the shore was getting farther away and the waves were getting larger—the boat pitched slowly, and had a slight roll. The sun was getting low and soon it would be dark.

Activity on the deck quieted down as the wind steadied from the east. The sails did not need much adjusting. The tired crew were hungry and thirsty. Bjorn had Ben sound the supper gong, and the crew came into the galley—talking, laughing, pushing, shoving in a friendly manner. Most of them had been on this ship several years. Only the cabin boy was new.

"All right boy—bring on the food, and we mean now."

"Well look at this little guy—he smells of land—he needs a little more salt," said Doffie, sprinkling a little salt on Ben's head.

Doffie managed to stay drunk on almost every voyage. He wasn't an obnoxious drunk, just a happy drunk. Small, round-faced, bald, and bearded, he was one of the Newfoundlanders that poverty had driven to the sea. Then in came Frank, large, redheaded, well-muscled, tough and mean.

"Just bring the food now, and I want my coffee now, and I want it hot."

No joking here. Frank was also a Newfoundlander, but bitter about life. Then Bailey—no one knew where he was from (Wales they said)—he didn't talk much. He stood only 5′6″, with broad shoulders, long arms, and a quizzical-but-friendly look on his face. He just came in and sat down. Bjorn cooked, Ben served. He learned quickly that carrying a hot plate on a pitching boat is not easy. There were ten ordinary seamen: five Newfoundlanders, one Englishman—Bailey, Jack from Maine, Manio from Italy, a Dane, and Borg from Norway.

The food was plain, but hot and filling. Conversation filled the air, as there was always excitement on the first day out. They finished supper, and then Ben had to serve the officers, who ate after the crew. The officers' mess was on the other side of the galley. They had a white table cloth, real china, and silver knives and forks. The Captain, Mr. Phillips, Mr. Rumans, Mr. Crockett, the second mate, and the third mate, Jones, ate together, but Rumans stayed on the bridge navigating while the others ate. First night out, supper was soup, fish, vegetables, meat, and fruit, which were all well prepared. Ben again was very busy waiting the table.

"Young man," said the Captain, "do you think you can manage this job for three months?"

"Yes, sir. I have wanted to go to sea all my life."

"You'll find it isn't as romantic as you might think. In fact it can be very boring. Do you like books?"

"Yes sir."

"Well, come up after supper, and look over the library."

"Thank you, sir. Thank you very much."

Washing dishes for two hours wasn't much fun, but the excitement of the day kept Ben going. The cook was working too, preparing for the next day, and they finished at about the same time, 10 o'clock. Darkness had set in.

Bjorn: "You have done vell, my boy, now you can rest—you go to your bunk because we must be up at 6:00 in the morning.

"Thank you, sir," said Ben and started down to the forecastle to the bunk he shared with three sailors.

Ten o'clock and the sun had been down three hours. They were now out in the open sea sailing south. The wind had freshened. The waves were a little larger, pitching and rolling more, but the steady wind kept them going at a good speed. No talking now—just the sound of the waves and the wind in the sails. Ben slowly made his way along the forward well deck; he stopped and looked up at the bridge where Mr. Rumans paced back and forth. Then he looked up at the sails catching the wind and passing its energy on to the masts which in turn pushed or pulled the ship quietly along its course. Looking up at the stars, he stopped and gazed.

The night was cool and clear. There was no moon, but the sky was lit up by the stars brighter than he had ever seen before. He could see the horizon; he could see the outline of the ship; and he could even see a slight shadow of his body on the deck. That bright one—that's Polaris, the north star almost directly behind and above the aft mast. The rest of its constellation was easy to see. Then he made out the Big Dipper or the Big Bear, each star shining brightly. Hey, there is Cassiopeia's chair! And look—the Swan, and the Milky Way directly overhead. His father had shown him the constellations many times, but never had they been this bright.

Ben thought that the stars at sea are different from the land stars. They had the same constellations, but at sea they were closer. He felt he was a part of them here in the universe, and he realized that he was on a small star or planet which shines and may be watched by someone on another solar system.

Stars have always charmed every civilization. The Phoenicians, the Egyptians, the Greeks, the Chinese; the Norsemen, the Tahitians. The stars charmed such great explorers as Columbus and Magellan. The seafarers seem to study the stars more, because they are able to see them better, and because they have time to look at them longer. And the sailors know the stars as friends: they are familiar—they move by the rules. You can count on them to help you get home. You push onward because your friends the stars are guiding.

Man, through the ages, connects the heavens with God. Here is the unknown before your eyes, and yet so useful that they help steer your boat. Navigators always depend on the stars whenever out of sight of land. It is easy to see how the heavens foster religion. It is also easy to see how they bring superstition—even now millions of people think their lives are guided by the stars. But Ulysses, Columbus, and Magellan all had one thing in common—knowledge, and great respect for the stars. It was the stars that told them to push on—that the unknown could be conquered. The sea is like emotions: it undulates up and down. However, the stars inspire resolve. Resolve to go on, to seek, not to turn back. Ben was transfixed. He looked at those stars every chance he got for the next two years.

* * *

The Baltimore clippers were famous for their speed. The *Nerrisa* made good use of the favorable east wind. By early morning, they were as far south as Hatteras. The wind had now shifted to the northeast and was much stronger. Some of the sails had to be taken in by the night watchman. Ben awoke in his upper bunk bed with the strong pitching of the ship, but on they went. Early the next morning there was a real Northeaster—a strong wind that often lasts three days, but it was favorable for the *Nerrisa* going south.

That morning, the whole crew was up on deck cleaning, chipping, painting, and holy-stoning the deck. The holy stone is a large flat rock with a long handle attached. Moved back and forth, it cleans and smooths the deck. The officers were on the bridge looking at the charts. They had been to Cuba before, but they deliberated on which route to take. They were going to Santiago, which is on the southeastern end of Cuba, to pick up sugar. The straight route was directly south into the open sea, and the first landfall would be San Salvador, which was the first land that Columbus saw in 1492. The safest route, though longer, would be along the coast, and if a hurricane approached they could make a run for the land and the sheltered ports along the coast—Charleston, Savannah or St. Augustine. Rumans and the captain discussed this—Rumans was for the coastal route—the Captain was for the high seas.

"The Baltimore clippers are seaworthy. We can pull down all sails, with the mizzenmasts' lower sail to guide us, and ride out any storm," said the Captain.

This was the last word; they proceeded southward.

By afternoon they had passed Portsmouth, Virginia, and were soon on the latitude of Hatteras, famous for its winds and treacherous seas. The wind was strong and still from the northeast, which took them speedily along with a slight tack. Another full day of good weather, and then in the night the stars disappeared. The waves were larger, but the clipper rode them well. The next morning, the skies were black and soon there was continuous, heavy rain. Rumans and the Captain both knew this was a bad sign. A hurricane might be coming. Rumans and the boatswain made a careful inspection of the entire ship and its cargo to be sure everything was secure.

Heavy rains continued all day and night. The captain knew that hurricanes usually travel toward the northwest. Therefore, he turned the ship almost due east, trying to skirt the storm. The sails were trimmed to the minimum, but even so there was quite a list to starboard. The wind shifted from northeast to east, and was now a gale. Time and again water poured over the prow, filling the forward well deck and draining out the scuppers. There was a rope to hold to from the fo'castle to amidships, where the galley was. The crew stayed amidships as it was best protected and the pitch and roll were the least there.

Ben and the cook stayed in the galley. Nobody wanted food. Even the oldest sailors lost their appetites, and many were sick. Ben fought the nausea; many would give up, but some inner pride told him to keep doing something and serving the crew, if they wanted anything. It was while they were all confined that Ben got to know the crew, and they learned that this thirteen year-old, thin, and lanky youth was acceptable to their fraternity. All but one accepted this lad—Frank the red-haired tough Newfoundlander.

"Ben," he cried, "this coffee is cold."

Ben rushed to his side, stuck his finger into the coffee, proclaiming, "So it is, sir."

Frank grabbed Ben's arm in a vice-like grip and twisted—"You little bugger, I'll teach you a thing or two."

Ben, in pain, did not cry out, and just then Bailey—six inches shorter than Frank—grabbed Frank's other arm and said, "Let him go, now."

The two stared at each other for almost a minute and then Frank released Ben, saying, "I was only kidding."

But everyone knew otherwise. Frank recognized Bailey's firm resolve and did not care to challenge it. Ben knew he had a friend.

The ship suddenly wallowed and began taking on water. The wind was screaming through the masts, and all but one tiny sail was down. This tiny sail and the rudder kept the ship pointed into the wind. The crew was immobilized. The officers all stayed on the bridge, which was drenched with spray. For twenty-four hours there was no letup.

And then there came a great calm, even though the sky was black. The crew rejoiced; food and drink was brought out, but Rumans and the Captain knew that this was the eye of the hurricane and there was more to come.

Soon the wind came from the west with equal fury as the east wind of the day before. Again they were immobilized to amidship, and the mountainous waves threatened to swamp the ship. They rode the storm out, and the next day there was only a light rain and choppy sea. They had been blown way off course. Mr. Rumans shot the sun, and found that they were not far from Bermuda. Now that the ship had weathered the storm, Mr. Rumans set a new course for San Salvador. A whole week passed, and they were about as far from Cuba as when they had set out. The sails were raised, and off they sailed again.

After six days, they sighted San Salvador off the Bahamas. It was October, the same month Columbus first sighted land, three hundred and fifty-five years earlier. After thirteen days at sea, the sight of land was, as Homer first put it in *The Odyssey*, a welcome sight. The weather was warm, the sea quiet, and the breeze gentle.

Santiago de Cuba in October, 1847, was the second largest city in Cuba. It had been founded in 1515 by Diego Velasquez. It was a Spanish colony, but there were many groups on the island that were fighting with one another. It had been active in the African slave trade. Many slaves were kept there on plantations, and later sent on to the United States. Creoles, Indians and Spanish immigrants, divided amongst themselves, were loosely governed by a Spanish government. Trade was free; sugar and tobacco were the big exports.

So Ben, you have been through a hurricane and are alive to tell the tale. You have adjusted to an entirely new way of life. You have done well so far, and here you are approaching a foreign country—a foreign culture. Very few thirteen-year-old boys get a chance for such experience. What will you do with these experiences in later life?

4

John and the Ministry; Cuba

October passed quickly to November, and then December. Every day John thought Ben would be coming home. Surely he would get over the emotional upset. Surely he'd realize that being away from home caused more problems than returning, swallowing his pride, and apologizing. John inquired around the city of Baltimore. He wrote letters to friends in Washington and Philadelphia and went to the local constable. No word. Ben disappeared without a trace, and there was nothing further John could do.

The children remained sad for some time, and they often asked their father when Ben was coming home. They were all in school now, stayed busy, and began to get along better with their stepmother. John was extremely busy—not only was he the principal of the Easton School, he was studying for the ministry. The theology school was in Baltimore. In those days it was possible for a student who was not able to attend school every day to come to the school once a week, get his assigned studies, and report back the following week. However, instead of a two-year, daily course, the program took three years. John had already completed two years, and now, in his final year, he had to write a thesis, and take comprehensive examinations. Ever since he arrived in this country, he had wanted to become a minister. When the opportunity arose in Baltimore, he doggedly went to the school near Bel-Air Academy every Saturday. It took the entire day—first he travelled by horseback to the ferry, then across the upper Chesapeake, and finally a long walk to the school. He left at six in the morning, and returned home about six at night. Often it was much later. He had done this for over two years, and even in snow and ice storms had not missed a single Saturday.

He was not sure why he did this. Reverend Davis at Bel-Air Academy had inspired him. In those days ministers were often teachers, as most churches could not support a full time minister. So the two professions were often combined. John was drawn to this combination, although he had kept the attraction in the back of his mind. However, when his wife, Susan, died in childbirth, John gained the impetus to go ahead with his suppressed ambition.

Susan's death came as a great shock. She had been perfectly well during the pregnancy, had been delivered of a blue baby, developed childbed fever, and died within six days. The child also died in a few days. John was left with seven children ranging in age from fifteen to four years. John grieved and kept brooding: "Why should this happen to me?" But he was a religious person. Instead of going into a depression, he consulted with his old friend, Reverend Davis at Bel-Air Academy. The Reverend told John about the opportunity of becoming a minister by the once-a-week attendance program at the theology school in Baltimore.

In those days there was better acceptance of such tragedies. Children often died young; scarlet fever, measles, and whooping cough often killed several children in a large family. Death of a mother in childbed was not unusual. There was grief among loved ones, but life went on. There was more of a philosophical attitude of acceptance than there is now. Still, John was very emotional when it happened. The funeral was a very distressing event for him and the children. Reverend Davis of Bel-Air had come to conduct the funeral at the small Episcopal church in Easton. Susan was laid to rest along with the newborn babe in the church cemetery. John's resolve to become a minister had become definite that very day, and grief was slowly pushed aside by a determination which grew stronger as time went on. So he would study his theology books every night after supper while the children did their homework. It was a joke among the children: "Papa is doing his homework." Christmas and New Year's came and went.

* * *

January 1st, 1848—James Knox Polk was President of the United States, elected in 1845 with Vice President George D. Dallas. They served until 1849, and were succeeded by Zachary Taylor and Millard Fillmore. During Polk's administration Texas gained its statehood. California was being settled by Mexicans, and Americans were pushing west. Collisions with Mexican troops followed, and Congress declared war on Mexico in 1846. General Zachary Taylor (who would become the next President) won one battle after another, the last in northern Mexico. Then General Winfield Scott took over, marched into Mexico City, and peace was made in 1848 by the Treaty of Guadalupe Hidalgo. The United States was given California, Utah, Arizona, and New Mexico. Mexico was paid $15,000,000 for this land, giving the United States the complete territorial form retained until the annexation of Alaska in 1867.

* * *

Slavery had put the South out of harmony with its surroundings, and the two sections, North and South, had drifted so far apart that they were practically two different countries. The South was agricultural; the North was beginning its industrial revolution. Over 100,000 immigrants entered the United States each year, with most of them heading to the North or the West. The abolitionists were organizing through the North, and Congress was almost paralyzed by long and bitter debates about slavery. Church denominations divided, and only the Catholics and Episcopalians, predominant in border states, retained their national character.

John was pulled by contrary forces concerning slavery. Maryland was on the border between North and South. It adjoined Virginia as well as Delaware and Pennsylvania. It finally sided with the North, but many people sided with the South. As a thinking man, teacher, and minister-to-be, John knew that slavery was wrong, but he was apolitical and more interested in teaching and in the ministry. So he stayed out of the debates unless directly questioned, in which case he would give an honest answer.

Social conditions were improving. There had been a general settling-down. The West was still wild, but law and order were more effective. Communities developed. Pure water, sewage, and roads improved. Almost all states developed good school systems—English was taught, and the universal use of English made the common schools most successful machines for converting the new material of immigration into American citizens. Good teachers were at a premium, and anyone who knew how to organize a school was much sought after. John had several offers from various parts of the country because teachers kept in touch with one another. But he was reasonably happy with his position as Principal of the Easton Academy. Still, there was part of him that said he one day might like to organize a whole school system. In any event, his main thrust now was to become a minister. He still was not sure why he felt driven towards this. He had been brought up as a Protestant in Ireland. He took his religion and belief in God for granted. At the University of Dublin, he had done some questioning of himself. He had a deep feeling of service to people, and he had seen good ministers help people—especially in times of distress. Something told him that he could help people spiritually as well as teach. So he studied the Bible carefully, thoughtfully. He learned about the history of religion, and he studied other religions such as Buddhism, Islam, and Hinduism.

* * *

As the *Nerrisa* sailed into the harbor of Santiago, Cuba, all hands were on deck or up on the masts ready to change the sails at a moment's notice. The boatswain was in the crow's nest; the Captain and Mr. Rumans were on the bridge. The heavy sails had to be raised, lowered, or turned quickly and correctly. The manpower was either on the masts or on deck pulling or letting loose rope at just the right time. Any heavy pulling required the hand winches. It was a tricky business coming into the dock and tying up. But the crew had done this many times before and each knew his job.

Only Benjamin and the cook were idle. They were standing on deck watching the land come closer, and Ben was looking at the blue-green water. The Chesapeake was never this color, the transparent turquoise water of the Caribbean. The docks were visible, but it was not nearly as big a port as Baltimore's. With only five or six ships present, there were several empty docks. Mr. Rumans was bringing in the ship slowly and gently. Many people were waiting on the docks, and as they approached, the crew threw out the forward and aft lines. These were tightened by the winches and the ship came to a stop. The sails were dropped, rolled, and wrapped, and the sailors threw down the gangplank.

The dock master came running up the gangplank with paper for signing. No one could leave the ship until their work was done and the hatches opened up.

"When can I begin loading the sugar?" Mr. Rumans asked the dock master.

"All in due time, sir. We are loading the other ships now."

"Our ship is already late due to the hurricane, and we would like to load as soon as possible."

"Yes, sir,—day after tomorrow for sure."

"What about tomorrow?"

"Not possible," said the dock master.

There were alway delays in Cuba—it was always tomorrow—mañana. Mr. Rumans knew this, and he also knew that the Cubans had to be pushed to get anything done. The ship was cleaned, the sails repaired, and the crew slept on deck, for the nights were warm and calm in the forecastle. The following morning they still had not begun loading, but the crew remained aboard in case they did.

Ben and Bjorn served an early breakfast and then walked into town's market to get fresh vegetables and fruit. Ben had never seen anything like it before. There were stores and markets in Baltimore, but here there were miles of little stalls selling all kinds of things, and crowds of people wandering about speaking Spanish. They came to the fruit and vegetable area. Bargaining was taken for granted, and Bjorn knew he could get a good price if he bought large quantities. They finally bought large boxes of cabbages, onions, pineapples, bananas, and potatoes, and they had to make the trip to the ship three times to carry it all back.

The crew remained idle most of the day. The Spanish had still shown no signs of beginning loading, however. The first mate gave the word in the late afternoon that they could go ashore, leaving just a few men to guard the ship from thieves. Most of the crew washed up and took off to town to the restaurants, saloons, and the red-light district. The Captain had lent Ben a book on the stars and navigation, and Ben read for several hours by the ship's lantern until he got sleepy and went to his bunk. The crew came back slowly, several together at intervals. Most had had something to drink and many had consumed too much. Ben had an excellent chance to observe the effects of alcohol on different people. He was awakened every time another group would return and finally stayed awake until they were all back and in their bunks. Doffie, who was always drunk, was no different. Frank was meaner than usual, and Ben knew to stay away from him. Some were laughing, some depressed and crying. Ben had never seen ten drunk men before. It was an opportunity for him to see the ways of the world. There had been no alcohol in his father's house, and although he had seen the occasional drunk person, this was a revelation. Here were ten people that he had known now for two weeks, and the way they changed with beer, whiskey, and rum amazed him. Most of the crew had become friendly with him because he really made an effort while serving meals, and he laughed at their kidding and talked right back to them. The cook had taken him under his wing, assuming the role of a father figure. Bailey was his physical protector; and the Captain recognized a bright boy and gave him books to read. Mr. Rumans, tough and mean as he was, kept after him, but they had a quiet admiration for one another. They both recognized that the other had the inner drive or ambition.

After only two weeks away from home, Ben, age thirteen, had learned what it takes some many years, if ever, to learn: how to get along with his fellow man. He was doing his job well, had kept his mind active, had been deeply impressed by the heavens and by the sea. He had been through a terrible storm and had seen a foreign country. But more important, he had learned to get along with people of all different types. He had been snatched out of one family where he was the eldest son to become a part of another family—the crew—in which he was the youngest.

Finally, on the second day of waiting for the loading to begin and with much prodding by Mr. Rumans, a large group of laborers began to bring large sacks of sugar up the gangplank. The laborers were of all colors, but mostly Spanish-speaking. The crew lowered the sugar into the hold, storing it so it would not shift in any storm while on the way to England. The loading took almost two full days, for the laborers often had to slow down and rest in the heat. Mr. Rumans and the boatswain were constantly after them to keep working, and Mr. Rumans later said that the job could have been done in half the time. Finally, they were ready to sail. The hatches were carefully battened down again, the sails raised, and the ship slowly moved out of the harbor. It was a very short run from Santiago to Port-au-Prince, Haiti.

Haiti, in contrast to Cuba, which was still a Spanish colony, was an independent nation governed by a black emperor. Discovered by Columbus in 1492, it had been a French colony in 1793 and remained so until the famous figure, Toussaint L'Ouverture, a former slave, drove out the invading Spanish forces and became governor, at first with the blessing of France. Napoleon sent Leclerc to stabilize the island. A black general, Dessaline, conquered the island with the insurgent blacks. Toussaint was captured and sent to France, where he died in prison. Haiti remained independent with black emperors ruling. The whites were expelled, and the population was mainly mulattoes with the French language and culture.

So Ben was able to see a completely different type of country, a land ruled by blacks in which there was no slavery. This made quite an impression on him and was to be a disturbing idea when he later had to make a decision about fighting in the American Civil War.

5
Two John Neelys

Mountains of sugar and coffee stacks everywhere. Huge piles of pineapples. They loaded them all at Port-au-Prince. It took three days. While Ben and the cook served the crew breakfast and lunch, most of the sailors caroused in town at night. So the afternoon and evenings were free. The cook, a great fisherman, invited Ben to go fishing along the beautiful sandy beach out from Port-au-Prince. Armed with long bamboo poles, strong lines, corks and hooks, the two set forth with cutup fish for bait. The beach began only a short walk from the boat. The weather was clear and warm. Soon they were on a fairly deserted beach, its translucent turquoise water gently lapping the shore. They took off their shoes, waded in to their knees, and began fishing. Ben had done some fishing on the Chesapeake, but rarely had he had the luck he was now delighting in: both were hauling in sizable catches, rapidly. They tossed back the small ones, but kept the many good-sized sheeps' heads, sea bass, and red fish. They also caught tropical fish whose splendid colors and strange shapes pleased Ben. Within a few hours, they each had a long string of fairly large edible fish, and trudged back to the boat, tired but from the kind of tiredness which warms the spirit.

There is something very satisfying about a successful afternoon of fishing. It's almost instinctive for man to fish. Clearly hunting is instinctive, although certain modern men inherit a stronger impulse to hunt than others. Man has been hunting at least 50,000 years and probably upwards of 200,000 years. With five generations to one-hundred years, fifty generations a thousand years and 2500 generations to 50,000 years, one can imagine that the good hunters survived, passing down their urge—and their skill—to hunt in their genes. In the old days, hunting was a necessity for one's basic survival, a Darwinian impulse of the survival of the fittest.

Today hunting is mostly a recreation. Still, no wonder we still enjoy hunting: with millennium after millennium of hunting behind us, the whole process must be instinctual. Question: Is this hunting urge passed down 2500 generations? Who knows? Fishing, one of the fundamental forms of hunting, surely has its roots in our ancient, our archetypal history. Apparently prehistoric man learned quickly that fish were both edible and relatively easy to catch with their crude nets or hooks. This was much easier than hunting large animals on land. So those who lived on and by the sea became, accordingly, more placid. They didn't have to be as aggressive as those inland who constantly had to stalk dangerous animals. These men had to be aggressive, quick, well-coordinated, and smart to survive. Those by the sea, on the other hand, were more reflective, more pensive, more restrained. Fights were to be avoided if possible. A sea environment changes man differently, yet the aggressive hunter survived in all cultures, a trait acquired by genes which helped the next few generations survive.

What about the two sexes? Are the genes of woman so different from those of man? Throughout the ages, man has been the hunter and woman the homemaker, cook, the mother. With the exception of such cultures as the Amazon women, such a system persisted since the start of civilization. Males are naturally more muscular, larger, more dominating. Women seem more placid, more retiring, more willing to remain in the background. Only in the twentieth century have women asserted themselves, competed with males, both mentally and physically. Women have not changed much physically, but in the later twentieth century we now find women lawyers, doctors, business persons, even laborers, policewomen, and firewomen. However, the genes of drive and aggression are still much stronger in the male; and we see many professional women still preferring the roles of mother and homemaker and stopping work for infant care.

Another change occurred in the twentieth century. The male, instead of going out hunting and expending great amounts of muscular effort, is now sedentary. Going to the office, conferences, engaging in telephone conversations, coming home late at night, eating a large, heavy meal—and then resting or sleeping soon after supper—all these factors contribute to man's sedentariness. Even laborers do not expend nearly as much energy as they used to, and their muscles seem weak, flabby. Smoking and over-eating lead to the hardening of the arteries, obesity, and culminate in the heart attack. Many women, on the other hand, remain much closer to their old role, the homemaker. Women don't seem as susceptible to heart attack as men, at least not until they are well past menopause. So although we do not have any accurate figures of the frequency of heart attacks in men one-hundred, two-hundred, or 2000 years ago, we do know that heart failure stands as the most common cause of death in males now.

The risk factors include lack of exercise, obesity, heavy smoking, nervous tension, and high blood pressure. Luckily there is now a better basis for health maintenance in the jogging and the fitness craze that's swept the nation.

The female is much less subject to heart attacks, although she is now experiencing a higher incidence of heart attacks because of her assuming more professional "male" roles. There may be sex-linked genes determining that women will not have the same heart problems men face. Again, it seems that inheritance and environment are so closely entwined that it will be very difficult to sort out which factors play which important roles.

Recent experiments in which rats were trained to solve a maze proved that rats, after much practice, could quickly solve the maze with few or no mistakes. The evidence suggests that the rats so treated with the brain substance of trained rats were able to solve the maze much faster than rats which were not injected with brain tissue of the trained rats. This suggests that environment can change the brain chemistry. Could the offspring of the rats trained to solve the maze solve it any sooner than ordinary rats? Are male genes more dominant than female genes? It's highly doubtful, of course. However, it's often easier to study the male line through the generations simply because we have more information about the males. Let us look at one such male line, and see if we can find his character qualities in the previous generations.

* * *

John Longstreet Neely, born in Atlanta, Georgia on December 13, 1947, is now thirty-seven in 1985. Parents: F. Levering Neely and Margaret Gardner Neely of Atlanta. On his grandfather's side we find Edgar Adams Neely of Atlanta and Emily Levering Eckfeldt Neely of Washington, D.C. and Philadelphia. On his mother's side we find Leroy Upson Gardner and Carabel McKenzie Gardner, both of whom originally hail from Meriden, Connecticut but resided most of their lives in Saranac Lake, New York. The great grandfather of John Longstreet Neely was Benjamin Neely and great great grandfather John Neely, teacher, organizer of schools, and minister.

The two John Neelys: one born in 1804 in Antrim County, Ireland, the other born in 1947 in Georgia. One-hundred forty-three years between their births, five generations apart. How much did the older John pass down to the younger one? How much did other genes and chromosomes affect the younger man? Who knows? No one, but let's speculate.

John Longstreet Neely is the fourth of six children. He has two older brothers, one older sister, and two younger sisters. He was not outstanding in the public elementary school. His grades hovered somewhere in the middle. He went to high school and the best private school in Atlanta, Westminster School. He played tennis regularly, improved rapidly, and earned a spot on the tennis team. He also enjoyed basketball, which he played on the varsity team. He was tall. During his high school years he gained a little self-confidence, a little drive or push. His competitive spirit emerged while playing tennis. His inner drive became a real asset in later years. Always in control of his competitive spirit, John received several sportsmanship awards and became one of the champion tennis players in the state. Upon graduation, he enrolled in one of the best southern colleges, the University of North Carolina.

He was a "B" student. He considered becoming a doctor, as he worked one summer in a hospital as a surgical assistant. However, by his second or third year in college, he decided he did not want to become a physician. His premedical subjects increased in difficulty, demanding persistent and constant study. He did not have the drive to become a doctor, so he changed his major from premedical, graduating with a degree in history. Self-confidence was there, but he did not have any clear purpose.

After graduation, he journeyed to Brazil with the Peace Corps. His sister had joined the Peace Corps in Africa, and this may have fired his imagination. During his two year stint in the Peace Corps, he contracted typhoid fever while swimming in the Amazon. Although he completely recovered, the Army rejected him because of the sickness.

Upon returning, he studied at the University of Georgia, taking a graduate degree in City Planning and Municipal Administration. Soon he married a classmate from the University of North Carolina. She worked while he continued schooling. His first job was in Milledgeville, Georgia, in a federally-funded program to improve the city. They settled into the town life of Milledgeville, both working until the birth of their first child, a girl, Jordan Neely.

Three years passed when John heard of a position in Savannah, Georgia, a job dealing with the economic development of the city. Although the federal government funded the job, local businessmen and government supported the position. After interviewing, he was offered the position. Once in Savannah, they bought a home in the suburbs and later one downtown. They restored the older downtown home with their own labor when time permitted. The job challenged John, whose expertise in the field grew with each day.

John adjusted to life well but was not "outstanding." However, somewhere along this time he seemed to change. His inner drive and steadfastness of purpose clearly emerged. He liked his job and its challenges. He worked steadily on his new house. Indeed, John continued any task until its completion. Soon their second child arrived, this time a boy, Thomas Neely. He was the sixth generation from John Neely, the emigrant from

Ireland in 1834, or one-hundred forty-eight years earlier. His wife Carolyn, was a great help, working at a job in a day-care rehabilitation center as a social worker. With the arrival of Thomas, of course, she stepped down from her job. John's self-confidence at times went a little too far, but never enough to cause trouble. The federal government did some psychological testing, concluding that John was somewhere "between self-confident and arrogant." This amused him immensely.

John had arrived in Savannah knowing very few people. He did have an opportunity to meet and deal with the business leaders, politicians, and some of the rather entrenched families. He and Carolyn decided that it might be a good idea to join a church. They had never been a very religious pair. Now that they had children and were part of a community, they decided to visit and possibly join one of the Presbyterian churches. They attended church several times and seriously considered joining. However, upon talking with the minister, they found that they would have to make a statement or vow that they believed in a number of concepts that they honestly did not believe in. Many a person would have gone on and made the vows, but John's personal beliefs were so strong that he felt he could not honestly join the church, so he stopped attending.

Here we have two Johns, one we will call John the 1st and the other John the 5th. How were they alike, and how did they differ? Can we trace any genes down that far, and can we draw any conclusions?

Physically they probably did look alike. John 5th is tall and handsome with blue eyes and brown hair. John 1st was also probably tall and handsome. We don't have pictures of him, but we do have pictures of Ben, John 1st's son, and he was large and had good features. Probably they both had certain physical characteristics that were passed down, as we know from genetics, there are dominant and recessive genes. The dominant characteristics are usually passed along unless two recessive genes occur in an offspring. What about the intelligence of each? It was also probably about the same. Both went to college and graduated. This might give an indication of similar I.Q's., but this could be just a misleading fact. If two very intelligent people have a child, the child is usually not as bright as either parent. This is a law of inheritance called the rule of the median. If two very tall people marry, the offspring will usually be shorter than either of the parents. Otherwise, we might develop into giants.

How about some of the other characteristics of the two Johns. John 1st must have had ambition. He made the decision to come to America. He came, found a job he liked and stuck to it. In his forties, he decided to become a minister which he did with considerable effort. John 5th went into the Peace Corps after college. This took a certain amount of courage or push as this meant two years of an entirely different life-style. John 5th was somewhat of a late bloomer. With the exception of sports, he was not a person who possessed drive early in life. His ambition seemed to be a slow and gradual process. John 1st may have had his ambition early in life. He was ready to change his whole way of life if the opportunity took his fancy. Thus he decided to become a minister when he was forty years old, and he moved from Easton, Maryland to Augusta, Georgia, at age forty-eight to organize the public schools there as well as be a minister. We are not sure how emotional John 1st was; certainly he wasn't so emotional that it interfered with his work. Toward the end of his life, we understand that he baptized several of his grandchildren, Ben's children—Edgar, John, and Hannah, and this was said to be one of the happiest days of his life. He was a very religious man. John 5th was not religious, but he was certainly a very honest man. Both had a rather favorable environment, and if they did have the same genes of ambition, they had a chance to enrich their lives. however, we must remember that John 5th got genes from sixteen different people and only one of these was John 1st. However, if the genes of aggression and drive are dominant, it is possible that John 1st passed them down the line to John 5th. [Carolyn says that Thomas Neely is an entirely different child from his older sister Jordan. She says she can see the seeds of ambition and determination as he gets hold of a tennis ball and won't let his sister take it away.]

John the fifth, we have great hopes for you. You are the one who definitely has ambition to succeed in this world. You seem to be on your way. But time will tell. If there are genes of ambition, you received a good share of them.

6

Ben at Sea; Frank Early-on

The loading of the ship in Port-au-Prince went a little more rapidly than it had in Santiago. Ben found it strange that a whole country, though small, could be almost entirely black and be under black rule. The Haitians were speaking French with some speaking a little English. Mr. Rumans seemed to be fluent in French and was able to converse easily with them. On the second day in port, the cook and Ben went fishing again a little further up the beach. They did well again, and as they walked along the beach, they came to a small clear stream coming out of the woods. They were somewhat hot and waded in the cool stream. They put the fish down in the stream and explored upstream. Soon they were in deep jungle with branches and vines hanging over the water. As they slowly made their way along, there was a sudden great fluttering of wings and the loud, angry cries of disturbed birds. Ben and the cook stopped and waited standing silently and still. The birds quieted down. They slowly walked on a little further and came to a small pond and an open place in the woods. Never had they seen so many birds of so many different types and colors. There were small bright green birds in one group; large pink cranes were standing in one edge of the marshy pond. Ben had known most of the ordinary birds in Maryland and could identify them, but here was an entirely new group—the birds of the tropical jungle. The cook was very interested in this colorful sight. Ben said he had seen a bird book in the Captain's library; perhaps they could identify some of the birds that evening. As it was getting late in the afternoon, they turned around and started down stream. As they did, a large black snake dropped out of a tree about three feet in front of them and swam across their path.

The cook shouted, "Stand still, Ben. They vill not bite vile in the vater. That is a vater moccasin."

They stood quietly in the water until the snake had slithered out of sight. They then went rapidly downstream, picked up their fish, and made their way back to the ship.

* * *

The wonders of nature—sea, waves, fish, trees, vines, ponds, birds, and snakes. It is usually at twelve or thirteen years that a boy begins to see and appreciate what an amazing world we live in. Darwin made his trip around the world, and came up with his theory of evolution. His theories have been enlarged and corrected, but still the general outlines stand. As a child he must have been fascinated with nature, especially animals. City children often do not get a chance to see and feel nature. These children have something missing in their lives. Show me a child that is curious, knows, likes, and is amazed by nature's ways in the plant and animal world, and there you will have a child with an inner peace—a peace that surpasses all understanding. He will see the harsh as well as the beautiful side of nature. He will be able to accept the unexpected, and good or bad, he will take it in his stride. From this acceptance of nature, and inner calm, there can develop a strong foundation for aspirations, goals, and aims.

* * *

The cook and Ben returned to the ship, cleaned the fish, and salted them down in the ice chests which they had filled with ice in Baltimore. Tomorrow they would sail for England with the seaport of Liverpool as their goal. It was now mid-October. With favorable winds, they should be there in two weeks, around the first of November. But it could take as long as thirty days. England would be getting cooler, but they would be sailing right up the Gulf Stream, and this would speed the journey.

The England to which the *Nerissa* would return in 1847 was well into the Victorian age, which extended from 1837 to 1901. It was an age remarkably free from sudden revolution and violent catastrophes such as France, Germany, Italy, and the United States experienced. Railways were just beginning to appear. Ships run by steam were just beginning in 1832, and fifteen years later would begin encroaching on sailing vessels. Public education (with only one quarter of the children in school in 1833) was beginning its rapid growth. Trade unions were beginning to spread. The British were colonizing the world. Slavery had been abolished in England in 1833.

Queen Victoria ascended to the throne in 1837. Melbourne became her Prime Minister, then later Peel, who began the first income tax. There was national prosperity in England, but Ireland had the appalling famine of 1845 with the failure of the potato crop. Eviction of bankrupt tenants following the famine caused Irish emigration to America in gigantic numbers. Russell became the new Prime Minister of England after Peel fell. His government from 1846 until 1852 was undistinguished, but did produce the Factory Act of 1847, which prohibited women and children from working more than 10 hours a day in the factory. The working class was organizing and working within the framework of the prosperous, capitalistic system. Dickens was writing his first novels. Tennyson's early poems were beginning to attract attention. The violent Victorian controversies between religion and science were beginning.

The good ship *Nerrisa* was now getting prepared for the long voyage to Liverpool. It was over 3,000 miles and food stores and water had to be sufficient if the voyage turned out to be longer than expected. The cook, Ben, and two of the seamen were busy bringing on water, fruit, and foodstuffs for the journey. Finally, the hatches were battened down carefully. The sails were made ready, and the ship sailed smoothly and majestically out of the harbor into the open sea of the Atlantic. The crew knew that the voyage was going to be long, and although those who manned the sails would be kept busy at times, there would be long periods of boredom unless the weather got bad. Ben and the cook would stay busy feeding the crew and the officers. They were up at 6:00 a.m. and did not stop work until about 10:00 p.m. Then to bed and another day over, only to repeat the same routine again. They had a little free time in the afternoon. Ben had chosen Dana's *Two Years Before the Mast* to read, as the Captain had recommended it. He was soon fascinated by its graphic descriptions of sailing vessels and whaling adventures.

* * *

Frank Henry Neely, you were born January 19, 1884 in Rome, Georgia, and you died in Atlanta, Georgia, in 1979, age 95. You are the one, Frank, about whom we would like to know more.

You were the son of Ben who ran away on that ship in 1847. You were born when Ben was fifty years old. You were born thirty-seven years after Ben left home. Many things happened to Ben during those years. And Frank, we see that you were only eight years old when your father died. You really did not have much chance to know him. You were probably in the third grade in the school system that would be organized by your father in Rome, Georgia. Your position in the large family may have been important. Your eight brothers and sisters were older than you. Robert Allen Neely, your oldest brother, was sixteen years old when you were born and twenty-four years old when Ben, your father, died. Your four older sisters and brothers were about one or two years apart in age, but you, Frank came a full five years later, and may have been considered an unexpected event. I can imagine the older sisters had a fine time looking after you.

Ben, your father, passes good genes onto you. Although Irish and emotional, he must have been intelligent. Did he have the genes of ambition to pass along to you? Very likely, if there are such genes. He became a teacher and organizer of schools very much like his father had been. He had a chance to know his father John, and must have been strongly influenced to follow in the profession of teaching and organizing schools. But you had no real chance to be inspired by your father.

It must have been a sad day when your father died at age fifty-eight. He left your mother a widow with nine children—four were over twenty years old and five were under twenty with your next older sister Mary thirteen years old. We are not sure just what the financial situation of the family was, but it was probably not good. Ben possibly had a reasonable salary as superintendent of schools, but when this was cut off, there was probably a difficult and trying time. The older children undoubtedly were able to help. In fact, they probably were working at the time Ben died. None of the children was married at the time—they probably lived at home, and pooled what money was brought in to support the whole family.

This was a time of great strain on your mother, Henrietta Eve Carmichael, who became the head of a large household. What kind of person was she? Were you, Frank, influenced by your mother to try to do something in this world? We don't know much about Henrietta. She also was one of a large family with eight brothers and sisters. Her father was Anderson Watkins Carmichael, born in 1821, who died in 1895 at the age of seventy-four. He was a prominent hardware merchant in Augusta, Georgia, before the War Between the States. Henrietta, the second child, born in 1845, married Ben in June, 1867, almost immediately after the Civil War ended. Times must have been bad then as the War was just over. Although Sherman had not gone through Augusta, it must have suffered greatly and must have been slow to recover. So, Henrietta had lived through the War, married Ben when she was twenty-two and had begun her large family. Her parents were descended from the famous Longstreets and also the Eves, whom we will discuss later.

So, we have a slight background for Frank. This may give us a little insight as to why he would become the best known of his clan, but we will have to examine Frank's life more closely before we can make any judgment. You are the one, Frank who had the strongest drive and ambition to do something in this world. You are the only one who could be called great—at least in your community, and at one time in the nation. How did you, the son of a superintendent of schools in a small Georgia community, become a leader in your time, an industrial and merchandising giant, a man recognized as one who did more than just make a fortune? Fortunes were made easily in your time, but people who went out into the community with visions were few and far between. People who could make other people do things beyond the call of their own duty and beyond their own selfish interest were indeed rare then as they are now.

How did it happen that you were perhaps the greatest Neely? We plan to look as carefully as we can at you, Frank. We have a great deal of information about you, but we want to know what made you tick. We want to know if you just happened to get an unusual inheritance from your parents, or did you get your drive or ambition after you arrived in this world? Was it a combination of the right genes and the good environment that just happened to suit your abilities? Great men do not seem to strive to be great. They just keep a steady course; they know when to change their course. They somehow seem to know what the future will bring and how to adjust for it. They can shoot the sun or the stars to know if they are on the right course. If they arrive at greatness, it is not because this has been their goal—it is more something that comes to them without any conscious effort.

So it was with you, Frank—you did not try to become great; it just seemed to happen as time went on, and slowly you added deeds. When enough accumulated, there was no question but that in the eyes of your own world, your community, you had changed some things for the better.

* * *

Lets turn back to Ben, Frank's father—what happened to him on the voyage to England at age thirteen?

The clipper ship *Nerrisa,* now fully loaded, began the long journey in early November, 1847. The sea was calm, the weather warm, the winds seemed favorable as the course was set north by northeast. They were not to see land for about three weeks, when they would come into the Irish Sea. The crew kept busy, cleaning, painting, scrubbing decks, and repairing sails. They were in the Gulf Stream, which should carry them along on their way, but the winds were variable. At times there would be a great calm, then a choppy sea, and then several storms. With the heavy load, the ship rode the waves better, but progress was slow. Ben and the cook remained quite busy as usual feeding the group, but Ben had enough time off to read, and the captain was very helpful. He was given *Robinson Crusoe* and *Nicholas Nickleby,* and read them avidly.

The crew generally was friendly, and he had the quiet Bailey as well as the cook as his protectors. Frank, the Newfoundlander, was the only one who was mean and unfriendly. The others enjoyed kidding him. The language of the crew was crude and often obscene; never had Ben heard so much cursing. Almost every sentence that they spoke had some obscenity and in their imprecations, they made all sorts of references to the Lord and to Jesus. The language came as quite a shock to Ben. He had been brought up in a highly protected environment and had, of course, heard some cursing in school, but nothing like this. His father, John, was a very religious man and was at the moment studying for the ministry. The children, his brothers and sisters, had all attended Sunday school and church in Easton. There was no question about whether they wanted to go. They all just bathed and dressed and went for the whole morning. Ben might have been bored, but he had not questioned anything his father wanted him to do. So here he was at thirteen, thrown into the midst of a group of apparently very irreligious people. There were some Catholics among them, and a few of them would go to Mass when they were in port. But most of the crew were cynics who felt that they had to make their own way, and the Lord didn't have much relevance in their lives. Although somewhat of a shock at first, Ben soon became used to it and very soon was talking very much like them, but he still couldn't curse the way they did. Thrown into a somewhat polluted river of life, Ben had to learn to become a part of it and yet rise above it.

7
Randolph Carmichael Neely

Meanwhile, it was not a very happy Thanksgiving at the John Neely household in Easton, Maryland in 1847. No word had arrived from Ben, who had now been gone for over a month. There was the family dinner, and several of the neighbors had been invited, but it wasn't the usual happy event. And soon Christmas was coming. Everyone thought that perhaps Christmas would bring Ben home, but little did they know how far away he was.

John had a tendency to brood about his son. It was worse not to know than to know, even if the news was bad. But the duties at school, plus the studying he had to do, kept him occupied most of the time. He was now in the last year of his three-year course to become a minister. The teachers at the theology school were impressed with his determination. Most of the other students were younger than he. He was in his late forties and obviously mature, and now absolutely certain that he wanted to be a minister. Perhaps Reverend Davis was his inspiration. Here was a man who combined teaching with the ministry, and the two professions seemed to complement each other. The pleasure of teaching was to see and inspire young minds, and later to see the long-term benefits. The ministry gave one the feeling of immediate accomplishment, of service to humanity. To officiate at christenings, weddings, funerals, and to give a weekly sermon would not seem to be satisfying, but these things appealed to John. Also, there was a certain amount of counseling or comforting, and John had a warm, outgoing personality. Many times he felt so tired that he wished he had not taken on this additional task, but the low burning inner drive to attain his goal kept him going. He had studied science as well as the classics while at Trinity College in Dublin, and he saw no problem between science and religion. He did not believe in the Bible literally, as he was brought up in the Anglican Church in Ireland, and even in those days that church was the most liberal in Protestantism. He felt that he could interpret the Bible in the broader sense. He was somewhat inwardly torn concerning slavery. He knew that it was wrong for a human being to be kept in bondage, but it was so common and accepted in the South that it was a way of life. The force of the abolitionists was becoming a very real factor in the North, and the movement came under the sponsorship of most of the churches in the North. John attended a few abolitionist meetings, but could not become very enthusiastic. He was just too busy.

The children were also busy in school in late 1847. Euphemia, the eldest, was now fifteen. Mary, the third child, had died in 1845 at age nine of whooping cough. Little John was nine years old. Elizabeth was seven, Maria, six, and Annie, four. Isabella, John's wife of only a few months, was falling into the role of new mother and wife very well, although the episode with Ben had been traumatic to all.

* * *

Randolph Carmichael Neely, born June 25, 1915, in Atlanta, Georgia, and died March 1980 at age sixty-five in Toronto, Ontario—how did you fit into this picture of Neelys? Your life is finished on this earth. Will you be much remembered? Will any of us be remembered for long?

We would like to look at your life and see if we can find whether you had any special genes from your forefathers. We would also like to know if perhaps environment played the bigger role in your life. You lived to be three score and five—a respectable age. Your life span covered a time of amazing changes in life-style, in knowledge, in science, in communication, and in transportation. But there were not any great changes in the exercise of compassion in this world. In spite of these great advances from horse-and-buggy to jet travel, people were fighting and killing one another when you were born, and were still doing so when you died in 1980.

You were only a very young child during the first World War, but you were a young man and an active participant in World War II. You were retired when you died, but you had been president of a part of a very large corporation before you retired. To attain this end, you must have had more than your share of personal ambition. You and your uncle Frank Neely are perhaps the only two Neelys who became presidents of large corporations. Frank participated in many activities for the community so that he was, and still is, recognized as a great man. But what about you, Randolph? Were you loved by your fellow man, but more important did you love your fellow man? How did you get to be president? We will probably never know, but we can make some conjectures, because we know your background very well. We can trace your rise from a small child to your death. What made you tick? If you were not great, why not? Was there a flaw? Did you pick up some bad genes? Let us take a look at your life.

You were Ben Neely's grandson. John Neely was your great grandfather. Edgar Neely, Sr. was your father. We don't really know enough about Ben yet to know if his genes had much effect on you. You didn't even know your grandfather—he died twenty-two years before you were born. Edgar Neely, Sr., your father, certainly passed on to you some good Neely genes. We will look at him later. Your mother, Emily Levering Eckfeldt Neely, brought down a whole new set of Yankee genes from Washington, D.C., and Philadelphia. Such a new mixture of genes may have been beneficial. Later, we also will try to see what she was like, and what her ancestors accomplished in this world.

You had two older brothers, Randolph—Edgar Adams Neely Jr. and Frederick Levering Neely. You were the youngest, and they say that position in the family is important. It is interesting that Frank Neely, your uncle, was also the baby of his family. They often say that the youngest child is the spoiled one. But this is not necessarily so, and even if it is true, who knows but that spoiling is good for the ego. However, with boys, the youngest may not be compatible with the older children. Boys particularly have very changing interest at certain ages. A boy of fifteen who is just learning to drive a car doesn't have much in common with a child of ten. And this was true with you, Randolph. Edgar, the oldest was five years your senior and had his own interests. He got along better with Levering, the middle child. You, Randolph, also got along better with Levering, but even two-and-one-half years makes a lot of difference in interest.

As a young preschooler, you were healthy, doted on by your parents, but too young for the games and bicycling of the older boys. At school, you had your own friends to play with. The older boys, especially Edgar, had a tendency to tease you and sometimes to frighten you, especially in the dark. But soon you were fighting back and did not hesitate to strike with a stick or whatever was handy. The fighting in the back seat of the Buick touring car was a great concern to your mother, and your father would often stop the car, threaten all three children, or even spank one or two.

In the summer, the family would go to Highlands, North Carolina. Your dad was an attorney for the Southern Railway and had passes on the train for the whole family. They would go to Cornelia, Georgia, by Southern Railway, transfer to the Tallulah Falls Railway (jokingly called the Total Failure Railway), proceed to Dillard, Georgia, and then get into a Model-T Ford and drive eighteen miles up a muddy road to the town of Highlands, the highest incorporated town east of the Rockies. The trip was usually memorable. The car would often get stuck in the mud and have to be pulled out by mules. It would often steam up, and flat tires were common. Finally, the family would arrive at the new world—a beautiful little town in the mountains, 10 to 15 degrees cooler than Atlanta, with beautiful mountain laurel and rhododendron. The family would stay in one of the old hotels, and later they would rent a house, and finally, in later years, would buy one. They would stay for the summer and come back home in the fall.

You loved this place, Randolph. You made friends there that remained friends the rest of your life. There was a large New Orleans contingent of summer people, and you fitted right in with these nice, friendly people. Highlands became a sort of second home for you, first as a young child and later as a teenager, and you loved to go back and reminisce about it the rest of your life. You were the one of the boys who loved Highlands the most because you had roots there.

Sometime, when the boys were young, about twelve to seven years old, your father Edgar bought a house and farm in North Georgia. This was in peach country, and he had an orchard, a barn, a hired man, and a pony for the children. This was another community of your youth, and although you enjoyed it at the farm, you always preferred Highlands. You, Edgar, and Levering would bring your friends to the farm. It was a good place for the boys—they got along with each other better, but you, Randolph, would fly off the handle if teased too much and even threw a knife at Edgar one time, when provoked. Your mother Emily could not understand why boys would fight so much and would get very upset if Big Edgar was not there to cope. But the farm only lasted a few years, and soon the family was going back to Highlands.

For your schooling, you attended the public schools in the elementary grades, usually walking to school or being taken by your father in his car on his way to work. You did reasonably well in school; if you enjoyed a subject, you would make a good grade, but if not, you would still get by. You had a good I.Q., but you did not show any great drive in the lower grades.

You were sent to Episcopal High School in Alexandria, Virginia—an elite prep school to which the good families of Atlanta sent their sons. It was an excellent school, and you were able to cope very well; you made fairly good grades and graduated. This school was where you changed from a boy to a man. The school had a good effect on you; it gave you self-confidence and made you into a pleasant and courteous person who impressed older people. Throughout your life you had to fend for yourself; you had to be tough to survive among your peers. You probably found you had to be somewhat selfish to get along in this world. You began to have an inner drive to get ahead, and you were bright enough to pick people and opportunities that would help you.

From Episcopal High you could have entered one of the good Eastern schools such as Princeton or Yale, but you decided that you would go to the University of Georgia where many of your friends were going. In that time, Georgia was known as a fun school. It was relatively easy to get in. You could get by with a minimum of study, but you could have a great time, especially if you had an automobile, and of course you did. You stayed three-and-one-half years and could easily have graduated, but World War II was beginning, and you decided that fighting a war would be better than studies at college. After your sophomore year in college, you had gone to Russia on a ship with your brother Levering. It was one of the big experiences of your life. You were the crew's mess boy—serving the crew just as your grandfather had in 1847.

This trip to Russia had a maturing effect on you, as you visited several different countries—Denmark, Finland, Poland, and finally Leningrad, Russia. The crew was a pretty tough bunch, but you managed to get along well enough. Your brother Levering had an easier job. He was assistant to the purser and wore a white uniform, did some typing, and held the door for the passengers coming to their meals.

You and Levering became much closer on this trip. You were experiencing things together, seeing new sights, and dealing with people together. This was in 1934, before World War II had begun. Russia was trying to spread Communism throughout the world. The ship's crew were good candidates for communism, and the Russians entertained the crew very royally to show how good communism was. They took most of the crew to a big rally in a large hall, gave them food and beer, and then asked the crew to tell how bad things were in America. Little did they know that Randolph and Levering were the children of a successful lawyer and that they thought capitalism was fine. When it came time to talk, they both tried to get the other one to say something. The whole hall full of people was waiting for them to talk, but the long silence was finally broken by another member of the crew who was glad to tell how bad things were in America. The trip took six weeks, and that six weeks created a bond between you and Levering for the rest of your lives.

You went back to school, but toward the end of your senior year you volunteered for the Army's officers' training school and were accepted and sent to California. You did well at the school, were stationed in Washington State for a period of months, and then promoted to Captain in the Infantry with orders to go the European theatre.

En route, you were in New York for a short time and there visited your brother Levering, who was interning at Bellevue Hospital. Levering's fiancee, Margaret Gardner, got you a date with her roommate, Kay Curme. You and Kay had a whirlwind romance, and as you were still waiting to go overseas were married in White Plains, New York, had a brief honeymoon, and then off you went to war in 1942, just as grandfather, Ben Neely, had gone off to fight in the Civil War in 1864.

8

Ben Crossing the Atlantic; Neely Plumb

The *Nerrisa's* journey across the Atlantic was slow by sail boat. One would think that travelling to England with the Gulf Stream from the Caribbean would be easy sailing. The Gulf Stream travels from south of Florida north-eastward along the coast of North America up close to Newfoundland, and then across the Atlantic. It warms Ireland and England which would have almost Arctic weather without the Gulf Stream and the warm winds that usually come with it. However, it only flows about four miles per hour and gets slower as it travels across the Atlantic. The direct trip from Port-au-Prince, Haiti, to England is a little further east than the main Gulf Stream, but with tacking and small storms, the *Nerrisa* went in and out of it. It was now well into November, and the weather was changing fairly rapidly. A small storm would sometimes force the *Nerrisa* considerably off course, and although it was warm at first, the wintry blasts from the northwest would push the ship southeast. However, slow progress was made.

The crew, though fairly busy when on duty, had a good bit of leisure time, and there was considerable sitting and talking in the crew's mess room by those off duty. Often there would be a card game going on, but mostly it was talk—talk about women, about places, and often about Mr. Rumans. They all respected Mr. Rumans, the first mate, but they did not like him. They knew he was tough, but also that he was usually very fair. They also knew that he and the Captain had guns and were not afraid to use them if necessary. The laws of the sea were still strong concerning mutiny, and if a member of the crew did something suggesting insubordination at sea, he would be put into irons for the rest of the trip. This had happened on the boat some months before, and Mr. Rumans' firm stand was well known to the crew. They did not like him—he was too aloof. He still had his Welsh accent. They all knew that he knew what he was doing—that he was an excellent sailor, and they trusted him. But they really enjoyed cursing him behind his back. Ben recognized that he was a man of purpose. Exactly what purpose, Ben was not sure, but there was a quiet firmness in his manner that Ben respected. Some of "the First's" characteristics rubbed off on Ben, and he was to use them in later years.

When Ben was not crossed, he was quiet and rational, but he was Irish in his emotions. He could be suddenly angry as he had been with his stepmother, but he could also be very sad and would easily cry in an emotional crisis. He was either high or low, but he tried not to show it. He was now beginning to have nightmares and would suddenly cry out in his sleep from the upper berth, disturbing the three other crew members. Because he was not so busy, he began to brood a little about being so far from home and going to an unknown country. This volatility of emotions was going to plague him in later life as he would sometimes make a decision when angry or sad. His father, John, was also emotional at times, but he did not let this interfere with his decisions.

It was almost a month since Ben had left home. He still thought that he could not go home again, but he was beginning to miss the security of his family life. Although he was now accepted as a part of the crew, there was something missing which he could not identify. He had a tendency to sulk and brood. The cook noticed it, understood, and tried to kid him out of it.

But it had to be lifted by itself, and as the good ship *Nerrisa* progressed northeast, he slowly came out of his depression, talked more with the crew, and began the long and painful change from boyhood to becoming a man. He was developing his muscles with the exercise he got around the ship. He could not see into the future. If he could, he would not have believed it. For he would one day return to Baltimore, rejoin the family, return to school, and would move with the family to Augusta, Georgia. His father, John, would become the head master of the Richmond Academy for young men. Ben would graduate from the Richmond Academy, work his way through college at the University of Charleston, and then go to Normal School at the University of Georgia where he would learn to teach. Meanwhile, his father—now a minister as well as a teacher—would begin the public school system in Augusta, Georgia.

About that time, the rift in America between the North and the South was becoming more polarized, and although Ben—like his father—knew that slavery was wrong, they both fell into the sentiment that the North was telling the South what to do, and both of them, being independent, resented it. Little did Ben know as he stood on the deck of the *Nerrisa* that he would one day fight in the Civil War, be wounded three times, but not maimed, and then get married and follow in his father's footsteps by teaching and organizing schools first in Augusta and later in Rome, Georgia. And if anyone should have told him that he would one day have nine children, five boys and four girls, he would have laughed. And also to mention that he would have at least twenty-four grandchildren and innumerable great-grandchildren—that would have astounded him. That one person should so increase the population of the United States and even the world—this was not a concept easy to grasp at his age.

Ben, little did you know that you would be passing along genes of intelligence, emotionalism, and ambition. Little did you know that some of your male progeny would have the right ingredients to be great or near great, whereas others would be finishing off the last days of their lives with six to twelve beers a day and merely existing.

However, one of your seeds would join with an ovum of your wife, Henrietta Eve Carmichael Neely, to produce a brilliant person, one with the ability to get along and influence people, one who exuded that magic ability to make people respect and love him. Somehow, this child of yours, Frank Henry Neely, did not get your excess emotionalism. This may have been the secret of his success. The wild swings of mood were somehow left out of Frank, thus enabling him to make more logical decisions.

The *Nerrisa* held to its course, no matter how buffeted and pushed around in the wintry, rough Atlantic. The Captain was on the bridge frequently taking his eight-hour shift on duty. He and Mr. Rumans had to shoot the stars and the sun frequently to keep on course, and they taught Ben the rudiments of navigation—how a sextant and a compass provided all the information needed to arrive at a point on the globe. The Captain, Mr. Phillips, continued to be very helpful to Ben. He offered him the run of his library, and for Ben this was a feast; his father had always had a large library. Somehow this ship's library was more intriguing, with books from several different countries; some were even in French, Italian, and Spanish. Ben learned a little French in school, and with the aid of a French-English dictionary, he read *Tartaran of Tarascan,* appreciating the wild humor of its author. There were a few books by Dickens which Ben read avidly. He tried to read poetry, but Tennyson was about the only one he could understand. He especially liked "Ulysses" and he wondered if his trip would take as long as the Odyssey—twenty years. He shuddered when he thought about it. But how could he go home with that woman still there? His hackles rose just to think about her.

What other genes would you be passing along, Ben, from you and from your father John? Often a characteristic will skip a generation. What about musical genes—either a great appreciation of music or the ability to play an instrument well? Leonard Bernstein's talent seemed suddenly to come from the sky and not from his non-musical parents. Again, we are not sure about the role of environment on fostering and encouraging music. Bernstein came from people who actually discouraged time spent on music, and still he became a great musician. Perhaps his grandfather or great-grandparent had genes of musical ability. Often there will be musical families who enjoy playing or singing together. Most of the Neely men had good voices, but few could carry a tune.

There was one of your grandchildren, Ben, however, who devoted his life to music. Perhaps he got his genes from his father rather than your daughter Hannah. This was in Neely Plumb and we think his full name was Benjamin Neely Plumb, named after you. This child was born in Augusta, Georgia, in November 1912, the youngest of four children—two older sisters and one older brother, William Plumb. These two boys were fairly close in age and interest, and they both learned to play the saxophone while in high school. They were both intelligent and both had enough drive to work their way through five years of engineering school at Georgia Tech in Atlanta. Their father had died young from a heart attack, leaving his wife Hannah with very little funds and four children from sixteen to ten years of age. She, being resourceful, and good cook, had begun a tearoom (as a restaurant was called in those days) in the center of Augusta. She continued to operate the tearoom for about twenty-five years until the children were grown and moved out of her house.

You, Neely Plumb, were somewhat of a character even as a child and young man. You enjoyed playing jokes on people, such as dropping water bombs from the second floor of the tea room onto unlucky passersby, one of your favorite pastimes until the police gave you a good scare. But you had a good sense of humor as a young person. You soon found that the ability to play a saxophone was a lucrative talent. You played with small bands in high school. When you were at Georgia Tech as a student, you began to play with the good jazz bands in the city. Then, after finishing college, you went with a big national band instead of trying to find a job in your field—engineering. True, it paid extremely well, much more than you could get in an engineering job during the persistent depression in 1935. You stayed with music, went to Hollywood, played for the movies, and you had your own jazz band at one time. After a stint in the army you went back to Hollywood, got into the record business and were highly successful financially. You did, unfortunately, acquire a Hollywood accent and seemed to be a different person from the humorous young man in the thirties. You married, had several children, and one of them became a television star in one of the soap operas.

You, Neely, must have had the musical ability and the business drive to run a very successful company. How much drive did you get from your grandfather Ben? Certainly you received a good intelligence quotient from the Neely side as well as the Plumb side of the family. But we don't know yet where the fascination with music came from. Your brother William stuck to a career in engineering, although he—like you—was a good musician and made good money working in jazz bands while in college in Atlanta. But William, although a very pleasant person, very gentle and loving, didn't get those genes of ambition that you received. Aside from your business success we don't know much else about you. You and Randolph Neely had the most success in business, but neither of you could really be compared to your Uncle Frank Neely, who in addition to business success was very active in the community and accomplished those things to make a wide variety of people consider him a great man. We shall see what those deeds were presently.

* * *

Liverpool was the largest port in England in 1847. It had first become an important port in the thirteenth century under King John. The port of Chester, formerly quite large, had begun to decline due to the silting in the River Dee, causing the port to be too shallow. Liverpool was on the Mersey River, and it had a sea lake at its mouth, making it well adapted as a harbor for vessels. The true rise of the commerce of Liverpool dates from the Restoration. Its trade was first with Ireland, France, and Spain, later with America and the West Indies. The Liverpool merchants entered the slave trade in the early 1700's, and the rapid growth of the port was due to this trade of slaves brought from Africa and sent out to America and the West Indies. However, slavery was abolished in England in 1807. Other trade was carried on, and the port became a great commercial center as manufacturing industries in the area flourished. There were miles of docks and warehouses in 1847, and new docks were being constructed all the time.

With the rapid growth as well as the many rich merchants, art, music, and education developed. It had a large Roman Catholic population, many from Ireland. A college and a Philharmonic Society were created, and many literary institutions were springing up. In 1847, Liverpool stood as a large bustling city based on trade and commerce, but it was developing along cultural lines also and was ahead of Ben's home of Baltimore in its fostering of the arts.

The *Nerrisa* sighted its first land after an eighteen-day journey in late November. The weather was cold, but not bitter, and remained fairly good as the ship sailed into the Irish Sea between Ireland and England and made its way toward Liverpool. Everyone of the crew was quite excited as the ship neared port. Mr. Rumans, "The First," knew the area well and was able to bring the boat right into the proper dock. The plan was to have a slight overhauling here, as there were ship yards with good repair service. After the ship had finished unloading, it would go into dry dock for a while, be inspected and repaired, and have some of its sails replaced or sewn. This meant that the ship would be there at least ten days. Some of the crew remained on board, but most took off and went into the city. There were usually a few crewmen who decided to remain in Liverpool and take another ship, as the jobs on boats were fairly easy to get. However, most of the crew on the *Nerrisa* planned to sail back to the States and perhaps change jobs there. Ben and the cook continued to serve meals in the morning and at lunch, but after that they would wander around the docks looking at the various ships from all over the world. They would also go into the city and walk about looking at the sights, eating at small restaurants, and generally whiling away the time.

And here was Ben, thirteen years old, soon to be fourteen, with his voice just beginning to change, seeing a large English port and city. He was pouring into his mind a whole new world, not realizing what this trip would mean to him later.

9
Frank Neely: Birth

When Frank was born in Rome, Georgia, in 1884, the whole household was happy. Ben had moved the family from Augusta to Rome, Georgia in 1882, and was extremely busy organizing the elementary school and founding the high school. Most elementary schools were private before this time, and high schools did not really exist. Ben had to work morning, noon, and night. He had to meet with the city fathers, the parents, and the potential teachers. He had to convince the town council that they should raise taxes to buy land and build the schools. Often he felt like giving up, but he held on to his temper and was obliging to everyone. The elementary schools were moving along well. But a high school was Ben's dream, and he set about methodically to achieve this purpose. Soon he had the plans, and the high school was begun. There was a small ground breaking ceremony and the mayor and city council decided to name it the Neely High School. This name remained until 1976 when the school was torn down and a plaque was put up to show where the school had been. There was another ceremony at the dedication of the plaque, and several of Ben's grandchildren came to see Ben's memory honored.

Ben was fifty years old in 1884 at the time of the founding of the high school, and he was the superintendent of all the Rome, Georgia schools. Henrietta Eve Carmichael Neely, his wife born in 1845, and married at twenty-two, was now thirty-nine years old and had already given birth to eight healthy children, all born in Augusta before they moved to Rome in 1882. The eldest boy, Robert Allen—called Allen—was sixteen years old at the time. Annie was fifteen, John was fourteen; Elizabeth, twelve, Ben, junior was ten; Hannah was eight; Edgar, six, and Mary, five.

It was at this time that the youngest child, the baby of the family, Frank Henry Neely, was born on a cold day in January. He was the first child born in Rome. The older children had come very fast with only a year or two between them, but it had been five years since Mary, the next youngest, was born. It was a closely knit family. Ben, who had come up in a large family where great importance was put on schooling and learning, carried on the tradition of his father, and all of the children were required to do their homework carefully and thoroughly. Ben would check their studies almost every night. Ben felt that the more people learned, the better the world would be.

The night Frank was born it was snowing lightly, and the wind was blowing, making noises on the window. They lived in a big, old, two-story house with six bedrooms upstairs, and with eight children, several had to sleep in the same room. The girls were old enough to help with the cooking and the boys did the chores. Allen and John worked at a grocery store on Saturday and earned a little money. The family was fairly well off, but not rich by any means. When Henrietta Eve went into labor that night, she knew it would not be long. She told Ben that her time had come, and he saddled his horse and galloped over to get Dr. Tom Harbin, the doctor who was a family practitioner looking after them. The doctor brought along two nurses—one to help him, and one to stay with Henrietta afterwards. There was a fireplace in every room in the house, and the fire was burning brightly in the bedroom where Henrietta was to have the baby. All of the children were very anxious to see if it would be a boy or a girl because there were four of each at the time, and there was considerable joking about which sex would have the upper hand. Dr. Harbin arrived with the nurses. They were none too soon; in about ten minutes a fine bouncing boy with black hair and blue eyes arrived with a loud cry. At about that time cries of joy came up from all the children downstairs.

"With those lungs, it has got to be a boy," cried John and the word came down the stairs that he was right.

The boys whooped with joy, and the girls shouted their disappointment.

Frank Henry Neely, child of the Covenant, you burst forth into this world with some great genes from both sides. Were you destined from that moment to be a great man? Was it all chance or luck? Or was it that special milieu or favorable environment that pushed you on to greatness? By greatness we do not mean simply business success; we mean your doing something for your fellow man, having new ideas, seeing the future better than others, influencing others to do things, helping a good cause, while having some modesty about it. Here you were, Frank, confronting the world with a lusty cry of anger at the sudden change from water to air. But you adjusted quickly and that may have been your secret: you could always adjust quickly to any situation. Also, somehow, you always got what you wanted; this time it was food, and you got it quickly.

The crib was ready, And after Frank was washed, fed, and dressed, the children were allowed to come up and see him. The boys welcomed him to the group, but the girls, Annie, Elizabeth Eve, Hannah, and Mary were overjoyed. They all had their dolls, but now they had a living doll. As time went on, the girls would vie with one another to look after him. So in the first six months, and really the first few years, Frank had every attention. The psychologists and students of child behavior say that the first few months are very important—that the infant learns a great deal during those months, which can have a great influence on his later life. Some infants who live in poor circumstances, where the parents cannot pay attention to and stimulate the child adequately, will not develop their intellect as rapidly as children who are cared for, picked up, talked to and fondled. Certainly, Frank had considerable attention as an infant and on into childhood. Never could four girls be more attentive in carrying him around, talking to him, singing, and in every way stimulating his little brain to go to work. It is thus possible, if the infant behaviorists are correct, that this unusual amount of attention by the girls began him on his road to accomplishing things. On the other hand, many would say he might have been spoiled in this early period of life. Anything he wanted would be instantly provided. This might have led him to assume that he was due all this attention. One would think that this might lead to selfishness. But this was not the case. He thrived on the attention, realized that he was loved, but for some reason did not become selfish. Some children who find out early that they have to fend for themselves and fight for what they want become selfish. It is the survival of the fittest, and these children know they have to be selfish to get along in the world.

But with all this attention, Frank just became a little smarter. He learned early to get what he wanted with a smile, rather than a cry. He walked early, talked early, and was reading and writing a little at four years of age. He did not get those Irish genes of emotionalism. He would laugh easily, cry, and get it over quickly, but he never sulked. He had a hidden temper as time went on, but any outbursts were usually justified. He did not get his father's mood swings. His personality was forming, adjusting to all situations. They all soon realized that he was extremely bright and could use his intelligence to live better in this world.

Henrietta Eve Carmichael Neely, we have neglected you. We have discussed Ben in great detail and tried to analyze his origins, his background, his childhood and his aspirations, but we have scarcely mentioned you. You poured genes of equal quantity and quality into Frank and the other eight children. Many of your genes were probably dominant, and when we speak of the shape of the Neely mouth, it may really be the Carmichael mouth. Any great qualities that Frank had may well have been from your genes and the environment that you, his mother, provided. Frank really did not know his father very well. He was eight years old when his father died, and although his father had been delighted with Frank's arrival, he had eight other children to look after. They were becoming teenagers, and there were the perennial problems. However, the atmosphere was always encouraging to scholarship and learning.

But you, Henrietta Eve, how did you fit into this picture? We know that you were beloved by one and all of your children. As the older children grew, especially the girls helped you look after the younger ones. Between feeding them and looking after the household, you did not have much free time. There was some time at night for relaxing after the children got to bed, and you did quite a little reading yourself. You had been born in Augusta, Georgia, and had gone to the public schools there which were run by John Neely, Ben's father. You finished high school in 1861, just before the Civil War. Your father, Anderson Carmichael, who was born in 1821 and died in 1895, was a prominent hardware merchant in Augusta before the Civil War. He also was in the milling business at one time. He had a large family and you, Henrietta Eve, had four brothers and four sisters. Your mother was a Longstreet—Elizabeth Eve—before her marriage in 1844 to Anderson Carmichael. Your mother was the daughter of Gilbert Longstreet, and he was the brother of James Longstreet, the famous Civil War general. His battlefield decision at Gettysburg was said to have been one of the reasons that Lee and the Southern forces lost that battle, but this is controversial. All we know is that you, Henrietta, were brought up in the Southern tradition—one of a large family in Augusta. Your father was originally from Ireland, but came over to this country and settled in Augusta as a young man. You were the oldest daughter. The family was well known in Augusta, and your father, Anderson Carmichael, was a good businessman and good provider for the family. He lost most of his savings during the Civil War, but was able to carry on the business during the War and expand after the War.

Ben, just back from the War, went back into teaching and organizing the schools with his father. You met Ben at church. The courtship was brief, with marriage coming in 1867. You had been through the War in Augusta. It was a time of great excitement at first, but later became a time of great sadness; so many of the young men of the town were killed.

500,000 men were killed in this war, many more than in World War I or World War II. It wasn't a very large country at the time—about fifty million people. Hence, one out of every hundred people died because of battle. More specifically, it meant one out of fifty men throughout the country and one out of every ten young men was killed. Some regions were much harder hit than others. In Augusta, Georgia, it was close to one out of five young men. So Ben was lucky to get home. He had been wounded three times, but not maimed. Times were hard when you two got married. Ben had his job teaching or organizing the schools when he came back, helping his father at first, but soon making a name for himself. Henrietta Eve, you were swept off your feet by the handsome-but-older young man just returned from war. And you were married by none other than John. And as we see, you had good genes and you provided a good environment for your children.

You and Ben were very well settled into the life in Augusta during the fifteen years from 1867 to 1882, and you had already had eight of your nine children. Ben was quite a success in organizing and developing the schools there in Augusta. His father John had begun the organization of Richmond Academy and later the public schools, but his church had grown and he was spending much more time as a minister than as a teacher and organizer. Ben was taking over the running of the school system and was the acting superintendent. He had to report to the town council, and they expected Ben to do whatever they suggested. Ben, however, was still emotional, and he suggested that they begin a school for the blacks or have integration. This caused a great deal of friction and controversy. Finally, at one very tempestuous meeting with the council, Ben cried out "I hereby resign." The room suddenly became quiet, and the head of the council looked at the fellow members, and then said, "All right, we accept your resignation."

So Ben had to go home that night and tell Henrietta that he had quit his job, right when he should be in the best possible position. You can imagine how Henrietta felt—with her eight young children. But she was quite calm when he announced it, letting him talk it out well into the evening. She was extremely supportive. Ben had been approached by some people from Rome, Georgia, whom he had met at a convention in Atlanta. He immediately travelled to Atlanta and from there to Rome, Georgia, a town of 25,000 people, much smaller than Augusta, where 60,000 lived. The city fathers in Rome liked him and offered him the job. So you, Henrietta Eve, had to pull yourself and eight children away from a nice house and a good position in the community and move to a small town in northwest Georgia. But you did not complain. If this was what Ben wanted, you were with him all the way. Ben went almost immediately to Rome and began working very hard. You came along a little later with the children, by train to Atlanta, then transferred to another train, and on to Rome to start a new life. The children were not too happy about leaving their friends, but adjusted quickly to the new way of life. Allen, the eldest, was thirteen and Mary, the youngest, was three years old.

So here's to you, Henrietta Eve, mother of nine fine children. We don't know much about you, but we can imagine. You were a wonderful wife and mother. The only report we have is that Frank loved his mother very much. I am sure that all the children loved their mother very much. As a mother, you somehow provided the right amount of guidance and direction without spoiling any of them. They all are a great credit to you. You didn't neglect any of them. You provided them with excellent genes from a college president (Augusta Longstreet of Emory University) to a general in the Civil War, James Longstreet. But most important, you provided them with a good environment so that by loving and respecting you, they came to emulate you. They all had honesty, decency, and compassion drummed into them so that it was a way of life.

And finally, Henrietta Eve, you produced a great man, your youngest, and whether he realized it or not, you were responsible for much of his greatness.

10
Ben to London; Frank and Virginia Neely

After the unloading was finished, the company representative, the first mate, and the captain inspected the *Nerrisa*. It had been through a hurricane and some other rough weather. After searching all parts of the ship, they found that the aft mast had loosened at its base deep in the hold. This meant a major repair job, as a loose mast can break a hole in the bottom of the ship. Thus the ship would be in port in dry dock for at least a month. The crew was advised that they could be off while the boat was being repaired. The cook and Ben could also be off. A number of the crew would stay close by and take temporary jobs at the docks where there was usually plenty of work. Ben had nothing to do and did not know how he would spend this time. The captain, Mr. Phillips, invited Ben to come to London with him. He had a sister living in London with her husband and their two children, both about Ben's age. Ben was delighted with the invitation and accepted immediately.

London in 1847 had a population of about two and a half million, ranking as one of the largest cities in the world. London also harbored the beginnings of reforms in the police system as well as public schools. There were new churches, better water supply, and sewage, with major works in the hands of the Metropolitan Board of Works. "Socially, commercially, financially, it was the hub of the Kingdom." The railroads were pioneered in England in 1826, and by 1847, the steam locomotives were increasing rapidly for both freight and passenger service. Forty-five hundred miles of track were laid down in England in the 1840's, and there was regular service between Liverpool and London.

Ben had read and heard a great deal about London. The invitation by

Mr. Phillips, could not have pleased him more. He had heard of the London fire, London Bridge, and Westminster Abbey, as well as the Thames River and Buckingham Palace. The captain's sister, Mrs. Mallory, was married to an Englishman and had a son and a daughter. They lived in the suburbs in a middle-class house. Mr. Mallory worked in town as an accountant for a large importing company.

The captain was going to have at least two weeks off. He had some work to do in the city at the company office concerning his next trip, and what he would load and unload. Communications were naturally slow, but several ships came straight from Baltimore to Liverpool, and often the orders took only two weeks from port to port. The company was The Baltimore and Ohio Railway, which owned the *Nerissa* as well as about ten other Baltimore clippers. So the office in London was fairly busy, as the cargo had to be arranged well in advance, and there were frequently changes in the routing. The captain learned that the *Nerrisa's* orders were to proceed to Glasgow, on to St. Johns, Newfoundland for salmon, then to Halifax, Nova Scotia, to Boston and New York, and finally back to Baltimore. This trip could take from six weeks to three months. It was now November 25th, and they were to sail during the latter half of December.

The captain, whose wife and daughter had died a few years previously, recognized in Ben a bright responsible child and had developed a fatherly liking for him. He was thus delighted to take Ben to his sister's house for a short stay. The Mallory children—Keith, fifteen, and Janet, twelve—were also intelligent, studious children, and they were quite entranced to have an American in their household for a visit. It took Ben a while to easily understand them, but it wasn't long before he had something of an English accent himself.

Keith and Janet were excellent guides through London, and there was a great deal to see. Ben had taken a course in English history, which proved to be quite helpful. The children would go into town by horse drawn trolley in the morning, visiting the Library, the British Museum, and the Tower of London. Walking by the river, they watched the large and small ships ply the water. Many steam paddle-wheelers already traveled the Thames. These ships were still a curiosity, but several were taking passengers and freight. Steamships had crossed the Atlantic. Most of these were steamers with sails. The clipper sailing boats were so fast and efficient that they were still very popular in the middle of the nineteenth century. The British built steamships rapidly, and the Americans soon followed with their own in the 1850's. Most of these ships were of the paddle-wheel type; they were efficient in calm water, but the paddles would come out of the water in rough seas. Propeller ships were not developed until about 1870, and these made steamships so superior to sailing vessels that the wind-powered ships died out quickly. So in 1847 there were both types of ships competing with one another, as the people still did not trust the steamship completely. Ben had an excellent opportunity to see all the different types of ships in Liverpool and London.

Ben fitted into the household very well. He was always courteous and helpful. After dinner, the family would often have a musical evening. Mrs. Mallory played the harpsichord, Mr. Mallory, the flute, Keith, the violin, and Janet, the recorder. They would all sing from time to time as well. Ben enjoyed these sessions because he had played the violin at home, and he had a fair voice, although it was beginning to change. They played Scottish and Irish jigs and would often do a little dancing and singing. They all enjoyed the simple songs, but Mr. Mallory was an accomplished musician and would get them to play short parts of the classics such as Haydn, Mozart, Bach, and Beethoven. London had many chamber groups as well as a symphony orchestra at the Albert Hall, and the whole family, including the captain and Ben, went to the concerts. For Ben, the experience of hearing a huge orchestra play was a new one.

Ben liked music, was reasonably talented, but was not about to devote his life to music. His father John was somewhat musical, but was mainly interested in church music. John enjoyed singing, but could not carry a tune very well. Ben, fairly good at keeping a tune, must have gotten this talent from his mother.

* * *

When Ben's son Frank was nine months old, he was walking. The four sisters were with him almost constantly, encouraging his every effort. He was breast fed and quite chubby the first year. He seemed precocious in all his activities. By age one he was saying quite a few words, and was into everything as soon as he became mobile. Ben had bought a fairly good-sized house after the first year in Rome. It had a large yard with a garden, plenty of room for the children to play, and sufficient chores to keep the boys busy in summer and winter. Ben, like his father John, was involved in the school system, and after supper there was a two-hour period of no talk or play during which Ben insisted that the children do their homework carefully. Frank, not yet in school, was given coloring books and crayons and would work at this until he got sleepy and was put to bed by his mother. Ben, like his father John, was a firm father and had developed schools—he felt that learning was of the utmost importance, and therefore accumulated a good library. So before Frank, there were two generations of teaching. But would Frank follow his father and grandfather? Teaching, though satisfying, is not very financially rewarding. Ben and John had earned enough to care for their families. Ben hoped to send all of his male children to college. Ben's plan for Allen, just seventeen at the time, was for him to attend Emory-at-Oxford, where one of his great-uncles—Augustus Longstreet, the brother of Gilbert Longstreet who was Henrietta Eve's grandfather—was the president of the college.

Frank grew rapidly in those preschool years. The older boys as well as his sisters took an interest in him, taught him to play with a ball, and even how to swim. It was a close family; although there were minor fights and jealousies, there was a great spirit of camaraderie, humor, and mutual help. Frank remained the center of attention, and he thrived on it. He was always able to come out with some funny remarks which endeared him to one and all. His father, Ben, also thought a great deal of him and rode him on his knee almost every night. By the time Frank got to the first grade, he was reading simple stories, could write a little, and seemed to understand most simple math problems. Thus, he became the most outstanding student in the first grade.

How many of us can remember one or two children in the first or second grade who seemed to stand out, be a little smarter than the rest, and go on to be outstanding throughout life? I certainly can—at least one—John Pollard Turman, nicknamed Booshie, who could do no wrong in the first and second grades. He seemed to make all A's, was the best athlete, and got along with his fellow students very well. Booshie continued to be outstanding in junior high and senior high school. He seemed to sail right through school. Similarly, he made good grades at college, was a good athlete, and became what was called a "big man on the campus." He seemed to have either a burning ambition to excel or a natural gift to solve the problems of life before they came up. Possibly it was the combination of his natural gifts and ambition. Booshie went on to law school and began practicing law in Atlanta. He entered community life and became president of a large corporation. He continued to work in the community with his college alumni group. Looking back, we can see that he was outstanding even as a first grader. His teachers said, "That boy will go far."

What of Frank—could that have been said of him as well? Probably. What about his sisters with the same genes, can we judge them in the first grade? In those days the girls did not have the same direction to their ambitions. Their main ambition was to have a large family. Also, opportunities other than this were few. So those superlative genes in the girls had their greatest possibilities in future generations by being passed along to the girls' sons.

In the current more modern age, what happens to the talented and ambitious female? One such female is Virginia McKenzie Neely Jackson—sister of John Neely, daughter of Margaret Gardner Neely and F. Levering Neely, and great-granddaughter of Ben. What course did she follow?

Virginia, you were a December child. It snowed lightly the night you were born in 1944. World War II was still in progress. V Day would come six months later, and the surrender of Japan some months after that with Hiroshima. Will you remember this, Virginia—that you were born close to the moment in history when man could devastate a whole city with one bomb? The future history of mankind may be described around that moment—BNB and ANB—before nuclear bombs and after nuclear bombs. Your father was not in the war, Virginia, because he had developed tuberculosis while interning. He practiced internal medicine at the time while most other young doctors in the city were off in military service.

You came rather fast, Virginia; we were able to get to the hospital in time, but you did not tarry long before arriving in this world. You were the second child, and there were to be four more children to follow you. You were perfectly bald at birth and very slow to grow hair. You were a very serene baby with very little crying. You adjusted to anything, but there was a firmness there, even in those early years. It showed itself most clearly when you got sick at the age of five, had a high fever, and three days later could not move your left arm (and you were left-handed). The pediatrician was called, came to the house, and then dropped the bomb on your parents. "Polio." In those days it was about as bad news as there was for a child. This was before the days of the Salk vaccine, and we had all seen the devastated bodies of polio victims. Off to Grady Hospital's Infectious Disease Unit where a spinal tap (no crying or even a whimper) confirmed the diagnosis and, after two weeks, on to Warm Springs for six months with weekly visits from the parents. Did you offer any complaints? Never. The left arm improved some so you could write, and fortunately no other paralysis developed.

So you came home and went to the first grade. It was with quiet determination—all A's in the first few grades. An outstanding girl—and what later? OK—private high school, college in the North, and into the Peace Corps in Africa for two years. You returned home, found your own job in Arkansas, took another degree at the University, married, had a child, and are now in a challenging and interesting job. So it can happen to girls as well as boys. And you have a super bright child—now four-and-one-half—Keefe. Where will you go? By the way, Keefe, we almost forgot, your mother had cancer of the thyroid found in 1973, had a major operation, and survived with no signs of recurrence. But, Keefe, keep in mind, your mother Virginia went through the whole ordeal of cancer without a complaint or a tear. Keep this in mind, Keefe.

11

Ben: Life in London; Death in Rome

Ben had a delightful time in London. He had two personal guides, Keith and Janet. It was perfectly safe for the young to travel around in the city alone. Keith and Janet were in school, but with the Christmas holidays, they were free until the first of the year. They took him to the British Museum and Library which were established in 1753 when the government acquired the art collection and library of Sir Hans Sloane. The present building was just finished in 1847. Extremely popular with adults and children, the museum had a fine collection of ancient and medieval artifacts and art. The three children spent several days there as it really brought history to life, and Ben, though a good student, found that Keith and Janet were much ahead of him in many subjects. They also visited the Tower of London, which housed the royal jewels as well as the collection of all arms and armor through the ages. It took them several days to explore this.

The captain told Ben that they would have to travel back to Liverpool in about a week. Ben remembered his father had studied at Oxford after graduating from the University of Dublin. He asked the Mallory family if he could possibly take a trip to Oxford. Mr. Mallory said that Keith and Ben could travel there by train and come back by boat as Oxford is on the Thames. This suited the two young adventurers fine, and the next day they rode the train third class up to Oxford. It was a beautiful town and an equally breathtaking campus with its many spires. They simply walked about. Most of the students had left for the Christmas holidays. Ben remembered that his father had been to Merton College, Oxford, and they visited there. It was now about the 15th of December, and the next day they took the boat back to London after spending the night in a nearby youth hostel.

The captain received word: they needed to get back to the ship; the repairs were almost finished, and it would soon be time to load. They said good-bye to the Mallorys and took the train back to Liverpool. The ship was almost ready, and they would be loading several steam engines to be sent to America. The plan was to pick up more cargo in Glasgow. The crew slowly came back, and Ben and Mr. Borg went to the market and bought provisions for the approaching trip back to America. The weather turned cold, and they had fires on the docks to keep them warm as well as wood heaters on board ship. The loading was finally done, and they set sail for Glasgow on December 21, 1847. They knew they would probably be aboard ship for Christmas, and Ben and Mr. Borg planned a big Christmas feast for one and all.

* * *

In 1890, when Frank began the first grade, at age six, at Rome, Georgia, he was still the pet of the family. The next older child was Mary, who was in the fifth grade. Allen, the oldest, was twenty-two. He had gone to college in Athens, Georgia for two years and was now working in Atlanta in a bank. He was a restless and adventurous young man, so the bank did not suit him at all. He had studied Spanish in college, and when he heard of a job with a coffee plantation in Bocas del Toro, Republic of Colombia, South America, he applied, was accepted, and was to leave for South America in a month.

Annie, now twenty-one, taught in the grammar school in Cedartown, a small town about thirty miles from Rome. She lived with a family there, coming home on weekends. John was twenty, had finished high school two years earlier, and was going to college at Emory-at-Oxford. He had become interested in biology and science since the loss of his leg in a train accident. Elizabeth Eve was seventeen, just graduated from Neely High School, with honors. She had not decided what she wanted to do and was now staying at home helping her mother.

Ben, Jr. was sixteen and had another year of high school. He was now the oldest boy at home and was taking the responsibility very seriously. He had to be sure that there was plenty of wood chopped for the winter, little repairs done on the house, the horses fed and walked, and the cow milked every morning. He was very good at mathematics and figures. Named for his father, he felt that he was in charge of the younger children. Hannah, almost fourteen, was going into the second year of high school. She was just blooming into a young woman who was vivacious, pretty, and full of good humor.

Edgar, age twelve, was the emotional child. He was in the last year of grammar school and would go to junior high school the following year. His voice was beginning to change and he was growing rapidly. He had red hair and the Irish temperament to go with it. He would get angry easily, but if anything upsetting happened, he would weep openly. When his dog died, he could be heard crying in his bed for a long time. He was also extremely friendly, very humorous, and well liked by his classmates. He got along well with adults, was able to talk with them with ease, and smiled disarmingly. Mary, ten, was also red-haired and also very emotional.

Ben, Sr. was very proud of his family. It was a tradition to do well in school, so he saw to it that they did. He had founded the high school eight years previously, and was the principal of the high school as well as superintendent of schools. He hired and fired the teachers, had many administrative duties, and enjoyed teaching history to the seniors. His purpose: to provide a school that would be both practical for those not going to college as well as preparatory for those students going on to college. He hoped that all of his children would go to college, including the girls; with the exception of teacher training at a normal school, this was almost unheard of in those days. (Annie had gone to a normal school in Atlanta.) All of the children were well indoctrinated with the virtues of higher education, and the children in high school were making plans for the future.

Ben, Sr. had resigned or had been fired from the job of superintendent of schools in Augusta because he thought that Negroes, like whites, should be allowed to go to school. He did not want to change jobs again, but he still thought that blacks should be educated. He decided when he began work in Rome to approach the problem a little differently when dealing with the mayor and town council. The black children did not have schools at all when Ben came to Rome, and the blacks often had large families and were running around the town while the whites were in school. So Ben approached the town council with the plan to keep them busy in school and off of the streets. After some discussion, the idea of educating the blacks passed the town council, was put into effect, and had been working well during the last few years. There were a few bright, older, black children who were taken into high school after a few years in grammar school, and there were no major problems. Several black graduates of Neely High School went on to college. Years later they came back to see Frank Neely, thanking him for his father's efforts.

Ben approached the zenith of his career in 1890 when Frank first started grammar school. A respected member of the community, he was also a good member in the Episcopal Church, and was a fine father and husband. Of the nine children, there were still six at home. Henrietta Eve was constantly busy with the household and the children. Keeping the family fed and clothed occupied her, but the girls as well as the boys were very helpful. She, like her husband, read a great deal at night while the children did their homework or slept. Frank, at six, kept the family amused with all of his bright sayings. He was reading simple books before he went to school, and his teacher talked to his father about how they would keep him busy while the other children in his class were learning to read. They decided to place him in the second grade because he was already large for his age. Even in the second grade he was a much better reader than most of his classmates, and was given special reading assignments to keep him busy. He seemed to breeze through school with no trouble and was able to get along with the other students. He was the kind of student who gave great satisfaction to his teachers but was not considered the teacher's pet by his classmates. He was also very good at sports that they played, including baseball and football. His brother and his father would often take him fishing in the rivers and lakes in the area, and he would enjoy fishing for the rest of his life.

Ben, Sr.'s day consisted of getting up fairly early, dressing, reading a little while Henrietta was getting breakfast, and chatting with the family. They usually had a large breakfast: eggs, grits, ham, bacon or sausage, coffee for the adults, and milk for the children. Ben, Jr. had already been up and milked the cow by breakfast. Ben, Sr. would walk to work, as it was only about a half-mile to the high school and his office. The children would usually walk with him. They had a horse and buggy but would use it only if the weather was bad, and Henrietta would come along to bring the horse and buggy home.

Ben would usually start the day with a short meeting with the teachers in the high school to discuss how certain problem children were doing and how to help the situation. Discipline was quite strict, but Ben did not believe in corporal punishment; he could usually remedy most situations with a chat in his office with the erring student. He had a secretary who would write his letters and notices by longhand, as the typewriter, although in use by many large companies, was not yet commonly used in small offices.

He typically scheduled several appointments with parents, teachers, and people in the community; then he would walk home for lunch. The children usually brought their lunches to school. He would return to school, teach his history class, and then visit the grammar school to see how it was doing. He often had meetings with the town council and had to defend his ever increasing budget. He would get home late in the afternoon, work in his garden a little, and then have a large supper with the family.

Henrietta, his wife, was forty-five in 1890. She had been the mother of nine children; the youngest was Frank, age six. She was serene, an excellent mother to all, and was just beginning the change of life, but the symptoms did not affect her emotions or her care of the children. She and her husband were considered the ideal couple by the community. On

Sundays when everyone dressed up and went to church, there was much admiration by the parishioners. Time slipped by and Frank, having finished the second grade, had a good summer at home with camping trips with his older brothers, and fishing and swimming in the river. He was ready to begin the third grade in the fall.

Late fall, 1891: Ben, Sr. began to feel the pain. He had been healthy all his life. He had had no serious illnesses except for the wounds which he had received twenty-five years earlier in the Civil War. He went through three episodes of being wounded, but he recovered, with only a few scars.

The pain was in his lower back on the right side. He did not pay much attention to it at first, thinking he had strained his back while gardening. But after a month the pain got worse and he was having trouble sleeping at night. It eased in the day. Finally, he went to Dr. Harbin, who thought it might be a kidney stone and prescribed some kidney medication and pain medication. Ben was able to sleep a little better, but the pain continued and at times worsened, and later there was a show of blood in his urine. Dr. Harbin prescribed a different medication as he continued to believe that this was a kidney stone. But during the next few months Ben became worse and, although he went to work every day, he was beginning to lose weight, was tired all the time, and lost his appetite.

Henrietta worried. She tried to coax him to eat more, to get more rest, but it was now obvious he was a sick man. He continued to work throughout the spring and summer, but by the fall of 1892 he was quite thin, in pain most of the time, and Dr. Harbin told him frankly that he probably had cancer and not long to live. The next few months were very distressing to both Ben and the family. He was able to get around a little until about the first of December when the pain was so severe that Dr. Harbin prescribed morphine. Ben had never been a drinking man, but Dr. Harbin recommended whiskey in the day to relieve pain and morphine at night, first by mouth and later by injection. The last two weeks of his life were miserable, and fortunately he went into a coma for about five days, developed pneumonia, and died December 19, 1892.

The whole family gathered. Although it was a sad, sad day for all, the death affected eight year old Frank the most.

12
Christmas and the Family

Christmas 1847 at the John Neely house in Easton, Maryland was not a happy time. Everyone missed Ben very much. Father John still thought Ben might be coming home for Christmas. Euphemia, the oldest girl at fifteen, was the most unhappy as she felt she was the cause of Ben's running away. Mary was eleven, John, Jr. nine, Elizabeth seven, Maria six, Annie four, and little Julia was three. It was a very large family for Isabella, John's second wife, to take on. They had been well brought up by their real mother, Susan Henry Neely, and would usually do what they were told. But it was still quite difficult to adjust to a strange mother who had just come into their lives two-and-one-half months earlier. Things were improving, however, and they were slowly accepting Isabella as their mother, just as she was becoming much more relaxed with the children. A few picnics helped to break the ice.

John was close to the end of his studies for the ministry, and he was very busy as principal of the Easton Academy. The school had grown remarkably since he had come there. Students from Maryland, Virginia, and Delaware were attending the academy. A preparatory school for college, it also included a military department, with many of its graduates moving on to West Point. John attended educational meetings in Washington, D.C., and he had met people from all over the country. At one of these meetings he met some people from Augusta, Georgia who approached him about helping to develop Richmond Academy, the first high school in Augusta.

Preparations for Christmas were beginning, and despite Ben's absence, the children were getting excited. They had been released from school for the holidays. There was a fine new snow on the ground. The ponds had frozen over. John took most of the children at home out into the woods to chop down a Christmas tree, and they were very enthusiastic over this tradition. Even Julia, the baby, was brought along, carried by Euphemia. They didn't have far to walk as the house was on the edge of town, and there were woods all around. John, Jr. was the only remaining son, and he helped his father while the girls watched. Finally they got the tree chopped down, and pulled it through the snow to home. They dried it off, took it into the house, and began decorating it. Cookies had been made; chains of popcorn and ribbons were attached. One large candle was put on top.

"T'was the night before Christmas and all through the house, not a creature was stirring, not even a mouse" ("A Visit From St. Nicholas," Clement Clarke Moore, December 1823). The children went to bed happily in anticipation of the next day. John and Isabella made sure there were presents for all on Christmas morning. John made a big fire in the fireplace, and the children came in and began opening presents. As there were so many girls, there were a great many dolls, dresses for dolls, and doll houses. Isabella and Euphemia were preparing Christmas dinner in the kitchen. John had invited his two cousins for the dinner as they lived in Baltimore and did not have any children themselves. Finally, the dinner was ready, and they all gathered around the table, except Julia who was in a high chair. John, Sr. said the blessing: "Thank you Lord for all our many blessings—for house, for home, for warmth; for schools, for churches; for friends and for family. Also, Lord, we thank you for friends and family not here." Here his voice cracked. The children whose heads had been bowed glanced up, and they knew that John was thinking about their own mother who had died two years ago, and about Ben who was not there. Euphemia and Mary let out muffled sobs. John, regaining control, finished the blessing firmly, clearly: "And thank you Lord for the things to come because we have faith—faith that one day our prodigal son will come home and be with us at Christmas."

Little did any of them dream that Ben was 3,000 miles away aboard the *Nerrisa* on its way from Liverpool to Glasgow. Ben and Bjorn had finished Christmas dinner about six hours earlier. They had cooked two turkeys and a goose, serving the crew and officers a sumptuous meal with potatoes, cranberries, apple pie, beer, and wine. The weather was quite cold, but the winds were favorable. The sea was calm. They expected to get to Glasgow the following day and would stay several days for loading—this time woolens and looms to be used in the States. So when John and his family were sitting down to Christmas dinner, Ben was about to go to his upper berth in the forecastle after a busy but happy day. It was not until he got into bed that he really thought about Christmas at home—then a surge of homesickness came over him; he fought back the tears.

Christmas is a great time for families to get together, to renew old ties. Children will make any effort to get home to join in the family customs and traditions. Ben would have his own large family Christmases in Augusta and later in Rome. Little did he know then that thirty-six years later in 1883 in Rome, he would be carving turkey for his family. The oldest son, Allen, was sixteen at the time, and the youngest, Frank Henry Neely, was eleven months old. It was a happy time for the family. All of the children except Mary and Frank were in school. Nine children had been born in the course of fifteen years; all were healthy. Ben said the blessing just as father John had done many years before.

But Ben had nothing to spoil his Christmas in 1884. He had been at his job for two years now, was well accepted in the community, and was working very hard and enjoying it. Henrietta, as always, was extremely busy with the family affairs but still had time to read and do some painting. Ben and Henrietta loved each other very much, and this love spilled over on the family. The children realized how devoted their parents were, and that made them feel safe and serene, as well as loving toward the world and their fellow man. Even Frank, not yet a year old, sensed this warm and loving relationship within the family. Also, a large family has a tendency to cure any selfishness that any one child might begin to develop. The working of each to help the other simply smothers out any selfishness, causing one to realize that there is no place for selfishness. Perhaps there will be a change in our society since families are now so small. Perhaps this smallness will give way to greater selfishness. But with the large family environment works, children learn not to be selfish.

Whither away, oh family of Benjamin Neely. Here you are in 1884, all of your nine children at home. Will they ever all be at home for Christmas again? Or will they soon begin to drift away? Their final destinations might surprise you, Ben, but let us take a quick glance at the future, forward from that Christmas day.

Allen, the oldest, will end up in Colombia, South America on a plantation and die there of yellow fever. Annie will marry a hardware merchant in Cedartown thirty miles away from Rome and produce a large family (of seven children). John will become a doctor, have two children, and practice in South Georgia; Elizabeth Eve will die in childbirth with her first child at age twenty-seven. Ben, Jr. will become an insurance salesman in Atlanta and have four children. Hannah will also have four children and live in Augusta; Edgar will have three children in Atlanta. Mary will have four children and live in Alaska a long time, while Frank will have only one child and will live in Atlanta. So, Ben, you will have twenty-three grandchildren; but unfortunately, you will not live to see a single one, and they will never get to know you.

Ben and the family would have a happy Christmas in 1884 and would have several more with the whole family. Thus, the Christmas dinner of John and of Ben 36 years later were very much alike: large happy families with many children—Christmas trees and presents for all—a time to count blessings. However, between John's Christmas in Easton, Maryland, in 1847 and Ben's Christmas in 1884, the Civil War exploded. The best young men would die in the war. Reconstruction was still going on, but the economy was still relatively poor. Sherman had sent some of his troops through Rome; quite a bit of the town burned, but it was almost rebuilt now. Everyone had helped each other with house-building. It was now sixteen years since the end of the war. The Federal troops had been pulled out, although there were small groups of Federal soldiers coming through at times. The blacks were free, but their lot had not improved very much. Now the blacks worked on the farms, earning low wages. There were a few small factories in which they also worked. They lived in the same little houses, but now they paid rent of perhaps $1.50 per month while receiving wages of only $10.00 to $15.00 each month. Ben convinced the city fathers that black children should be in school. Later, Ben also had the black adolescents in the Neely High School, but they were few in numbers because most of them dropped out of school, working to support their families.

Christmas day in 1884 was a high point in Ben and Henrietta's life as well as the children's.

The next generation down produced only one large family; this was the family of Annie Neely and Augustus Young in Cedartown. Annie married Gus in June 1897. She very soon began their large family of five girls and two boys. Her husband owned and operated the only hardware store in Cedartown. He was well-known throughout the county and surrounding counties for his honesty and fairness in dealing with people. Annie was known as a lovely, gracious lady as well as a wonderful mother. So here was another large family, and they had a similar Christmas family party in 1915 with all of the children gathered around. Life was beginning to change a little. The horse and buggy was still used as the main mode of transportation for short distances. There were a few automobiles in Cedartown, but they were a rarity and a great curiosity. Annie encouraged all of the children to do well in school in the tradition of her father. She passed along some very good genes, but the boys did not seem to get the genes of ambition. Several of the girls did get those genes, but in those days, there wasn't much opportunity for a female to do much in the world except teach. Several of the girls went to college and several did become teachers. But no one in this family had the fire and drive of Frank Neely. However, in the next generation, there was some resurgence of ambition, as we shall see. Does ambition skip a generation at times? No one really knows.

There weren't many large families in the fourth generation down from John, Sr. The largest was probably the Levering Neely family, which had six children. The son of Edgar Neely, Sr., Levering and Margot Gardner Neely produced six children—three boys and three girls. What about Christmas at the Levering Neely house when the children were about fifteen to five years old? This was in 1958. The changes in life-style over the years from 1915 to 1958 were probably the greatest changes the world had ever seen. The advent of the automobile began these changes. It was slow at first—everyone distrusted them, but in the 1920's, when Henry Ford began producing his Model-T at a reasonable price for a family of modest means, the explosive change was on. Then electric lights, good running water, telephones, radio, and air travel came hurtling into their lives, changing them completely.

But did the tradition and customs of Christmas change much? Was there still that warm inward glow of all on the morning that the children got up and opened their presents? Yes, certainly in the Levering and Margaret Neely family everyone looks back on that Christmas morning when the children came rushing out to open their presents. Perhaps the custom of the big Christmas dinner was changing—sometimes it might be held on Christmas Eve or on Christmas night, but there was usually the large turkey and big dinner just as there had been with John Neely's family in 1847.

And to follow down one more generation—John Longstreet Neely—this being the fifth generation from John, Sr.—had his first family Christmas with Carolyn and the two children, Jordan and Thomas, in their new home in Savannah, Georgia in 1982 with little Thomas, eleven months old, just as his great-grandfather's brother Frank had been brought to the table for Christmas dinner ninety-eight years before in 1884.

How long will customs and traditions of Christmas continue in this country? No one knows—but certainly the families are getting much smaller. Now, one or two children are typical, while large families are almost a thing of the past. What will this do to the genes and chromosomes? Will it be helpful to have so much attention for one or two children or will they miss the benefits of the large family so much that they will become selfish or spoiled? Of course it is also quite possible that this extra attention will help the child. It is much like the argument of private schooling versus public schools. At the private schools, the child gets the extra attention and is forced to study and learn more. In the public schools, the child does not get so much attention, but he mingles with all kinds of people, and this in itself can be a good education for later life. The answers are not yet known.

* * *

Ben's Christmas had been aboard ship. The *Nerissa* plowed on toward Glasgow. The Captain and Mr. Rumans arranged the trip so that they would arrive in the middle of the day. As they approached, Ben was standing on the forward deck and leaning out. Here was another new world for him to see.

13
John Carmichael Neely, Physician

John Carmichael Neely, physician, born November 11, 1870, third child and second son of Benjamin Neely—what manner of man were you? Born in Augusta, Georgia, you moved to Rome with the family when you were twelve. You were named for your grandfather John, started into the sixth grade grammar school in Rome, and did well, transferring to your father's high school in 1884, the year Frank, the baby, was born.

You developed a fascination for railroads. Rome had a large railroad yard where they did repair work on engines and railroad cars. Rome was a small railroad center on the main Southern Railway line from Atlanta to Cincinnati, with spur lines from Rome to northern Alabama, Copper Hill, Tennessee, and Knoxville. While in high school, you would go down to the railroad yards every day after school and watch them repair the engines and switch the cars around in the yards. You spent so much time there that soon you were known around the railroad yard by most of the workers.

The repairmen and switchmen occasionally needed a little help; you gave your time freely, enjoying every moment. Working around the railroad was heaven to you, and as you advanced into high school, your interest in railroads did not waver; it became stronger as you came to know more about them. You watched as they took an engine apart and put it back together. You knew the name of every engine that came through Rome.

You had taken the train ride from Rome to Atlanta and back, and this was one of the great experiences of your young life. You worked in the grocery store on weekends, saving your money to take a longer trip. You continued to have such a strong interest in railroads that you decided that you would one day be a railroad engineer and possibly travel across the country; the railroad to the West coast had been completed.

Your father, Ben, at first discouraged this ambition. However, Ben soon saw how determined you were and decided to let you follow up with your interest as much as you desired. Ben had hoped you would go to college, learn to teach, and possibly follow in his footsteps. His eldest son, Allen, was somewhat erratic and had not done well in school.

In 1887, while you were a junior in high school at age seventeen, something happened which changed your plans and your life.

You still went to the railroad yards almost every day after school. The chief switchman, Mr. Murphy, liked you and had let you help with the switching of trains. It was all done by hand in those days, and it meant lifting a heavy piece of metal weight and rotating it—the switching was effected, and the two rails were thus changed to another track. You had helped Mr. Murphy many times, and as the switchmen would often have to run rapidly to change the switch, Mr. Murphy let you do it by yourself at times. You were a husky lad and could lift the switching mechanism easily.

One day in October, 1887, they had to switch a freight train onto a side track while the express train from Cincinnati to Atlanta came through. Mr. Murphy was standing at one switch and John at another, as both needed changing to let the express come through. Mr. Murphy had a flag he would raise when he wanted John to change the switch. John stood ready beside the tracks. The freight train had been put on the siding, and it was time for John to fix the switch so the express would go through. Mr. Murphy raised the flag: it was time to throw the switch; the express train was a little early and coming through the yards rather fast. John tried to raise the weight to throw the switch, but it was somehow stuck and John had to put one foot across the tracks to get better purchase on the weight to be lifted. He was finally able to raise the weight and throw the switch to the correct position. In doing so, his right foot was caught between the track of the switch, and in spite of jerking his foot as hard as he could, it remained stuck!

The train bore down.

He made one desperate effort to reach the weight of the switch and loosen it a little; but he could not reach it. By now the train was only fifty feet from him and the engineer saw him and slammed on the brakes. John lay down with his body away from the track, but he could not extricate the foot.

The train took 100 yards to stop; Mr. Murphy and the engineer rushed to where John was lying on the ground. The wheel had cut off his foot cleanly above the ankle. It was bleeding but not badly. Mr. Murphy took off his shirt, wrapped up the stump, and kept pressure on it; meanwhile he called for help. Soon there was a large crowd of people around. John was as white as a sheet, but he was not crying in spite of the pain.

They brought a litter from the station, put him on it, took him to the station, and then someone brought a horse and wagon. They galloped to the Harbin Hospital's Emergency Room. Dr. Harbin was there. He gave John an injection of morphine and took him into the operating room immediately.

In 1842, Dr. Crawford W. Long in Atlanta discovered that medical students who smelled ether became insensitive to pain—they would often have ether parties, and some students would fall down the stairs and break a leg with no apparent pain. In 1842, Dr. Long removed a tumor from a patient under ether, and the patient felt no pain. But Dr. Long did not report his findings in the literature. Meanwhile, Dr. Martin, a dentist at the Massachusetts General Hospital, performed dental surgery on a patient under ether with no apparent pain. Dr. Martin did write up the case, and he is credited for the discovery of the value of ether as an aesthetic. Thus in 1887, about forty years after this discovery, ether was in most hospitals; nurses were putting patients to sleep with what was called drop-ether. A cloth mask was applied over the face, and ether was dropped on the cloth, just enough to keep the patient asleep.

When they wheeled John into the operating room, a nurse immediately put him to sleep with drop-ether. The operation took about one and one-half hours. The arteries and veins were tied to stop bleeding and the bone sawed off evenly. Of course, the big danger was infection after surgery. John awoke and was sent up to a big ward. He was a very observant and friendly boy, and he soon knew all of the nurses, doctors, and other patients and their ailments. So here was the beginning of a new interest which he would pursue the rest of his life.

John had much pain the first few days after surgery. The wound looked clean, was dressed and bound with a huge bandage. However, on about the seventh day after surgery, John had a nightmare, fell out of bed striking the stump on the floor, and causing a bad bruise. The next day he began to run a fever and Dr. Harbin found that infection had set in. No antibiotics were available in those days. As Dr. Harbin watched the infection begin to spread with red streaks going up the leg, he knew that if he didn't do something John would lose his life. He called in Ben and Henrietta and told them the bad news—that he would have to operate again, removing more of the leg, this time with the amputation above the knee.

In the operating room John was put to sleep once again, and the amputation performed. This time, he did well—the wound healed slowly, but properly. They kept him in bed for two weeks; they thought this would give the wound the best chance to heal.

During this time, John had a chance to really observe the other patients, nurses, and doctors. He saw patients with broken bones come in and have the leg or arm put into a cast. He saw men come in with heart attacks.

Some recovered, some died. He watched as men with pneumonia seemed to go through a crisis and miraculously recover. He saw people with large tumors who, after surgery, were sometimes cured completely. He observed old people suffering from strokes who were unable to talk or move one side of their bodies. He watched as Dr. Harbin and the other doctors did their best to remedy those illnesses. Sick children would come in and die, and he saw the grief amongst the bereaved relatives. There he saw life in the raw. Instead of being repulsed, he became fascinated with the drama of the hospital.

After two weeks in bed, he was allowed to get up and was given two crutches. He would use a crutch for the rest of his life and learned to get around quickly and with ease. He later had a mahogany artificial leg made, but he never really used it. He preferred the crutches, and later just one crutch.

After Dr. Harbin saw the wound was healing well, he let John go home. He had been in the hospital almost six weeks, but now he had a new ambition. He knew he could never be a railroad engineer; now he was determined to become a doctor. His father encouraged him. He had one more year of high school to complete, after which he would go to Athens, Georgia to the University for two years, and then to the medical school in Augusta, finishing by 1893. His father helped him financially at Georgia and later at medical school, but John worked every chance he got, thus earning money to pay his own way.

In 1891 his father became ill. John was in his second year of medical school, with two more years still ahead. When visiting home, he often went back to the hospital and made rounds with Dr. Harbin. Dr. Harbin would let him come into the operating room and observe. He also let John help with delivering babies, and he often would act as Dr. Harbin's assistant. He even delivered one baby alone when Dr. Harbin could not get there in time. John was in his last year of medical school at Augusta when his father died in 1892. It had been a long and painful illness, and John had come home from school as frequently as he could to see if he could help his father and comfort his mother.

Allen, the oldest son, had already gone to South America with the coffee company. The family received letters from him time to time, but they were often several months late and communication was very unsatisfactory. Ben, Jr., just nineteen, had not gone to college, but had begun work in an insurance agency, the New York Life Insurance Company and had been sent to Atlanta by the company. He was the business man of the family, and he was the one who helped the other boys, Edgar and Frank, to get an education after their father died. He also helped John with that final year of medical school although John was working at night and on weekends in order to support himself. Following Ben, Sr.'s death, the family was very disorganized, but Ben, Jr. took over, saw that his mother had enough money, that the other children continued their schooling, and that the family stayed together as much as possible.

After John finished medical school, he did not have enough money to open an office on his own. Upon hearing of a job as doctor for a large saw mill community in south Georgia, near Bainbridge, he went down by train, was interviewed, and offered the job. It entailed caring for about 500 saw mill workers and their families; it paid a salary which, to John, sounded very good. He talked it over with his mother, Ben, Jr., and the rest of the family and decided to take it. This meant general practice in the community. He had to treat many injuries, deliver babies, and care for the sick and dying, all of which kept him busy from morning until night and often into the night. Having only one leg did not hamper him; he had learned to get around remarkably well. He used only the one crutch by now, and was so skilled at using it that he could run quite fast when he had to. His arms and torso were very well developed. And with the help of the one crutch, he was able to mount his horse very easily and make his rural calls.

After about four years at the saw mill, he heard of a position at Iron City, a mining town on the border of Georgia and Alabama. It was a better paying position practicing medicine, with his own office and the possibility of charging the patients for his services. One and one-half years later he met Mary Godbee in Bainbridge, and she persuaded him to begin practice in Bainbridge, a larger town. He married her in 1902 when he was thirty-two. Their first son, John, was born in 1904, and the second child, Bess, was born in 1906. In about 1912 John bought his first automobile to make his rounds. It was one of the first automobiles in Bainbridge.

The family got along well in Bainbridge. John had a very busy practice; he was well respected in the community.

He had been a pipe smoker for many years and during his rounds on horseback and later in the car, he always had the pipe in his mouth. In 1912 he noticed a small sore on his tongue. It would not heal and seemed to grow. He first went to Atlanta to a specialist who took a biopsy and told him it was cancer. After this he had a series of operations on his tongue, but the growth kept coming back. There was a well-known cancer center in New York; accompanied by his wife, he went there for further treatment. He was advised to have a major operation which would include removal of glands in his neck. He decided to have this done; the next day he died on the operating table at the age of fifty-four.

Young John and Bess had been left with their negro mammy, Aunt Jane, and a maid named Jeanie. Bess was out playing with her dog Snip when the news came by telegram. Jeanie finally found her and cried out, "Fo' God Miss Bess, fo' God, your paw is dead." Jeannie's weeping and wailing could be heard all over the neighborhood, and soon Bess joined in the crying. Sixty-eight years later Bess Neely—who had established a successful career in interior decorating in Atlanta, Georgia—could clearly recall the cries of their maid Jeanie when she came to bring the shocking, unexpected news. And Bess could still remember her dog, Snip. Seventy years later.

14
Ben's Voyage Back; Frank as a Child

The *Nerrisa* experienced a cold but calm passage from Liverpool to Glasgow, with a Christmas dinner aboard ship. Ben and Bjorn worked hard, giving the crew a feast. It was a jolly time for all, with good food and free ale and rum. There was much toasting with loud and hilarious talk. Ben, who still would not try any alcohol, was the only sad one. The memory of the many happy Christmases at home caused a lump in his throat. But he was careful not to show it.

The following day, the 26th of December, 1847, they came into the Glasgow harbor. It was a large, protected harbor with many ships. The harbor master had several little paddle wheel boats which would pull the sailing vessels to the right dock. One came up, threw a line to the *Nerrisa,* and towed her into the right spot. It would take several days to load.

The crew readied the ship for the long journey across the Atlantic. They replaced the sails, checked all ropes and winches, and got everything shipshape for the long voyage to St. Johns, Newfoundland, and then the States. Ben and the cook were also very busy buying enough food for the trip. They had to feed the crew and did not have much time to go into town except to shop for provisions. The trip to Newfoundland might take two to four weeks. They bought mutton, large amounts of cheese, bread, and oatmeal; also they loaded up on coffee, tea, and ale.

Glasgow was the largest port on the Atlantic side of Scotland, as Edinburgh was on the east coast. The people were extremely friendly, but their accent was difficult to understand. Ben was amazed to see all the kilts, to hear the bagpipes. Ben had never had a Southern accent, having come from Maryland, but he enjoyed listening to the Scottish brogue, and he had a knack for picking up the different sounds, mimicking them. This ability would be useful to him when he was in the Civil War; he could change accents at will and became a spy for the South, at times going behind the enemy lines.

Soon the *Nerrisa,* loaded, was ready to start the long voyage across the Atlantic. The plan was to drop off clothing and potatoes in Newfoundland and pick up a load of salmon from the Newfoundland fishing boats. The weather had turned bitter cold with winds and storms to be expected. The north Atlantic might have a few icebergs, but the season for icebergs was more in the spring and summer when thawing caused the ice to break up. They would still have to be on constant watch in the day and slow down in the night.

The crew manning the sails had to wrap up very well with heavy clothes, gloves, and caps because of the freezing weather conditions. There was ice on the rigging and a constant danger of falling. However, the crew was quite experienced about just what might be dangerous.

Ben felt quite happy about going back to the United States, but he felt sad when he realized that he could not go home again, as he still thought that he had done something that would not be forgiven by his father. However, the ship had now become his home and the crew his family, and he was beginning to feel fairly secure in his job.

It was a long, rough journey across to Newfoundland. The ship, famous for being fast as well as seaworthy, was blown off course several times, and Mr. Rumans and the captain stayed very busy shooting the sun in the day and the stars at night—that is, when the stars were out and they could see them. There were days when they did not make any progress; the whole journey took about three weeks. The crew was tired, uncomfortable, and everyone was happy at the first signs of land: seagulls coming out to meet the ship.

St. Johns, the capital of Newfoundland, has a fine protected harbor but a very small harbor entrance. In the seventeenth and eighteenth centuries there was a big chain that could block the harbor so no enemy ship could get in. The navigation going through the entrance had to be carefully planned; Mr. Rumans and the captain had done it many times before, and they knew exactly how to sail in with the tide to avoid any danger of shipwreck.

St. Johns had been a small village in the sixteenth and seventeenth centuries, but had grown by 1848 to a population of about 40,000 people. The land was fairly fertile, but the growing season was short. With excellent fishing both in the bay and in the ocean close by, there were numerous fishing boats at the dock. The *Nerrisa* was to pick up many barrels of salmon and other fish and take them to Boston and New York, remaining in port only about three or four days. Several of the *Nerrisa's* seamen came from Newfoundland. They had not been home for several months now, and they planned on staying to join their families and then taking another ship. However, it was not difficult to sign on good seamen to replace them, and Mr. Rumans interviewed several before choosing the ones he wanted.

Ben, meanwhile, remained fairly busy in the galley, but he did have a little time off. He and the cook did a little more fishing, again coming back with a good catch. The weather remained quite cold, but not bitterly cold; the port was well protected from the wind by the mountains.

* * *

Newfoundland (pronounced New-found-lánd, contrary to how most people pronounce it) was discovered by John Cabot in 1497, sailing under the English flag. It became extremely important because of an abundance of cod fish in the area. A rivalry developed among the fishermen of France, Portugal, and England. There were practically no settlements until the eighteenth century, when it became a British colony, but the French still tried settling in certain areas. Finally, the British sent in a large group of laborers to work in fish processing; and in 1814, 11,000 people came in, mainly to St. Johns where the great harbor was. These immigrants were mostly Irish Catholics, a divisive force against the Protestant middle class. It remained a poor county, but was a training ground for British seamen.

* * *

Frank Neely, age nine, had lived through his father's death fairly well. It was extremely traumatic at first, but after the funeral, things seemed to settle down. He was now in the third grade of the public school at Rome which his father had founded. Frank had been extremely fond of his father and felt the loss deeply, but with all the support of his mother, sisters, and brothers, he entered back into school life with great enthusiasm. His teachers, realizing that he was ahead of the class, gave him extra work to challenge him.

His third-grade teacher, Miss Fitzgerald, was a hard task master. The students disliked her, although she achieved good discipline and somehow made the children work hard. She insisted that the children learn how to spell, write a sentence, and do fractions. The latter was usually begun in the fourth grade, but she thought the children should be prepared for the fourth grade. When she gave an assignment, the children had to finish it before they went home in the afternoon, and there was a great deal of complaining when they had to stay late. In fact, the families of the children constantly complained about the children staying so late. Still, she would not give in, and when the children finished the third grade, they were well prepared for the fourth.

Frank took the third grade in stride. Having been taught to read by his sisters, he was well ahead of the class at the start. Also, he really enjoyed arithmetic. He knew the multiplication tables early in the third grade and was able to add and subtract; when he came to fractions, he quickly mastered them. Miss Fitzgerald recognized that Frank was well ahead of his class, so she gave him extra difficult assignments which he seemed to thrive on. He always did his homework right after supper and would never play with the other children until this was done. There was never a question of being the favored child of the class. He and Miss Fitzgerald had a mutual respect and admiration for each other, but they never showed it. So the other children in the class were not resentful, but they recognized that Frank was usually ahead of them. Very early in his career at school, it was obvious that he was very bright; but in addition to this, he got along well with his fellow students. He was able to laugh and make jokes with the other students and was well accepted by them.

* * *

Frank at age five was extremely observant of both his siblings and his parents. He could sense their moods, their wishes. He did not talk much at the dinner table, but listened to everything attentively. His memory was astounding. All of these things were helpful in his school work. The older brothers also coached him after he had been to school.

His next older brother, Edgar Adams Neely, was fourteen years old when Frank was eight and in the third grade. Edgar was red-haired, full of vitality and love of life; he was a high school student doing very well. He would go over Frank's special homework assignments at night after Frank had finished. Edgar drilled him, pointed out mistakes, and gave him a good review of the problems that had been brought home.

Thus, every one in the family made an effort to teach Frank what they thought was important. His mind was developing very rapidly, and he was able to pick and choose, and learn what he thought was important. He was soon able to read rapidly, write legibly. He also could think clearly, quickly. It was as if he was able to learn something from everyone in the family. Also, he was able to judge other people by comparing them with someone in the family.

He developed a strong sense of honesty at an early age; and this was to be a great asset in the years to come, when he was dealing with influential people who sometimes can bend the truth a little in their favor. But not Frank. Later, if any of his employees gave an indication of dishonesty, they were immediately discharged. How this trait developed in him is difficult to say, but it was probably from observing his large family. Frank had a smouldering fire of honesty in him, and any suggestion of dishonesty would set off an explosion. Other qualities developed while he was in grammar school: once he had decided what his course of action should be, he would not stop until he finished a job. These tough qualities, a great ability to get along with people, as well as having a quick, curious mind and a good sense of humor—all these things started him on a fruitful career. In addition to these characteristics, he inspired other people to do their best.

So when he finished grammar school at age twelve in 1896, he was quite mature for his age. He transferred to high school, and during the next four years changed from a boy to a man.

Frank's best subject was still mathematics; he was just beginning algebra and geometry. He had been coached by his sisters in algebra, so he was ahead of all the other students. In addition, algebra and geometry suited his quick and orderly mind. The mathematics teacher, Mr. Cromartie, who had been there about ten years, was notoriously tough; Frank made the highest grade ever given by him. Frank was also interested in science and did well in chemistry and biology. Toward the end of high school, he was still getting all A's and was first in his class.

He had heard about the Georgia School of Technology in Atlanta. The school, known for its excellence in technical studies, was a much better college than the University of Georgia. He matriculated at Tech in the summer of 1900, was accepted, and began school in the fall.

15
Robert Allen Neely, Colombia, South America

Robert Allen Neely: born August 13, 1868 and died at age thirty-two on November 19, 1900 in Bocas del Toro, Republic of Colombia, South America. Bocas del Toro is now in the Republic of Panama.

You were the oldest child of the Benjamin and Henrietta Eve Neely children. You were called Allen. You were born in Augusta, Georgia right after the Civil War. Times were hard. Augusta had been impoverished and reconstruction had not started. Your father, Ben, did have a job teaching in the school system that his father had begun. Ben and Henrietta had been given "the Cottage" on the Carmichael plantation near Augusta by Henrietta's father, Anderson Carmichael, a hardware merchant. Your grandfather, John, was delighted at your birth and would later baptize you.

You, Allen, are the enigma of the family. We have almost no concrete information about you. All of your younger brothers and sisters married and had children, and we have some handle on them—we have a fair amount of information about the lives of most of them, enough at least to make some inferences as to what kind of person each was. What we know about you, Allen, can be said in just one sentence—Robert Allen Neely was born in 1868 and died in 1900 in Bocas del Toro, Republic of Colombia, South America, of yellow fever. You lived to see almost a year of the twentieth century and then died at age 32 far away from home. There is no one alive who knew you. There is almost no one alive who had heard anything else about you, Allen, and so we will have to make some fairly wild conjectures.

First, you inherited good character traits. Your father Ben gave you intelligence and possibly some of his emotionalism, including sudden flare-ups that, in Ben's case, changed his whole life for a time. From your mother, Henrietta Eve Carmichael, you picked up genes from the Longstreets, the Eves, and the Carmichaels. You came into this world with an excellent genetic heritage. And you were the first born, which may be important. The oldest son in those days usually followed in his father's footsteps. But not you. You must have had a deep adventuresome spirit within you to leave family and friends and to go off to make your way in a country which at that time would be classed as very backward and underdeveloped. You must have been independent, that is for sure.

You were a child of the reconstruction in Augusta. Devastation covered Augusta; even though Sherman had not come through, there were Federal troops stationed in Augusta for about eleven years, from 1865 to 1877. There was still much enmity against the North and what they were doing. There were secret organizations such as the Ku Klux Klan and the Knights of the White Camellia. We don't know for sure whether Ben belonged to any of these organizations, but he certainly knew about them. From his later actions concerning helping the black race to gain education, we think he remained neutral or perhaps spoke out against these groups.

You, Allen, as a child must have had some idea as to what was going on. Children don't have prejudices built in; they pick up the ideas of their peers, and when you were in grammar school, your schoolmates must have been talking about the Federal troops and the blacks. So for the first nine years of your life, Allen, you were living in an occupied country. Surely you realized this, and you probably saw some violence, some intimidation by the Federal troops against the local citizenry; and then there must have been many problems with the newly freed slaves.

We know that you, Allen, were baptized by your grandfather, Reverend John Neely on June 9, 1871 when you were about three years old. At the same time, he also baptized your sister Annie and your brother John Carmichael Neely. This was one of the happiest days of his ministry, and he died less than a year later in March 1872. The children of Ben and Henrietta came very fast after you with eight born in eleven years; there was then a five year hiatus before Frank was born. Frank's personality was affected by his position in the family, and we think that your personality also must have been affected by your place in the family but perhaps in a different way.

The family became so large so quickly that you could not have had the individual attention that Frank must have received. Rather, we think perhaps you did not really get enough attention, yet very soon certain things were expected of you. You came from three generations of teachers also. So first, you were supposed to make very good grades in school.

Your father Ben had organized the first girls' high school in Augusta and was superintendent of both city and county schools.

We suspect that you did not make quite as good grades in school as the rest of the children in the family, and this must have subconsciously upset you, perhaps causing you to rebel in a small way. Ben, your father, although a good family man, felt strongly about his children making good grades. You were fourteen when your father resigned from his job in Augusta and moved the family to Rome to organize the schools there. This must have been hard on a teenager, to be pulled from one environment and put into an entirely new one.

However, we think you did finish high school in Rome, and then decided to get a job in Atlanta. This was in about 1886 when you were eighteen. We think you labored at several different jobs, including working in a bank, which was rather dull. You heard of your father's adventures. You made up your mind to find something else to do. After your father, Ben, died in 1892, you sent money home to your family.

In 1894 at age twenty-six you heard through the bank that there was a chance to work in Bocas del Toro, Republic of Colombia, South America; you jumped at the chance.

Bocas del Toro, a small seaport town on the Caribbean Sea in the northwest part of the Republic of Panama, in 1894 belonged to Colombia. In 1903 Panama revolted, seceding from Colombia. Panama was recognized by the United States and provided us with the ten-mile-wide Canal Zone. Colombia, itself, began as a republic in 1819 when the area was liberated from Spain after 300 years of Spanish rule. The history of the Republic of Colombia, since its beginning in 1819, had been stormy. There had been a long struggle between the conservatives and the liberals with the conservatives in power from 1880 until 1930. It was because of a long civil war, the War of a Thousand Days—1899 to 1903, that Panama was able to secede.

Most of the people of Colombia were of Spanish descent, although there was a slow mixing of the Indians with the 25,000 black slaves who were freed in 1851. The Roman Catholic Church had always been strong, but as early as 1860, a revolution developed. The government, headed by Mosquera, expropriated the church lands and adopted a constitution guaranteeing freedom of religious practice. Nevertheless, the Catholic religion dominated.

Panama in 1894, although a part of Colombia, desired independence. It was a long way from Bogota, the capital of Colombia. Also, it was separated from the densely populated part of Colombia by three high mountain ranges. Communications were bad. There were no good roads and Panama was vastly different from Colombia, which was almost landlocked from the Pacific by the mountains. So if a revolt were to take place, it was a long time before Colombia would hear about it and a long time before they could respond. When the revolt did come, Colombia was involved in several local border battles. The country simply did not have the time, money, or inclination to cross the mountains and try to hold onto Panama. So Panama became free by treaty, not warfare.

Panama had well-protected seaports on both the Atlantic and Pacific Oceans. It was tropical with a long rainy season. It had a fairly high range of mountains running down the center of the country and large areas of fertile volcanic soil which were good for growing bananas in the lowlands and coffee on the sides of the mountain ranges. The warm, moist, and rainy climate of Panama was perfect for both of these plants.

The history of coffee dates back to Arabia over a thousand years ago. A shepherd, according to legend, found his sheep became extremely frisky and could not sleep at night after eating the leaves and berries of the plants which were growing halfway up the mountain. The shepherd tried the berry himself and found that he too could not sleep that night. He took the plant back to his village, and soon coffee, as a drink, was made and the idea spread rapidly over the world.

Colombia and Brazil were excellent places to grow coffee. In the nineteenth century, Americans were beginning to drink coffee in place of tea, which had fallen from popularity after the Boston Tea Party. There was a rapidly expanding demand for more coffee in 1894. Bananas were also becoming much more popular.

These two plants lured Allen Neely to Bocas del Toro, Colombia. Rafael Rodrigez, a relatively young man of Spanish descent, came to Atlanta in early 1894 looking for a large loan to expand his coffee and banana plantation in Bocas del Toro. He had been referred from Savannah to the Trust Company of Georgia where Allen worked. Rodrigez brought his credentials with him. He had been shipping both of these products to Savannah, and the merchants of Savannah found him trustworthy. But the banks in Savannah did not have the large amount of money to lend. They sent him on to the Trust Company in Atlanta. Young Allen was in the loan department, where he took down all preliminary information for several of the vice-presidents and even the president before such a big loan could be approved. Allen talked to Rodrigez and took him out to lunch and dinner several times. In addition, Allen had taken Spanish in high school and had kept up his Spanish with some friends from Cuba who were working at the bank.

The Trust Company was very much impressed with young Mr. Rodrigez and approved his loan for doubling his plantation operations. Mr. Rodrigez was also very much impressed with young Mr. Neely; on the last day when the deal was closed, he offered Allen a position in Colombia. He would be the business manager for the whole plantation with its three hundred employees and large warehouses. He would make the shipping arrangements for the coffee and bananas. Allen immediately accepted. Banking life bored him, he wasn't making much money, and he had a strong spirit of adventure. Rodrigez told Allen to wind up his affairs in two weeks, take a ship from Savannah to Colon on the Caribbean side of the Panama isthmus, and he would be met there.

His brothers were supporting his mother Henrietta fairly well, and with the big increase in salary, he could send home more money. So there was really very little that he had to do to get ready—arrange to leave the bank, buy some clothes for the tropical climate, and tell his family and friends good-bye.

In 1894 John was twenty-four, Ben, Jr. was twenty, Edgar was sixteen, and Frank was ten. Ben, Sr. had been dead for two years. John and Ben were working; Edgar was finishing high school, and Frank was still in grammar school. The girls were also teaching or working at something. They still had the house in Rome. Allen came home from Atlanta and broke the news about his plans. Everyone was very excited but had mixed emotions about his going. They felt it was a great opportunity, but to go such a long way meant that it might be a year or two before he got back. So in the spring of 1894 Allen took the train to Savannah, boarded a steam ship, which was propeller driven, and began the eighteen-day journey to the Panama city of Colon. There was much talk of building a canal across Panama at the time, but no definite plans had been made. A Frenchman had devised a plan which looked feasible. The United States was negotiating with Colombia at the time, but Colombia was in such an unstable state politically that no agreement could be made.

Bocas del Toro was a medium-sized village of about 4,000 people on the island of Toro. Toro nestles in a large lagoon close to the mainland. It had considerable flat lands with one mountain seemingly coming up right out of the sea on one side while the other face sloped into the tropical jungle and large open fields. The town was situated at the base of the mountain. Mr. Rodrigez's plantation was the largest on the island, and he was now about to take over another plantation.

The town itself was small, but had a large Catholic Church with priests and nuns. The latter took care of a small primitive hospital next to the church. The people of the town were a mixture of Spanish, Indians, and negros. Spanish was the official language, but many spoke English.

Allen arrived in Colon, was met by Mr. Rodrigez personally, and they traveled by coach to the mainland close to Bocas del Toro; they took a ferry from there. Allen liked his job immediately, was given a small house of his own, and worked very hard organizing the overseers, the laborers, and arranging the transportation for the coffee and bananas. The weather was extremely hot throughout the year, but was the hottest in the summer.

Sometimes during the rainy season all work would have to be stopped for weeks as small floods were a problem.

One of the secretaries was a beautiful dark-skinned mestizano named Francesca. Allen began taking her out to the beaches, and it was not long before she moved in with him. Mr. Rodriguez, married and father of six young children, made no objections to their living arrangements. In fact, he thought that this might help him to keep Allen; he was delighted with the job Allen was doing.

Time went by rapidly. Allen did go home once after three years, but things had changed so much that he was happy to return to Bocas del Toro.

In 1900, right after the turn of the century and at the end of the rainy season, Allen was badly bitten by mosquitoes, developed a high fever, and became delirious. He was taken to the Catholic hospital, rapidly became jaundiced and went into a coma. Death from yellow fever came at age 32.

Allen's life was snuffed out just as he was beginning to live. He had just found himself. Who knows what he would have achieved had he lived a full life. He never lived to see an electric light, an automobile, or an airplane. All of his siblings lived well into the twentieth century, but he died right at the beginning of an era which would change the world.

16
Ben to Boston; Death of a Sailor

John Neely (the 1st) was admitted to the ministry in Easton, Maryland in 1848. He had studied faithfully while still continuing his position as superintendent and director of the Bel Air Academy in Easton. He had taken a series of written and oral exams, passing them easily. He was now forty-four years old. He had had a lifelong ambition to be a minister, spurred on by the untimely death of his first wife. He was admitted to Deacon's Orders in Christ Church, St. Peter's Parish, Easton, Talbot County, Maryland by the Right Reverend William Rollinson Whittingham, Bishop of Maryland, on May 7, 1848. His presentor was Reverend Henry M. Mason, D.D., pastor of St. Peter's parish. Reverend Spencer, D.D., of St. Michael's parish, Talbot County, and Reverend Robert Goldsborough of Hillsborough in Caroline County were present at the ordination. Four years later his second wife, Isabella, would die of consumption in March of 1851. Eighteen months after her death he would marry Rebecca Sommerville Dorsey on September 7, 1852 in West River, Anne Anundel County, Maryland. It was in 1852 that John moved to Augusta, Georgia to begin the first Girls High in Augusta as well as be superintendent of the schools there. He brought the entire family there, including his third wife Rebecca and all of the living children by Susan Henry, his first wife. Euphemia was twenty at that time; Ben was eighteen; John, Jr. was fourteen; Maria was eleven, and Annie Wyatt nine. Mary had died in 1845 at age nine; Elizabeth Dorsey had died in 1850 of scarlet fever at age ten, and Julia had died in November 1844 two days after her birth. So out of the eight children, he had lost three girls. This was not unusual in those days. John was greatly saddened by the death of his children, but becoming a minister had helped him accept such tragedies. He was able to call these deprivations the will of God.

When Ben disappeared in 1847, John was saddened and depressed. Fortunately, he was extremely busy at school as well as with his nightly studies for the upcoming exams in April of 1848. Euphemia, now sixteen, had faith that Ben would some day return, and her hope somehow spilled over to her father.

* * *

So where was Ben in February 1848? He was still on the *Nerrisa* in St. John's, Newfoundland, and the ship was preparing to sail to Boston via Halifax, Nova Scotia. The weather was bitter cold, and the sailing was delayed several times by snow and sleet storms. Finally, in February, they had a beautiful, sunny day with a favorable wind and tide. They sailed.

There were fairly large rolling swells as they came out into the open sea, but the Baltimore clipper *Nerrisa* was quite seaworthy and took the waves well as the ship sailed south toward Halifax, Nova Scotia. The tough seaman, Frank, stayed in Newfoundland because that was his home. Also Doffie, the alcoholic seaman, stayed in St. John's; he too was a Newfoundlander. Mr. Rumans had signed on two new seamen to take their places. One was Rolf, a huge but gentle man with a beard, the other Charley Fitzhugh, another Newfoundlander of Scotch-Irish origin. He had been a sailor most of his life—he was older, about fifty-five, and rather fat. He had children who were sailors.

Ben now felt so at home on the ship that he could banter with the sailors on an equal basis. He seemed to have grown in the short few months he had been gone from home and was now as tall as some of the sailors. Ben knew he did not want to be a sailor for the rest of his life, but the ship, the crew, and the officers were now his home. Bjorn was a great help, and the captain readily provided advice and encouragement. Ben knew that within the next month or two the ship would be returning to Baltimore, and he must make up his mind as to whether he could go back home again. He stayed awake many a night trying to decide. It was an agonizing decision, but he kept it to himself.

Nova Scotia, or New Scotland, is one of the smallest of the Canadian Provinces. It seems to stick out into the Atlantic Ocean; it is bounded on the west by New Brunswick. Its northeastern border is Cape Breton Island. This is a part of Nova Scotia separated by the narrow Strait of Canso. North of Nova Scotia lies Prince Edward Island. The city of Halifax on the southern coast was settled by the British—with 4,000 settlers—in 1749. It has remained British ever since. The French were expelled in 1750, never to regain control of the area again.

After eight fairly rough and cold nights, the *Nerrisa* sailed into the quieter waters of the bay on which Halifax was located. The waves were much smaller and the wind and tide took them into the dock with an easy landing. The ropes secured the boat to the dock and most of the crew went into town.

Halifax was the largest port in Nova Scotia. Fisheries were the major activity with shellfish, haddock, and cod the biggest catches. Ship building and lumbering were the other industries which made Halifax a large city even then. The *Nerrisa* would leave off some woolens from Scotland and take on a load of fish bound for Boston. This loading, plus preparing for the next leg of the trip, kept most of them pretty busy during the last few days. Bjorn and Ben did get a chance to do a little fishing through the ice on one of the fresh water lakes nearby. A small shed was pushed out on the ice; they cut a hole in the thick ice, and the little house was right over the hole. There were seats on the sides and there was a stove with a wood fire to keep them warm. They spent the best part of one night and came out in the morning with a fine string of fish for each, which they took back to the boat and put into the ice room kept cold with lake ice they had taken on in Newfoundland.

It was now well into February. The weather had moderated slightly, but it was still bitter cold at night. Mr. Rumans and the captain had sailed these waters many times in all kinds of weather and did not hesitate to set sail in the winter months even though the spray would often freeze on the sails, make them heavy, and sometimes break a spar which would come crashing down on the deck; God help anyone that was beneath it. One evening, the new sailor, Fitzhugh, crossed the forward-well deck to go on duty on the bridge when a huge spar from one of the upper sails crashed down. The spar struck Fitzhugh on the right temple. Knocked out for a moment, he was lying on the cold, icy deck. Ben saw what happened from the door in the forecastle. He cried out for help, and soon they had lifted him to the galley area. There was a small room with two bunks used as an infirmary. They laid him down gently on one of the bunks. He had a large, bleeding gash on his forehead with swelling around it. Mr. Rumans came down from the bridge, cleaned the wound and dressed it.

Fitzhugh came to but was confused. Rumans acted as the doctor, and the cook and Ben were delegated to look after him. He lay on the bunk groaning but did not appear to have any other problem beside the head injury. Bjorn and Ben took turns staying with him in the other bunk. By the next morning he appeared worse. He could not move his left arm or leg, but he could answer a few questions. He had a tendency to drop off to sleep if left alone for long. Rumans would come down twice a day to examine him; Ben and Bjorn continued to nurse him as best they could. They would feed him with a spoon and clean him up when needed. On the following day he was in a deeper coma. Ben stayed with him most of the time as Bjorn was busy preparing meals for the crew. On the third day

Ben noticed that Fitzhugh felt very warm and was beginning to cough. He was now in a deep coma. Mr. Rumans, who had seen many head injuries, told Ben that Fitzhugh did not have long to live. The following night his breathing became more rapid and noisy; then it became weaker and finally stopped. Ben ran up to the bridge to tell Mr. Rumans. Rumans raced down, only to confirm the worst: the poor fellow was gone.

Rumans said, "Ben you have looked after him well, but he had a bad head injury and must have developed pneumonia. We will leave him here for the night. You go back to your room in the forecastle, but be especially careful when you cross the forward-well deck as it is still slippery."

This was the first time Rumans had complimented anyone.

Ben just said, "Yes, sir, thank you, sir," but he felt that a word from Rumans really meant something.

This was the first time Ben had seen someone die right before his eyes. Although he did not know Fitzhugh well—only for about eight days—it shocked him: just three days before, he was talking and joking with Fitzhugh; now he was dead. Ben got into his bunk and cried quietly for a short time.

From life to death, just because a spar had broken—could the accident have been prevented? Ben wondered. For a young person of fourteen to have seen someone he knew die before his eyes was tough. Women often died in childbirth as had Ben's mother, Susan, but he had not been there at the time. He had gone to the funeral of his mother and watched her lowered into the grave. For a loved one to die was one thing, but simply to watch death slowly come to a person he scarcely knew was different. Here was a person Ben had known for only a week but had seen daily at mealtime. Death haunted Ben's consciousness. A simple accident—could anyone have prevented it? Who knows?

Death troubled Ben. Heaven and hell were well known to Ben as ideas, but now at fourteen, he was beginning to wonder: would Fitzhugh go to heaven or to hell? He was sure his own mother ascended to heaven; but watching a man die had had a sobering effect on him. He began to question whether there really is a life after death. He began to reflect that he himself would one day die and also that his life could be snuffed out by an accident. He would see death in great profusion later—in thirteen more years he would be a soldier and a spy in the Civil War. Then he would see so many dead and dying that he would have to become hardened. Now he had had personal experiences with only two deaths—those of his mother and of Fitzhugh.

They gave Fitzhugh a sailor's burial at sea. He was wrapped in a shroud and after a short prayer by the captain, he was dropped into the sea. The finality of this also made Ben think—if there is a God watching over us all of the time, why would he let something like this happen? Had Fitzhugh done something wrong?

Death in wartime is explainable. But a simple accident causing death baffles. The Catholics seem to have an answer: "It is God's will and we can't question what he does," they say. But does this satisfactorily explain death? Recently there was a terrible bus accident in France and fifty children were killed. One of the mothers of a child that was killed asked: "Where was God the morning that it happened?" (*Atlanta Journal,* June 2, 1984.) Or when a child gets an incurable cancer, the family and every one asks: "Why, why did it happen to this child?" So Fitzhugh's death set Ben to thinking and questioning. Going through this experience may have been the event that turned him from a boy into a man.

* * *

Boston, Massachusetts was settled in 1630 by Puritan Englishmen of the Massachusetts Bay Company. The Puritans had split from the Church of England, but within fifty years, England had sent a royal governor, and he brought the Church of England with him. So there were both the congregationist churches of the Puritans and the Anglican Churches. It was an unusual coincidence that Ben should be coming to Boston at a time when he was trying to understand what religion was all about.

Boston was the center of the beginning of the Revolution. The Boston Massacre occurred in 1770. This was when the British troops fired on a crowd of civilian hecklers and killed several people. Then there was the famous Boston Tea Party of 1773 in which colonists disguised as Indians dumped three shiploads of tea into the Boston Harbor. Then there was the exchange of shots at Lexington and Concord in April 1775, and the Revolution was on. Less than 100 years later (85 to be exact) the Civil War would begin and Ben would be in it. Boston in 1848 was one of the most important cities in the United States. Its industries, including shipbuilding, fishing, manufacturing of woven cloth and machinery, made it a booming, growing city. Its harbor was excellent and quite large, with hundreds of ships loading and unloading all of the time. Across the Charles River in New Town, later called Cambridge, a college was founded in 1636 to provide the colony with religious scholars and ministers. In 1848 Harvard College was renowned throughout the world.

Ben was on deck as the *Nerrisa* sailed into the large and protected Bay of Boston. He had studied about Boston and the Revolution in school. He knew about the ride of Paul Revere. Here was the real center of the United States; there was Ben. What would he learn there?

17

Frank to Georgia Tech

Frank finished high school at the Neely School in 1902. He excelled in high school—he seemed to go through the courses and make excellent grades even though he apparently did very little homework. He was especially good at mathematics. He would often make one-hundred on the exams. This was most unusual as the Neely School was known for being difficult. Ben Neely had been dead nine years. The family still had the house, but the children had begun to spread out.

Allen had already died in South America in 1900. Ben was working and living in Atlanta and sending money home. John was practicing medicine in south Georgia. Edgar, age twenty-four, was working in Atlanta as a court reporter, saving money to go off to law school. Annie had married Augustin Young in 1897, moved to Cedartown, Georgia, and begun her family which would be the largest of any of the siblings. Elizabeth Eve had married Joe Jones in 1900 and had died in childbirth one year later. Hannah, twenty-six, was married to William Plumb and had moved to Augusta. Mary, twenty-three, was about to marry Reginald Grinell. So when Frank finished high school in 1902, he planned on going to Georgia Tech to study engineering. Henrietta Eve would be almost alone in the big house and was making plans to go to a smaller house or possibly move to Atlanta. However, she decided to stay another year or so in Rome, and the children would come and visit often.

When Frank arrived at the Georgia Institute of Technology, it already had a good reputation. The University of Georgia was in Athens and even in those days was known as an easy school of only medium academic standing. Georgia Tech, on the other hand, was small, had excellent teachers, and was known to have a fine training program for engineers. The first two years were basic sciences—mathematics, including trigonometry and calculus and also physics with special emphasis on mechanics, forces, motion, and gravity. The laws of thermodynamics were stressed, but the greatest amount of study was on electricity. In the early 1900's, of course, electricity was lighting much of the city; the light bulb, invented in 1880 by Edison, began replacing gas and candle light. There was a huge revolution of energy. Water power with turbines to produce electricity and the transmission of electricity by wires were booming. Chemistry also was a basic course, and Frank had to take all of these subjects. He was a good student with a quick mind, and during those first two years, he did well.

He stayed in the dormitory the first two years, but after finishing the sophomore year, he began looking around to find a boarding house where he would have more freedom. Mr. and Mrs. Arthur Schlesinger lived in a big house on West Peachtree Street. They had five children—one boy and four girls. Mr. Schlesinger had a small candy factory which had done reasonably well, but not outstandingly. In 1896 Mr. Schlesinger began a novel type of candy, soft taffy with a core of peanut butter. Each piece of candy was individually wrapped in brightly colored paper, and about fifteen pieces were put in a box. They were called "Charleys" and the name was printed in bold letters across the box. Each box sold for five cents. The school children learned about them and loved them. They became extremely popular in Atlanta and outlying districts. The candy factory had to be expanded to fill the demand.

New machinery to make the candy and an assembly line to turn out the fully wrapped, finished product were necessary. Mr. Schlesinger had gone to Georgia Tech many years before, but he had kept up an association with the school; he went to alumni meetings and knew many of the professors. He naturally consulted one of the professors of mechanical engineering for help in developing machinery for his growing factory.

Professor Bernard Little of the School of Mechanical Engineering consulted with him. The professor needed some help in the details of what was required, and he just happened to have a very bright student in one of his classes who needed a practical project in engineering. This student was none other than Frank Neely, now in his third year at Tech. Frank went out to the factory, took notes, and had several sessions with Mr. Schlesinger and his plant manager, Mr. Elsas. Frank got all of the information, wrote it up, consulted with Professor Little, and came back with a plan. A contractor was brought in and work was started. Frank would come out from time to time, and when it was finished and operating, Mr. Schlesinger had a party at his house on West Peachtree Street. Mr. Schlesinger's children ranged in age from twenty-four to seventeen. The middle girl, Rachael, called Rae, was twenty and was just back from Smith College in Northhampton, Massachusetts. In those days, it was most unusual for girls to go to college; to go to a northern college was almost unheard of.

When Frank met her at the party there was an immediate, mutual attraction as if a light had been turned on within each, by the other. Both were twenty years of age. Frank was tall, dark-haired, handsome. Rae was pretty, vivacious, and attractive. But it was their bright minds that each recognized in the other which seemed to bind them together from the start. Frank found that she had been to Smith College, and he was amazed. She found that he had designed the machinery for her father's factory and was equally amazed. They were soon talking about their lives, their aspirations, and the wonderful world they lived in. Before the evening was over, Frank had asked Rae if he could come calling on Sunday afternoon. Rae, after asking permission of her mother, said yes.

The Schlesingers were Jews and were a part of the relatively small Jewish community in Atlanta in 1904. There were a number of well-known Jewish families who were outstanding in the business world, and there were many Jewish shopkeepers and small businessmen. Most had come down from New York soon after arriving in America. One family would get established in some business in Atlanta, and then they would write their relatives or friends in New York to come on down. Often there would be several families staying together, until they got established.

There was a certain amount of prejudice against the Jews, but it was not evident until some problems arose. The Gentiles were jealous of the way they could start a business. They seemed to be able to find a need and fill it. They could sell things and had new ideas of business dealings which were unknown to the Gentiles. So the richer Gentiles did not let them into the private clubs but would do business with them when they could fill a need. The harsher prejudice was among the lower classes; they resented the Jewish people for the jobs that they took away. They were more open and cruel in their prejudice. Even in school, the children, having learned from their parents that Jewish people were a little different, would sometimes torment and taunt the Jewish children. Still, in general, the Jews and Gentiles in Atlanta got along fairly well with one another.

The Schlesingers were the so-called Reformed Jews. This division in Judaism began in the eighteenth century in Germany, by Moses Mendelson. This meant that they did not have to be as strict about the laws of dress, eating, and customs as the Orthodox Jews. The two groups got along well with one another, respected each other, and would still stick together. However, their actions and reactions in the world were slightly different. The Schlesingers had come from Germany where the reform movement had begun.

Mixed marriages between Gentiles and either reformed or Orthodox Jews were almost unheard of, but Mr. and Mrs. Schlesinger had been so impressed with this young man, that they readily agreed to this beginning courtship. Rae had been very choosy as to whom she would go out with, and she seemed delighted to have Frank as her suitor.

Frank was brought up in the Protestant Church. His grandfather, John, worked as an Episcopal minister as well as a teacher. Frank's parents—Ben and Henrietta—went to church regularly, and Frank went with them. He had not been an overly religious person. He had not gone to church while a student at Tech, mainly because he was so busy and had to study a great deal. Now that he was in the third year at Tech, he had a little more free time but was now out of the habit of going to church.

He personally had no prejudice against the Jewish people. He had seen a little prejudice against Jewish boys at school in Rome, but he had respect for their intelligence and quick minds. He, of course, knew about the prejudice against blacks. It was now forty years since the end of the Civil War, but there were no enforced laws protecting the rights of the blacks. Only in the larger cities and towns was there any schooling for blacks; the schooling which existed for blacks was separate from the white children's schools.

Atlanta had some separate schools, but this was mostly elementary, and the law for school attendance was not enforced. There was one small, black high school which was more of a technical school in Atlanta; almost no blacks went on to college from it. Atlanta University was talked about but consisted only of a normal school or a small school to teach blacks to become teachers. Later, Atlanta University would become a great school, but in 1904 it was just beginning.

Ben Neely Sr. resigned from the school system in Augusta because they would not permit blacks to attend school, but in Rome he had convinced the town council that admitting blacks to elementary and high school would be beneficial to the community; some of the black students had graduated from high school and later gone to northern colleges. So Frank had a tradition of racial liberalism behind him, and although he never got to know his father well, he heard the stories of how his father had helped the blacks and how racial discrimination was a bad attitude. Somehow this broad liberal attitude applied to the Jewish race, and he saw no reason why he should not go out with a Jewish girl if he wanted to.

As he continued into his junior year in college, he became more and more involved with the Schlesinger family. Frank had found a place in a boarding house which was not far from the Schlesinger home on West Peachtree Street. He at first saw Rae about once a week, but later saw her more often, as she worked late at the factory, and he had to visit it often. Mr. and Mrs. Schlesinger were extremely impressed with Frank and had no reservations about the attentions he was paying their daughter.

Frank wrote many papers on various subjects while in his junior year.

He could understand the problems, but often had some trouble expressing his thoughts and describing operations of machinery on paper. He once showed one of his papers on electric motors to Rae. Rae had been to Smith College for two years majoring in English and English literature. She read the paper, asked a few questions, and then made several suggestions about the wording and expression. Frank took these suggestions, made the changes, and turned in the paper. He received an A on the paper with the comment that the precise and clear English expression had been extremely good and to keep up the good work. From then on Frank was to get help from Rae in almost all of his endeavors. She never told him what to write or what to do. She was never a "bossy" woman. She would make suggestions which he would take at times and reject at other times; here were two bright minds which would work together for the rest of their lives.

The junior year at Tech went into the senior year; Frank and Rae were becoming closer all the time. There was no question of marriage at this time. Marrying in college was almost never done; it was taken for granted that a young man had to have a good job and be able to support a wife before even considering marriage. In his senior year at Georgia Tech Frank was still doing an outstanding job in his studies, earning the highest grades in his class (with some help from Rae along the way). It was the custom in those days for representatives from some of the big companies to come to Georgia Tech in the spring and interview the brighter students. Frank was interviewed by several companies, and the man from the Westinghouse Company was impressed and offered Frank a job in Pittsburgh where they were working on the early electric locomotives. This was the kind of job that was very interesting to Frank. He had a special interest in electric motors, and here was a chance to become involved in a field that was to revolutionize train engines.

He talked it over with Rae, the Schlesingers, and his brothers Ben and Edgar. Everyone thought it was a great opportunity, but he did not want to leave Atlanta, and especially he did not want to leave Rae. They had by now talked about marriage with each other and with their families, but it was entirely too early to set a date as Frank was still in college. Rae encouraged him to take the job. It paid a good salary, and there was good railway service between the two cities. Frank would not have much expense living in Pittsburgh, and he could visit Atlanta frequently. Finally, he decided to take the position.

Graduation from Tech was coming up in June of 1904. Ben Jr. was helping to support Frank during his school years. Edgar had also helped. He worked as a court reporter, and had just been accepted to law school at Georgetown University in Washington, D.C. He had saved enough money to support himself through law school; he would also work part time in Washington as a court reporter. So the two brothers would be going north at about the same time, both beginning their new careers. They would visit each other frequently while in Washington and in Pittsburgh.

Frank began his work in the research engineering department of the Westinghouse Company along with several excellent electrical engineers from the Massachusetts Institute of Technology. They had to prepare the plans for the first electric locomotive, work out all of the details as to the source of electricity, the horsepower of the motor, the control of the motor, the ordering of some parts, and the actual manufacture of other parts. This was all done in great secrecy as it was known that some of the other companies were working on this also. There were reports that the English already had a small engine; Westinghouse wanted to have the first American electric locomotive and obtain a patent on it.

Frank remained extremely busy throughout the year. He did return frequently to Atlanta, and early the following year Frank and Rae began to seriously consider marriage. That was when some problems began to arise.

18

Leroy Upson Gardner

Let us now examine a great man who also infused his genes into the Neely family. Let us see if his genes mixed with Neely genes and made some new mutation of character, either good or bad. One child has eight great-grandparents, and to try to say that the characteristics of a child come from a single great-grandfather is obviously foolish when the child gets the genes from eight different people. Sometimes certain characteristics will emerge and be dominant and will seem to come from perhaps one side of the family. Obviously tracing genes through generations is mostly conjecture. However, we can perhaps pick one person, say a non-Neely grandfather of the six Levering and Margaret Neely children, and see if we can find some genes that he passed along.

Leroy Upson Gardner was born in 1888 in Meriden, Connecticut. His father, Irving Isaac Gardner, outlived his son, Leroy, the eldest of six children. His father was in the insurance business in Meriden. He was able to send his children to college. Roy, as he was called, was able to attend Yale and Yale Medical School. Roy graduated with the class of 1912 at the age of twenty-four, did his three-year internship in Boston, and after graduation married his childhood sweetheart, Carabel McKenzie. He had gone through grammar school and high school with her. The newlyweds had a flat in Boston and one year later Carabel became pregnant—with Margaret. This was in 1916; World War I had been going on for two years. The United States was not yet involved in the war, but it was becoming more evident that they would join the side of Great Britain and France.

The United States built up its army and needed doctors. Margaret was born February 8, 1917 just two months before the United States declared war on Germany. Three of our ships were sunk by German submarines. President Wilson and Congress decided we must enter the war.

After the United States' entry into the war on April 12th, 1917, Roy volunteered. He came home one night in an army uniform, his way of breaking the news to Carabel with their new baby only three months old. Roy was to go to Fort Dix in New Jersey for training and would be working in the hospital there. Carabel and the baby would have to go home to her parents in Meriden until he got settled and could find a place for them to stay. He would be able to come up to Meriden by train for some weekends. After a few months of training, he was put to work in the large army hospital in Fort Dix.

Roy worked extremely hard at the hospital, staying up late at night. He performed many autopsies; in those days tuberculosis was rampant. Often there was no suspicion of tuberculosis in the patient and very few precautions were taken. Roy began to notice that he was coughing and running a slight fever. He had an x-ray taken which showed several spots on his lungs. Being a laboratory man, he examined his own sputum, and after seeing the x-ray, he was not surprised to see the bright red tubercle bacilli under the microscope. So here he was with his wife and new child, and he was sick with active tuberculosis. You can imagine how difficult it was to go to Meriden and tell her that he was a sick man and that he would have to go to a sanatorium. Since he had developed tuberculosis while in the army, his disability was service-connected and he would receive three-fourths of his salary as a captain in the medical corps until he was cured.

He consulted with the best specialists in tuberculosis and was advised to go to Trudeau Sanatorium in Saranac Lake, New York, the best known sanatorium for tuberculosis in the country. Carabel and infant Margaret stayed with her parents in Meriden; Roy took the New York Central train from New York City to Saranac Lake. This was a small town of about 4,000 people, but it was already a big center for tuberculosis treatment.

Trudeau Sanatorium was started by Dr. Edward L. Trudeau in the last years of the 19th century in one small cottage. It grew rapidly. The theory was that the extreme cold was good for tuberculosis and each cottage had porches where the bed could be rolled out for sleeping at night, under many blankets. The temperature would go down to 40 degrees below zero Fahrenheit. The weather was never really warm; even in summer, the temperature rarely climbed above 75 degrees in the middle of the day. The discovery of complete bed rest as a treatment for tuberculosis was made when a patient with tuberculosis also was suffering from a severely broken leg. The patient was confined to his bed for several months; his condition greatly improved during this time, and the idea of complete bed rest for tuberculosis was born. So it was thought that the extreme cold plus complete bed rest would really help to cure tuberculosis. The sanatorium was producing a good number of cures, or at least great improvement, with this treatment. Soon the sanatorium had a long waiting list, and many of the houses in town were opened up to tuberculosis patients where they were treated by private doctors. By 1917 the sanatorium had attracted some of the finest doctors in the country, and Dr. Trudeau, himself a rich man, contributed funds for building a complete sanatorium. He also worked at raising contributions for research in tuberculosis, and the Trudeau Foundation was thus established.

Roy improved rapidly at Trudeau and after about six months sent for Carabel and Margaret and rented a flat in town. Dr. Trudeau was extremely impressed with this bright young man. The sanatorium needed a pathologist, someone to oversee the laboratory and take charge of the research program. Roy was superbly qualified. He had attended Yale Medical School and trained in pathology and laboratory work at Harvard for three years. Trudeau offered him the position and Roy was delighted to accept as he had come to like the challenge of living in an extremely cold climate. Also, he had come to know the fine group of doctors that Dr. Trudeau assembled.

So here was the beginning of a career in pathology and research. Dr. Leroy Gardner was to become famous all over the world for his research in lung diseases, especially silicosis or miner's disease, but also in asbestosis. He would travel all over the United States and the world and would show the effects of many different dusts on the lungs of both animals and man. He would predict what is now a great tragedy—the huge number of cases of cancer which are showing up in workers who have been exposed to asbestos. He wrote numerous papers on the various pneumoconioses. He contributed to textbooks of medicine, and was invited to write the article on silicosis for *The Encyclopedia Britannica*. He would be listed in *Who's Who*. So here we have a man who contributed greatly to scientific knowledge that benefited humanity.

But let us see how life was in Saranac Lake, New York, in the 1920s. There are several large fresh water lakes in Saranac Lake. These freeze solid in December and thaw in May. There is usually one day in which suddenly the ice disappears, sometimes over night. But there is snow on the ground from November until March or April. Saranac Lake is about 100 miles from Montreal, the closest big city.

Tuberculosis became the biggest business in the town with the great influx of people coming to the town to "cure." They would rent or buy houses and often come to cure and stay for life. Other sanatoriums sprang up as time went by. The U.S. Veterans opened a sanatorium; later, the Will Rogers Memorial Sanatorium was built for actors who developed the disease. It was now the Magic Mountain of the United States. Later Arizona and Denver became centers for curing tuberculosis, but Saranac Lake remained the best and most famous center until drugs were developed which would ultimately almost eradicate the disease in this country.

Tuberculosis, or phthisis as it was sometimes called, plagued the world for a very long time. Even some Egyptian mummies had tuberculosis. It was known throughout the middle ages, recognized by its symptoms of cough, bloody sputum, and slow wasting of the body. It did not become rampant until the industrial revolution in the nineteenth century, when crowding in factory villages became widespread. It would sometimes wipe out entire families and often struck the young and adolescent. It was probably the first cause of death of the young in the 19th century. It was still prevalent around the turn of the century and was really not under control even in the 1920s. Koch discovered the cause of the disease in Germany when he isolated the germ or bacteria. He was able to stain the bacillus from the sputum or from lung tissue obtained at autopsy with a dye which was acid fast. The red dye remained in the bacillus even after the slide was run through acid, and it stood out under the microscope bright red, whereas the background had lost the red dye with the acid wash. He showed that the disease was transmitted from person to person. Doctors were slow to understand this until the test for the germ was developed. Later, the tuberculin test was developed in France. This was the use of tubercle bacilli which were killed with heat and then made into a solution which was injected into the skin of a patient. If he had active tuberculosis or if he had been exposed to tuberculosis and developed a subclinical case of the disease, the skin would swell and turn red. This test and the sputum test were very helpful in controlling the disease. School children could be tested when they entered school. Anyone with a spot on the lung would be skin tested. If the test was negative (meaning no observed reaction) one could be sure it was not tuberculosis. However, if it were positive, it simply meant that it might be tuberculosis or that perhaps the patient had had tuberculosis in the past and had overcome it. Thus a positive test did not make the diagnosis, but would give an indication of the meaning of a spot on the lungs.

The x-ray of the chest was also an extremely useful tool in the detection, care, and treatment of tuberculosis. The x-ray of the lungs in tuberculosis was very characteristic. The disease was usually at the top or the apex of the lung. Frequently a cavity or hole in the lung could be seen by x-ray.

Thus the disease was beginning to get under some control in the 1920s, although the specific cure was not found until the 1940s. Patients found to have tuberculosis in the 1920s were taken away from their families, sent to sanatoriums for the cure of one to two years, and then sent back home. In the 1920s and 30s, as knowledge of contagion became more prevalent, there often was a stigma attached to anyone with tuberculosis. These patients were not allowed to stay in general hospitals, and they were often neglected, especially the poor. The disease would strike rich and poor, but it seemed to be much worse among the poor and undernourished, particularly in the black and Puerto Rican population. Both of these groups seemed to have very poor immunity.

Roy began his career in pathology and research in tuberculosis when the state of our knowledge about the disease had advanced to the point that there was beginning to be some control of this devastating disease. There was still a huge number of deaths from tuberculosis, but it was not the leading cause of death in the United States in the twenties; heart disease and cancer had now surpassed it in the yearly mortality figures. But the disease was still far from being well controlled, and although there was much solid information about the diagnosis, no definitive treatment except rest existed. If the lung did not have to work so hard, it had a chance to heal.

Dr. Baldwin was in charge of the laboratory when Roy came along, and he was the one who actually hired Roy. Dr. Baldwin was already a well known scientist and doctor in the field of tuberculosis. He was in his sixties at the time and had contributed much to the world literature on tuberculosis.

He and Roy had a fine research laboratory, and they discussed the direction their work should take. Roy observed that miners seemed to develop tuberculosis much more commonly than other people. It was well known that they breathed in a great deal of coal dust. Silicosis as a disease had been described and its association with tuberculosis had been established. The question was why this combination predisposed to tuberculosis. Roy said he would like to try to find out why this occurred.

From the simple question asked by Roy, there began a long series of experiments involving spraying various dusts into the lungs of guinea pigs, rats, and monkeys. One experiment would lead to another concerning the effects of dusts on the lungs of man. Roy managed to interest several of the big companies and persuaded them to contribute funds for the laboratory experiments.

Roy found that many lungs did not contain much dust, but he did find that pure silica caused a chemical reaction within the lung which in turn caused scarring throughout the lung. As the scarring progressed, large masses of scar tissue prevented the lungs from expanding and the patients became short of breath after slight exertion. Apparently, if at this point the miner was exposed to the tubercle bacillus, his low resistance would encourage the rapid spread of the disease, leading to eventual death. Thus it was found that the coal miner might get a lot of coal dust in his lungs, but that this did not cause much trouble unless he had also inhaled large amounts of silica.

Under Dr. Baldwin and Dr. Gardner the laboratory grew; each summer courses were held at the laboratory, known as the Trudeau School, for lung specialists, and the school became famous for its definitive experiments. Slowly Roy became well known in experimental lung disease circles. He was invited to talk at universities here and abroad and won prizes for his excellent work. He studied pathology of asbestosis both in animals and in man, and he found how it caused cancer years after the disease had developed in a person. Funding from various sources kept coming in. Dr. Baldwin retired and another young doctor was brought into the laboratory —Dr. Arthur Vorwald. By the early forties, Dr. Leroy Gardner was a famous man, one who had contributed greatly to the knowledge of dust diseases and their relation to tuberculosis.

What kind of a man was he? He was a very gentle, friendly soul, a hard worker with great determination and an intelligent mind. He was a good family man, loved by his wife and two daughters. He went to work early in the morning and came home about seven o'clock at night. He thus worked about twelve hours a day and would also work on Sundays for part of the day. He smoked a pipe and enjoyed working in his garden. He and his wife Carabel had a great interest in wild flowers; they would often go out in the woods on Sunday afternoon to collect and identify wild flowers.

Some time in 1946 he came home one evening, complained of a little indigestion at supper, and went to bed early. Carabel went to see how he was, heard him gasp, and he stopped breathing. Dr. Woodruff, the surgeon who lived two doors down, came over and said that he had had a heart attack and had died instantly.

So here was a life snuffed out prematurely. He was fifty-eight years old and possibly at the peak of his career. Some people say that is a good way to go. But here was a man who had done great things in this world and who probably would have done much more

What kind of genes did he pass along? Will his great-grandson, Thomas Gardner Neely, pick up some of those admirable characteristics of his great-grandfather's?

19
Ben Neely's Ancestors

Let us look briefly at the present and then go back in time a little further than we have before. One of the latest additions to the Neely family is Thomas Gardner Neely, age eighteen months, born in March of 1981. This child displays some evidence of over-activity; he does not seem to pause for long on any of his projects. He likes to climb up on a sofa arm and dive onto the sofa. He frequently misses or bounces off and lands on the floor. He may cry for a moment, but is soon back at the same occupation. Persistence is a good characteristic and may help him to go far in this world. He loves his sister Jordan, and he also loves to pull her hair. This causes much weeping and wailing, but the parents are modern and don't like to try to correct Thomas. They just say: "Jordan, we told you not to get your hair close to Thomas. You know he likes to pull your hair." She finally extricates herself and quiets down. Thomas, meanwhile, has some new activity.

This boy Thomas has a splendid variety of genes. His mother, Carolyn, has a quiet but strong determination; she trained as a social worker after a college career in history at the University of North Carolina. She has always managed to find a useful and interesting job, usually caring for people, and she really does care. She worked in a rehabilitation center before Thomas was born, and now she is into psychological testing and judging the skills and abilities of children in school.

Her parents—her father is a retired army officer and her mother is a college graduate—lived all over the world as he was a career officer after serving in World War II. Carolyn and her sister lived and went to the American Army schools abroad for many years. The family finally settled down in North Carolina.

Now, Thomas' father, John Longstreet Neely, was born in 1947. We have already discussed John in some detail. John's father, Levering, was born in 1912, and his father, Edgar was born in 1878. Edgar was the son of Ben Neely, Sr. who was born in 1834, and his father, John Neely was born in 1804 in Antrim County, Ireland and came to this country when he was about thirty years old. His father, Benjamin, was born in 1783 in Tyrone County, Ireland and died in Lisburn, Antrim County.

So we have about seven generations of Neelys over a time span of about two hundred years. We have a considerable amount of knowledge over the last four or five generations, but then our facts begin to get hazy, and we have to rely on conjecture.

Let us see if we can trace back a little further. Ben Neely, Sr. married Henrietta Eve Carmichael in 1867 in Augusta. What about Henrietta Eve Carmichael and her forbearers? She was the daughter of Anderson Watkins Carmichael and Elizabeth Eve Longstreet. Henrietta, was their second child and born in 1845. Anderson Carmichael, her father, was born in 1821. His father, John Carmichael, was a native of Ireland, but we have very little information on this.

What about these Eves and Longstreets—whence did they come? We see that Henrietta's mother was a Longstreet. Her name was Elizabeth Eve Longstreet, and she was the daughter of Gilbert Longstreet and Henrietta Eve. Gilbert Longstreet was the brother of Rebecca, Augustus, James, and William Longstreet. James Longstreet was the famous Confederate general. Gilbert Longstreet married first Betsy H. Leigh in 1812. When she died, he next married Martha Henrietta Eve, daughter of Oswald Eve.

Oswald Eve was the progenitor of a whole tribe of Eves. The Eve name, being short and pretty, pops up through the generations such as Rachael Eve Neely Parker and Elizabeth Eve Neely. So let us see what we can find out about Oswald Eve who was surely born in the middle of the 18th century. In fact, we know that he married Ann Moore in Christ Church in Philadelphia on June 2, 1744, so he must have been born in the early part of the seventeen hundreds.

We have a large number of isolated facts about him. Let us see if we can piece them together and find out what kind of a man he was. Remember from Thomas Neely to Oswald Eve we are going back eight generations and spanning about two hundred and fifty years.

Here's some historical references about Oswald Eve. First, he was appointed port warden for Philadelphia in 1766, and he held that post for many years. How big a job was it in those times? We don't know, but it must have carried a lot of responsibility. Second, on August 22, 1776 Congress ordered $224.55 paid to Oswald Eve for surveying the inlets to the north of Cape May. The date is very interesting. 1776—The Revolutionary War—had it begun by then? And $224.55—what would that be in dollar value now—certainly fifty times that amount, but it could be 100 times more, making it about $22,455 by present standards.

Third, in the history of Philadelphia, Oswald Eve was a member of the American Philosophical Society as well as a member of the American Society for the Promotion of Useful Knowledge. This was in 1768, before The Revolutionary War. Also, the history states that he was a lieutenant assigned to a company commanded by Captain Samuel Mifflin in 1758. His marriage had been in 1744. The history states that Oswald Eve's ship, the *Martha Vodroid,* was taken by the privateer, Captain Lehay in Delaware Bay. It further states that he signed the non-importation resolution. One final historical fact: On July 4th, 1775, Oswald Eve was appointed by Congress to a committee to view the Delaware River (the word "view" must have meant survey).

All of the isolated facts don't tell us very much. Or do they?

Other facts gleaned from old Philadelphia archives: (1) The secret committee of Congress bargained with Oswald Eve to manufacture gunpowder at $8.00 per hundred weight. (2) In May of 1778 Oswald Eve was referred to as a merchant and gunpowder maker late of Philadelphia. (3) In 1778 a complaint was made in Congress with respect to powder made in his mill. (4) Two of Oswald's sons were jailed in August of 1778, and their goods were sold for use of the State. (5) In 1781 Oswald Eve forfeited his estate, which included a house, two plantations, a powder mill, and two hundred and seventy acres of land. It was sold to Captain John Eve for $108,000, in conventional money, and 54 bushels of wheat.

Again, interesting information and undoubtedly true, but it still is very difficult to fit the pieces of the puzzle together.

Here we have a little interesting information about one of Oswald Eve, Jr.'s sons. He was Dr. Paul Fitzsimmons Eve, M.D. (1806–1877). He spent most of his professional life in Nashville, Tennessee. He was one of the founders of Vanderbilt Medical College. He studied medicine in Philadelphia and Paris. When twenty-three years of age, he joined the Polish Army. For distinguished service, he was awarded a badge of honor. Soon after the Medical College of Georgia was founded in 1818, Dr. Joseph Eve and Dr. Paul Eve were appointed to the faculty.

Oswald Eve, Jr., who keeps popping up in history, still stands as a puzzling figure. Oswald Eve, the younger, died in 1829. His will was probated in Richmond County, Georgia on September 24, 1829 and had been signed by him on July 24, 1829. So he must have died between these two dates. He left a large family of ten children; the youngest was Martha Henrietta who married Gilbert Longstreet (whose daughter, Elizabeth Eve married a Carmichael, and their daughter, Henrietta married Ben Neely). Here's what's presented in the Neely ancestry published in March of 1960 by Roy Hand:

> In 1745 Oswald Eve, the elder, was a sea captain commanding the ship *George* and was so prosperous that from 1756 to 1760 he was part owner of no less than twenty-five different vessels. During part of this time, he was a shipping merchant of Philadelphia and in 1756 was a lieutenant in Captain Samuel Mifflen's company of Philadelphia Associators. In 1766 he became a member of the Society for the Relief of the Poor, Aged, and Infirm, Masters of Ships, their Widows and Children.

So we see that Captain Eve was a seafaring man, a very prosperous man, a philanthropic man—all before the Revolutionary War. Captain Eve met with misfortunes in business, left his family in Philadelphia on May 1, 1768 and with his sons John and Oswald, Jr. went to the West Indies where he engaged in business principally in Montego Bay, Jamaica. After an absence of five years, he returned to his family in Philadelphia.

John Eve returned to Philadelphia before his father and brother, and in 1770 married Jane Campbell, but he apparently returned to the West Indies.

Mrs. Emma Eve Smith, Oswald Eve, Jr.'s daughter, who was eighty at the time of the following observation, provides further insight:

> My father, Oswald Eve was born in Philadelphia of English parents in 1754 and married a lady in Charleston, S.C. by the name of Aphra Ann Pritchett whose parents were of Irish descent. She owned a house in the city and a plantation on Cooper River and many Negroes. My grandfather Eve was captain of a vessel named *Roebuck,* and his son, Oswald, followed in his father's occupation until his marriage when he abandoned the sea and became a planter.
>
> He called his place *Middleton* and resided there until he had six daughters. After selling this place he moved to Ninety-Six, S.C., and lived there many years. Five more children were born while there.
>
> In 1800 father came to Georgia and purchased *Goodale* near Augusta and a fine plantation on the Savannah River. Here he built a large house called *Forest Hall* for a winter residence and a cottage nearby for a summer retreat.
>
> Father was educated in Philadelphia, but was said to be very dull and slow in learning. After commencement, however, he grasped knowledge with surprising ability and became prominent in many positions of trust. His mind was very comprehensive. He brought up his children with the fear of God and by precept and example, taught them the ways of life. Himself, the patriarch of his household, he conducted church services at home when no place of worship was within reach, and when accessible, going through heat and cold, wet or dry weather to worship. He was particularly faithful to his duty to his numerous slaves, having these around him at family worship every morning and evening.
>
> He built for his plantation a church on Butler's Creek which he had dedicated by four ministers and which is still in use by the Negroes. Father

was an elder in the Presbyterian Church in Augusta, his contribution to which entitled him to the ownership of three pews.

Father loved the Indians, a company of Cherokees. They would come, and he would lend them a house. . . Father's health declined in a year or two and he closed a long and successful life in 1829, having previously buried my mother who died in 1821. He was seventy-five years old.

Of my mother, I can safely say she was a true help-mate from the Lord. She was wealthy when she married but did not raise her children in indolence on that account but taught them the strictest economy, order, and management. They were made to go to the kitchen and learn to cook everything, to take charge of the dairy by the week, giving them profits of all the butter they could sell while employed. Each one had his own bureau or trunk, and the property of each was strictly respected. Thus she instilled justice in little things as well as in great. Our home was the seat of hospitality. Every Saturday the house was filled with a happy group of children, grandchildren, and strangers who my married sisters would bring with them from Augusta. Our parents united with us in our sports; went with us to the balls and parties and encouraged and contributed to all our enjoyments. Mother was a very large person. She had a fair complexion and large blue eyes. She always wore silk when in full dress and a velvet mantle trimmed with white lace. Instead of a cap, she wore a muslin turban which stood high on her head. She was father's banker. He knew she would take care of what he made and he trusted her with everything. After suffering for a long time, she died in 1821 at peace with her God, at peace with the world and at peace with herself. This was her own language as she was departing.

Mrs. Emma Eve Smith also gives some descriptions of Oswald Eve, Sr., her grandfather. In addition, we have these historical references about Oswald Eve, Sr., the sea captain.

We are not sure when Oswald Eve, Sr. was born or when he died. His son, Oswald Eve, Jr. was born in 1754 and died in 1829. Oswald, Sr. must have been born in the early seventeen-hundreds and probably died in the late seventeen-hundreds.

We know that Oswald, Sr. was first a sea captain, then an owner of many ships, and later had a gunpowder plant, selling gunpowder to the U.S. Army. The journal clearly states that Oswald, Sr. left his large house and family in Philadelphia because of business troubles and went to the West Indies. This was in 1768. He took his sons John and Oswald, Jr. (age fourteen) with him. They remained there for over five years, returning in 1773, three years before The Revolutionary War began. Oswald Eve, Sr.'s granddaughter writes that when the war broke out, her grandparents went to Nassau and died at the house of Joseph Eve, Oswald, Jr's. brother, in Providence.

From the other historical references, we see that he, Oswald, Sr. was port warden in Philadelphia in 1766. We think this must have been a rather important position. Also, he married Anne Moore in 1744 in Philadelphia. In 1775, he was appointed to a committee to "view" the Delaware River. One year later he was appointed by Congress to survey the inlets to the north of Cape May.

Thus we see a man with many ups and downs. He apparently was quite a prominent citizen of Philadelphia with responsibilities in the city and to Congress. He went to the West Indies, retrieved his fortune, and returned to America in 1773. He was doing surveying as late as 1776 when the war began, but now we hear that he moved to Nassau where he remained until his death.

We still don't know much about you Oswald Eve, Sr. It seems a little unusual to leave your family in Philadelphia in 1768 and go to the West Indies with two young sons for five years. And then when you did return you stayed from 1773 to 1776, only to journey again to Nassau, this time with your wife, and this time until your death.

Now we have a great deal more information on you, Oswald Eve, Jr. You went to the West Indies with your father when you were fourteen. When you returned you were nineteen. You were twenty-two when The Revolutionary War began in 1776. Did you fight in the War? Were you the one jailed in 1778? But we do have this fine journal of your daughter's which tells about the later years after the war and about your life with your wife, Afra Ann Pritchett. You two had a huge family. After moving around a good bit, you finally settled near Augusta. You apparently lived a full and satisfying life. You died in 1829 at age seventy-five. A God-fearing man, you felt that religion should be a big part of our daily lives.

Seven generations later we have little Thomas Neely. Will he manifest some of your genes?

20
Edgar Adams Neely, Sr.

Children, once they become adults, frequently ask their parents: "What kind of person was my grandfather?" They may have a dim recollection of their grandparents. Occasionally a child will grow to adulthood and know his grandparents well, but this is the exception rather than the rule. I knew only one of my grandparents. My maternal grandmother, already elderly when I was in grammar school, died when I was in high school. She would visit us often but did not live with us. My other grandparents had died before I was born.

Ben Neely, Sr., my grandfather, died in 1892—twenty years before I was born. The others, Frederick Levering Eckfeldt (on my mother's side) and Henrietta Eve Carmichael Neely (Ben's wife), also died before I was born. My parents did not talk about their parents very much; now that they are gone, I have lost a wealth of information which they must have stored up but were never asked to relate—so that information was buried with them.

My son, John, age thirty-four, recently asked his mother about her family. He had never asked this question before, and it was an extremely difficult question to answer in just a few minutes. He had heard that his maternal grandfather was a well-known physician, but he had no idea that he was in *Who's Who* because of his pioneering research on the effects of various dusts on the lungs. Then he asked me about my father, and I told him I would write up what I knew and send it to him.

John was four years old when Edgar Adams Neely died at age seventy-three. John's maternal grandfather died before John was born. His maternal grandmother, Carabel McKenzie Gardner, died when John was about twenty-five; she had slowly been losing her faculties for about five years. His paternal grandmother, Emily Levering Eckfeldt Neely, died when

John was about nineteen. She was ill for a long time and John did not get to know her well.

So here we go, John—here is the story on one of your grandparents—Edgar Adams Neely, Sr. who was born on October 2, 1878 at Sunnyside, Richmond County near Augusta, Georgia. He was the seventh child and the fourth son of Ben Neely, Sr. and Henrietta Eve Carmichael Neely. His father was superintendent of schools in Augusta and was forty-four years old when Edgar was born.

In any event, Edgar was four years old when his father moved to Rome, Georgia, and Edgar must have started grammar school there in about 1884. His father, Ben, being a school teacher and a superintendent, must have insisted that his children do well in their studies. We think that Edgar did very well in school and probably excelled in writing and English as he was later destined to become a lawyer.

Edgar came up in the middle of a large family. Large families usually have a good effect on any individual child as the child is very soon part of a group, and this may be very helpful in learning to get along with people in later life. It promotes unselfishness. Some modern psychologists, however, take the position that an only child in a family gets much more loving attention. They contend that a child of a large family cannot get sufficient love and attention to really thrive. I think there are many advantages to being a child in a large family.

Edgar developed well within the large family. In the 1880's discipline of children was quite strict. The father ruled with an iron hand while the mother supplied the loving care. Edgar got along well with his brothers and sisters in later life, and we must assume that he loved his family and was loved by them. He was on the younger side of his siblings but did not have the unusual position of his brother, Frank, who was not only the baby but was five years younger than the next child, Mary.

Edgar was redheaded and fair-complected with many freckles. He was a nice looking child, but as he became older became more handsome. Edgar did well in elementary school. He was probably in his first year of high school when his father became ill. This illness was very upsetting to all of the children. Edgar, being redheaded as well as Irish, was extra emotional. Even later in life he was wont to weep easily if something stirred up his emotions.

The death of his father Ben in 1892 was very traumatic for Edgar. Here was a large, close family, whose nature was influenced by their father, Ben. Edgar was fourteen when his father died. The last month of Ben's life was spent at home as he was too ill to get out. He had become extremely thin, and Dr. Harbin had told Henrietta that he did not have long to live.

When the end finally came, the children, and especially Edgar, were devastated. Edgar would cry for hours. Allen, the eldest at twenty-four, worked in Atlanta at the bank. Annie, twenty-three, taught in Rome. John was twenty-two and in medical school. Ben, Jr. was eighteen, worked in Rome to help support the family. Hannah was sixteen and in her last year of high school. Mary was thirteen and in school, and Frank, the baby, was in the second grade.

Following the funeral, the older boys and girls got together with their mother and discussed what was to be done. They had the house which had been paid for, but Ben's income, never very much, had now been cut off. Allen, Annie, and Ben, Jr. were all employed and none were married, and they agreed to combine incomes to support the household. John was supporting himself in medical school in Augusta and was able to send home some money.

Edgar, only fourteen and in high school, was also working in the afternoons and on weekends in the grocery store, so he contributed to the household finances. Ben, Jr., although only eighteen, was working full time; he took charge of the household fund as Allen was in Atlanta and John in Augusta. Henrietta took in sewing from the neighborhood and was helped by all the girls at home. There was thus enough income for everyone to survive; as time went on, the financial picture improved. The family pulled more closely together by this common bond of working and living together with a common purpose.

Edgar, now fourteen in 1892, was one year older than his father was in 1847 when he pushed his stepmother down the stairs in Baltimore and went off to sea. Edgar had his father's strong emotions, his bright mind, his easy laugh, and his father's ability to get along with people. But Edgar had some other genes in him, and there was a love of theatrics which his father did not have. Also, he seemed to be very good in English and he liked to talk even before a crowd at school. This was a field that was just becoming popular at school. Edgar was in the school play and had an active role. It was about this time that Edgar found that girls were very attractive and charming and with his quick wit, ease in conversation, and improving good looks, girls were attracted to him as well. It was about this time he was said to have chased Miss Lucille Moses around a haystack five times. We don't know whether he caught her or not.

Edgar continued to do well in high school, graduating in 1895. It was taken for granted that he would go to work full time and help support the rest of the family as the older siblings had done. A good writer, he found a job with a cotton company in Rome as a male secretary. Typewriters were just beginning to appear in business places but had not come to Rome as yet. It was his job to write letters, write up bills, and do whatever the boss wanted him to do. He had to take dictation from his boss, writing as fast as he could and then rewriting the letter. He heard about shorthand and began studying it at night.

He kept this job about one year. Ben, Jr. had moved on to Atlanta as that was where the good paying jobs were, and he had found a good job in the office of the L & N Railroad in Atlanta where he did accounting. He suggested that Edgar come to Atlanta as he had heard of a job as a checker of railroad cars for the Georgia Railroad. This paid almost twice the salary that Edgar was receiving in Rome, and so he too left Rome for the big city of Atlanta. This left Henrietta, the girls, and Frank at home with no grown male. The family would frequently get together on the weekends; since three of the older males of the family had jobs in Atlanta, there was a general decision to move the whole family to Atlanta when possible.

Mary was just finishing high school; upon her graduation in 1896 they began making arrangements for the move. In about 1897 they made final arrangements for the move to Atlanta. Ben, Jr. was in charge as Allen had moved to Colombia, South America. The first child to get married was Annie; in 1897 she married Augustin Young, who was in the hardware business in Cedartown, Ga. Now Frank had just turned thirteen and would finish high school in Atlanta.

Edgar Adams Neely, you must have had the genes of ambition: you seized opportunity when you saw it. You changed jobs frequently while in Atlanta, always bettering the last. You would study at night; you learned shorthand, how to operate a telegraph with the Morse Code, and how to type. You finally became a court reporter taking shorthand in the court and typing it up later. Within the courtroom you saw the fine lawyers of Atlanta as they would argue cases. You realized that your deepest ambition was to become a lawyer.

Your genes and environment began to work together and your ambition slowly heated up to the boiling point. You were becoming more adept at the English language. You began to see that the word was mightier than the sword. As you took down those words of the lawyers, the plaintiffs, the defendants, the witnesses, and the judges, you became a part of it; you felt the drama. As a court reporter, you learned the art of a trial lawyer. It was here that you unconsciously learned some basics of forensics which would someday make you one of the finest trial lawyers in the South.

So, slowly, this resolve to become a lawyer came over you and you began to consider how you would do it. As the century changed to 1900 you were twenty-two and had just become a full-fledged court reporter, making a very good wage for one so young. You kept this job for almost three years, and you were able to save a considerable amount of money.

There was a young lawyer in town who had just begun to practice, Mr. Hughes Spalding. He came from a fine, old Atlanta family; they were Catholics. You and Hughes got to know each other and you talked to him about your ambition to become a lawyer. He had gone to Georgetown University in Washington, D.C., whose tuition was not high. You had done very well in high-school, and so with a letter from your friend, Hughes, your high school diploma, and several other letters of recommendation, you took the Southern Railway from Atlanta to Washington, went out to the school, and applied. The law school agreed to accept you in 1902. In those days you did not have to have a degree from a college to get into law school.

So you returned home and told the family the good news. Your happiness, however, was counterbalanced by two tragedies within the Neely family in 1900. Elizabeth Eve Neely, who had married Joseph B. Jones in June 1899, died one year later in June 1900, in childbirth, just as her grandmother had died in 1845. In December 1900 word came from Colombia that Allen had died of yellow fever. So now there were only seven siblings. Hannah had married William Plumb and moved to Augusta. Mary had married Reginald Grinell when she was twenty, and they had gone to Alaska in the gold rush which began in 1897 and was still going on in 1899 when they left Alaska.

So the only ones left in Atlanta at the turn of the century were Ben, Edgar, and Frank, along with their mother. And now, in 1902, Edgar would go to Washington, D.C. for almost three years. Edgar thrived in law school. Having had almost three years in the courtroom as a reporter, he knew the language of the legal profession, and he had learned not only a good bit of law, but he knew how to express himself. He made excellent grades, and the teachers at Georgetown Law School soon realized he was a superior student. He would make frequent trips back to Atlanta on the Southern Railway as Ben worked for the railway and had passes. Edgar was able to support himself completely in Washington; he had saved money, but he also worked nights typing up legal documents for law firms.

In 1902 when Edgar arrived in Washington, D. C., Theodore Roosevelt was in the White House. McKinley was assassinated in September 1901 and Vice President Teddy Roosevelt took over. He was only forty-two years old and was the youngest person to become president. This period in American history was the time of the so-called Progessive Movement. Roosevelt, the Republican, was its prime mover. Congress was conservative, but Roosevelt put through a variety of reforms and restraints on big business which included enforcement of antitrust laws against railroads, beef, oil, and tobacco monopolies. He also put through the Meat Inspection Act and the Pure Foods and Drugs Act. He was one of the early and great conservationists and brought millions of acres of exploitable land out of private hands and under government control for parks, national forests, and water power. The new laws required whole new courses for Edgar at Georgetown University.

During his senior year he met Emily Levering Eckfeldt. They were both twenty-six years old in 1904. She was the only daughter of a physician, Frederick Levering Eckfeldt, who had practiced in Philadelphia but had died relatively young, in 1902, of pneumonia. Emily and her mother, Kate Coogler Eckfeldt, had moved to Washington, D.C. in 1903 to be close to several members of Kate's family as well as a number of the Eckfeldts in Washington, D.C. The Eckfeldts, of Dutch descent, had been employed in the United States Mint for several generations, working in and directing the Mint.

One of Emily's good friends, Margaret Pokenhorn, met Edgar at a party for the young lawyers of Georgetown. She had recently been married herself and invited Emily and Edgar over for supper. Bright, vivacious, and very pretty, Emily was immediately taken with this handsome and charming man from the South. The romance progressed very rapidly, but Edgar felt that he should be established as a lawyer before considering marriage. He graduated with honors in 1905, began working back in Atlanta with one of the best law firms, and traveled back and forth to Washington, D.C. frequently as he and Emily were now engaged. They were married in Washington in 1907. Ben got passes on the railroad for most of the family: Henrietta Eve, their mother; Frank, now twenty-three; and Augustine and Annie Young. The wedding took place at the Episcopal Church in Washington, D.C. with a small reception at Emily's home, and the Eckfeldts, the Leverings, and the Neelys all had a splendid time.

So here you are, Edgar, standing right on the threshold of a new career complete with a new wife who would be a great asset to you as time went on. What will you do with your life? Will you live a full and fruitful life? Will you contribute to this world of yours? The world is your oyster at this point. Perhaps you will be the greatest of the Neelys. Let us see. Time will tell.

21
Levering—Earliest Memories

Earliest memories? What are some of the first things in your life that you remember? After seventy years, to look back in time and come up with anything before school age is very difficult. School is a landmark. School was a beginning. The first grade has many memories—blurred it's true. But with a little prodding, a little thinking, one thing comes back and then another.

But to dig back before school is tough. Still there are impressions before school, and often we can tie into some event, some simple occurrence which made an impression. Old picture albums help a lot, and you say: "Okay, I remember that!" If you had gone somewhere away from home, you can often remember that. But trying to recall something out of early childhood is difficult for most of us.

Three things stand out in my mind that occurred before the first grade at Tenth Street School in September 1918. The first: sitting on my father's lap while he was reading the paper at the breakfast table at 4 Barksdale Drive N. E., Atlanta, Georgia. The talk was about the headlines which told of the sinking of a big ship. I believe it was the *Lusitania,* now, but that date was May 7, 1915. I would have been two and one half years old. This all seems very unlikely, but I know it was about the sinking of a big ship. Perhaps there were other big ships sunk when America was actually in World War I. But I still have the strong impression that it was a major boat which sank. I remember the heavy acrid smell of my father's cigar.

The next thing I recall is my bicycle. I can date this very well. I was five years old. My bicycle was red. My older brother's bicycle was black. He had not received his bicycle until he was six years old, and when he heard and found out that I received a bicycle on my fifth birthday, he flew into a real tantrum. He was an expert at tantrums. They did not do any good, but well do I remember the fuss he made. Also, I remember learning to ride my bicycle and the joy of mastering it after many falls. I can still see that little red bicycle clearly as well as my brother's black one.

I remember my home in a vague way and a couple of nurses that we had, Irene and Jennie Zellner. They were sisters from Highlands, North Carolina, and they served us successively, perhaps taking turns looking after those three little boys who were so hard to handle.

The family went to Highlands every summer, even when I was an infant. The family went by train to Dillard, Georgia, and then by horse and buggy; later by a Model T Ford. I vaguely remember the Model T, the curvy mud roads, the getting stuck in the mud, and the rescue by a team of horses.

I do remember Cornelia and Alto, Georgia in 1917. Our family had some good friends and neighbors on Barksdale Drive where we lived. They had a farm near Cornelia and Alto, Georgia and one daughter, Katherine, who was about two years older than I. They were planning on staying the winter of 1916–17 to avoid the flu epidemic which was beginning to spread throughout the world. They suggested that I come to spend the winter and attend kindergarten while Katherine was going to the second grade.

I remember the house in the country very well, the big trees out in front, the hammock between the trees, and riding to school and sometimes walking. I can remember cows and milking; I can remember the barn where they milked the cows and kept the horses. I can vaguely remember the small, private school, but I don't think I went there long. I can hazily remember being shown the difference between boys and girls by Katherine. I was amazed as I had come up in a family of boys. My family bought a house near the Lindners very soon after World War I, and we went there often in the summer. Thus my memory of certain times may be wrong, and things could have happened later. Oh yes, I also remember dogs—one in particular named Joker. He was a black and white Llewellen Setter who was a very fine dog. I also remember a couple of goats that were given to us and how we tried to get them to pull a wagon.

All of these experiences were B.S. (before school), but things began to happen A.S. (after school) when I attended Tenth Street School—elementary or grammar school as it is alternatively called. Here I was back home from Alto—almost six and ready to start real school in the first grade. School began the day after Labor Day in September 1918. I won't say that I remembered the first day of school, but I do remember early in the first grade. Dad usually took us to school in his car on his way to work and we would usually walk home. As I drive up or down Piedmont Avenue now, I often think how many times I travelled this road as a child from age six to ten in the first through the fourth grades. That was five days a week for nine months (5x4x9 = 180 trips a year x4 years = 720 times) before I transferred to another school as the districts had changed. The first and second grades sort of merge together, but the second year was quite different from the first.

For some reason I don't remember my first grade teacher very well, but I do remember my second, third, and fourth grade teachers; I can see them in my mind's eye. I think we changed teachers several times in the first grade. My fellow students I do remember very well. I have a picture of my first grade class—there were two groups with different teachers. There were forty-eight students, and the class was divided into two sections.

In looking at this picture taken sixty-four years ago, I recognize most of the pupils as looking familiar, and I find that I can name about twenty-three of them. It is true that many of them remained in my classes for those four years; some went with me to Spring Street School in the fifth grade, and some rejoined me at O'Keele Junior High in the seventh, eighth, and ninth grades. Also, some were at Boys High School and Peacock School for Boys when I was there. Some went to college with me, and two of them came to my seventieth birthday party. So there have been many things to reinforce my memory over the years. But many of those students I have never seen again, at least for sixty years.

Still, I remember some students very well—what they were like and how they acted. Billy Fuller, one of my friends, was very jolly then and was pleasant and jolly the rest of his life until he died about five years ago. I remember his sister, Peggy, very well also. She must have been his twin as she was in the class also. She was also very pleasant but not pretty. Then there were those pretty girls that I held in awe, since I had learned about girls when five years old in Alto. Wilhelmina Doughtry—long blond hair with a ribbon in it. And that brunette—I have forgotten her name. I could talk to the boys, but I could not think of a thing to say to the girls.

K. T. McKinstry, Clatt Armstrong, Boisfeuillet Jones, Julius de Give, Tom Law, Sims Bray, Albert Stuhman, George Lowndes, Pollard Turman: I remember you all then; and sixty-four years later those I know now have enough resemblance to their old picture that I can pick them out easily. At least two of this group in those first and second grades were showing some qualities which would go with them the rest of their lives. Pollard (Booshie) Turman was outstanding in everything in those first few grades. He was not timid, nor was he too pushy, but he was the one that made good grades and excelled in athletics. He was curious about things and about people, and he still has that curiosity about people. He had a gentle father but a very strong-minded mother who was working for the League of Women Voters and Women's Liberation even then. So he had the genes of ambition from his mother. He became a lawyer, a president of a company, and was very active in community affairs. I could have told you in the first grade that he would go far in this world.

Then there was Boisfeuillet Jones—a fine French first name that was quickly shortened to Bo. He was quiet, dependable, studious—no nonsense in those days. He was not a fine athlete like Booshie, but he got along well enough. He became an outstanding citizen in the community with his continued work for Emory University, the Federal Government (he is said to have written the original Medicare law during the Kennedy and Johnson administrations); he now heads a large foundation giving away millions of dollars to worthy causes.

But I don't see any others who became outstanding later. Some may have moved away. Many, I am sure, are dead. But the seeds, some genes, some determination had already begun in Booshie and Bo in early childhood which persisted, grew, flourished, and bore fruit. Then there are several that I know that lived fairly ordinary lives making a living, but they did nothing outstanding later. What about the opposites of these ambitious children? Were there any that were either of low intelligence or low ambition who were somewhat of a failure in later years? Unfortunately, I am unable to see any of this category.

I still remember several of the girls especially well. Barbara Baker was a very pretty brunette, and by the second grade I could communicate a little with her, though I was still terribly shy. I remember a group of four children being taken to a movie downtown by our parents. We went to the Rialto Theater and saw *The Birth of a Nation*. It was a very stirring movie. Barbara comes to me as a patient now, and she says that toward the end of that movie she began to cry and I held her hand and said, "Don't cry, Barbara, I am here." I don't remember this part, but I do remember going and coming from the movie.

In the second year the flu epidemic was still going on and the school decided to use the second grade as an experiment. We would hold all classes in the outside under a shed, to ward off the rain, with open air on all sides no matter how cold it was. Large woolen sacks were provided, and we would slip into these at our desks after being wrapped up in sweaters and mittens. We did this for the whole second grade. Apparently the experiment was a failure as they did not do it again.

1919 was the year World War I ended. It had been a long four-year war for England and France, but America had been in the war only a year and a half. Woodrow Wilson, the idealist, did his best to keep us out of World War I. He was born in Stanton, Virginia in 1856 and was a child during the Civil War. He must have grown up in the bad times of reconstruction and seen the hard times of the post war years in the south. He spent those early years in Georgia and South Carolina. He entered Davidson College and then went to Princeton where he graduated in the middle of his class in 1875. He became president of Princeton, then governor of New Jersey, and then President of the U.S. in 1912. He was re-elected in 1916; one of his slogans helped: he promised he would keep us out of the war. However, with Germany continuing to sink ships in neutral water, he finally went to Congress to ask for a declaration of war, and Congress passed it with an overwhelming majority. The country was ill-prepared for war but geared up rapidly. The Treaty of Versailles was signed in June of 1919, about three months before I entered the second year of elementary school at Tenth Street. I had just turned seven years old. America suffered during World War I, but not nearly as much as England, France, and Germany. There were many young men who died in World War I, but there were less than fifty thousand deaths, about one fifth the number of Americans that died in World War II. The figures on the Civil War are so tremendous that it is difficult to believe—five hundred thousand men died in the Civil War.

As a child of seven I never realized that a war was going on until later. My father made speeches for liberty bonds, but as he had three children, he did not have to go to war. He was thirty-nine in 1917. I don't think this country ever had a serious shortage of food, although there were undoubtedly some shortages. A child of my age of course did not understand war, but my family and I as well as my friends were not affected much by the war. That was the winter that we had the outdoor classes to see if it would prevent the influenza. Fresh air was supposed to ward off the flu.

Influenza, which struck the world in 1918, was much more devastating in its widespread mortality than World War I. During the year of 1918 twenty million died throughout the world from influenza. This is of course an estimate, but it may even be low. For the United States we do have figures: there were twenty million people who became ill, and there were eight hundred and fifty thousand people who died of influenza across our country. There were probably one hundred and sixty million people in the United States at the time, and thus one out of eight people developed the illness, and one out of twenty of those died. The disease spread like an uncontrolled forest fire raging across the country, and there were many households that suffered loss of one or more members. This worldwide illness made World War I, with all its direct and indirect mortality, look small.

So how can the influenza virus be so deadly? Will it come back again to claim millions of people almost like an atomic bomb? Or do we have it under control? Will our vaccines stop it? We don't know for sure. There have been terrible epidemics throughout the years. There have been thirty-one pandemics recorded since 1510. Will modern medicine control these ravages? The influenza virus has now been identified and vaccines have been made and proved effective. But the virus keeps changing and new vaccines are needed almost every year. Antibiotics do not affect or kill the influenza virus, but most of the deaths come from bacterial infection which follows on top of the virus to kill people with pneumonia. Perhaps these antibiotics will prevent or cure this secondary pneumonia. But we really don't know.

So, here was a child of seven living through one of the most terrible wars of history and living through one of the most terrible epidemics in history, and he really did not know that anything unusual was happening in this world. Many years later he would study World War I in college and be fascinated. Many years later he would go to medical school and learn about viruses and epidemics of influenza. And just last week he took his yearly influenza vaccine.

22
Three Elizabeth Eve Neelys

Betsy

She Marches to a Different Drummer
Written by the family

Turn back Oh time -
Go back with us to August 1953,
A child is born, a blessed child.
Who would have dreamed then
What a great person
What an unusual woman
What a life would evolve
In those thirty-two years.

She began life slowly,
Looking at it, drinking it in,
Then as time went on,
She began to grow and blossom,
And soon we knew that
In spite of her gentle innocence,
She would always be
Wonderfully different.

This difference really began
When she discovered herself
And life in Colorado College.

Then we began to know
That Thoreau was right.
Here was the frontier woman
She found the West and
We knew
She would never come back East.

Who has a ten year old VW Beetle?
Who rode her bicycle alone
Through Wyoming 400 miles?
Who would camp out anywhere
Even in deep snow?
You guessed it—
Betsy (Elizabeth Eve Neely)

After college to the Philippines
With the Peace Corps two years
To a little town
High on the side of a mountain near Bontac
She taught the natives in their language—Igorot
And they loved her
They cried when she left.

Then off alone around the world,
Backpacking to Malaysia, Thailand,
Nepal, Greece, Italy, France,
England and home
No fear, no problems for her.
Around this entire globe,
All by herself.

Back to Utah
(can you believe it—Utah from Georgia?)
To get another degree
Again in Botany and
Work in the herbarium
She loved plants with
A deep and abiding love
Her specialty?
Endangered species of plants.

To her, clothes were irrelevant
She never had a skirt till thirty,
She always wore
Her good pants—Army fatigues.

Other activities?
Tennis, running, drawing,
Canning, gardening, baking bread,
Cross country skiing.
This girl was a
Rugged independent woman,
A real Amazon,
And yet beautiful,
Raven hair down to her waist,
A long tall drink of water (or wine)
Shapely yet muscular.

Oh, I almost forgot
Pinto beans.
They were a part of her life
She knew the whole life cycle
Of these beans
And ate them almost daily,
With and without relish.

What is this story
About Tierra del Fuego?
Mountain lions and
Endangered plants?
No big deal, she says.
So sweet Betsy
we love you.
Everybody loves you,
But not as much as we
And we are happy
That you now have
One to love you
And appreciate
This delightfully different,
Lovely woman,

Betsy,
You are the greatest.

* * *

Let us look at some Neely women. There have been three Elizabeth Eve Neelys—two still living and one deceased. The first Elizabeth Eve Neely was born in 1872 in Augusta, Georgia. She was Ben Neely's child—

the fourth child and the second girl; John was two years older and Ben, Jr. was two years younger than she. She died in 1900. The second Elizabeth Eve Neely is the daughter of the late John Neely, the doctor. She was born in Bainbridge, Georgia in 1906 and is living in Atlanta. The third Elizabeth Eve Neely was born in 1953. She is the daughter of F. Levering Neely and Margaret Gardner Neely. She is living in Utah. There were two Elizabeth Eves. One was the sister of Oswald Eve who died in 1822, and the other was the daughter of Oswald Eve. She died young at the age of ten in 1850 of scarlet fever.

Let us concentrate on the three Elizabeth Eve Neelys. We know a little more about them. Let us call them EE I, EE II, and EE III in order to distinguish them more easily. EE I, when you were born in 1872 your father Ben was thirty-eight and was organizing the schools in Augusta, Georgia. The Civil War had been over for only a short time and painful reconstruction was slowly getting under way. Allen, Annie, and John were your older siblings; all were less than two years apart. Allen was only four years old when you were born. So your mother, Henrietta, really had her hands full. Your brother, John, age one-and-one-half, was still in diapers, and the other two had only recently outgrown them. Of course, as the baby, you were the center of interest at least until the next baby, Ben, Jr., came along. Then there was a steady procession of children until there were eight little Neelys within eleven years. Five years after Mary was born, along came Frank.

But you can imagine that household in Augusta a few years before Frank was born. Eight children, from infant Mary to Allen, had to be fed, clothed, put to bed, and the older ones sent off to school in the morning. You, Elizabeth Eve Neely, were right in the middle of this gang. You had just turned seven when Mary was born. You were in the second year of school. You were a brunette with blue eyes. You were bright in school and studied hard as did all the children of a school master. There wasn't any question about it, all the children in school did their homework well, and most of those in school were well ahead of their class. You, Elizabeth, were no exception. You knew the alphabet perfectly. You were able to read simple books and you were especially good at spelling. You would continue to spell almost perfectly on through grade school. At seven you also were already a great help to your mother. Annie, the oldest girl, was ten, and she taught you how to look after the baby Mary. Hannah, the next girl down the line, was just three at the time. But you girls stuck together, and as you grew older you were always very close, each having taught the younger ones. You all loved your mother very much.

In 1882 all of the children were in school except Edgar who was four and Mary who was three. This was the year that your father, Ben, Sr. resigned from the school system in Augusta and took the job in Rome to develop the school system there. You, Elizabeth, were ten years old. You were doing very well in school, had many friends, and your father's decision to move away was like the end of the world to you at first. But your mother, Henrietta, was so calm about it that you children finally took it philosophically. So father went to Rome first, found a house, and then came back and things were packed up, and the family took the train to Atlanta, changed trains at the old Union station, and went on to Rome. The town was much smaller than Augusta, and you could not help being homesick for a while. But once you got into the spirit of adjusting to the newness of it all—a new house, a new school, new friends—it became an exciting and interesting time.

You, Elizabeth, fitted into the new school very well, and quickly, established yourself as one of the best students in the fourth grade. You were the musical one, could keep a tune, sing, and dance; you learned to play the recorder first and later the flute. You had a *joie de vivre,* you loved your sisters and brothers, and as you grew and became a young woman, everything you did was right. You went through high school with very good grades, finishing at seventeen. Your father wanted you to go to college or normal school like your older sister, Annie, and off you went to Shorter College which was close by. You went there two years, graduated, and began teaching at age nineteen in 1891. That was the year that your father became ill, and when he died in 1892 your world disintegrated. You were not twenty, a pretty, intelligent, young woman. Fortunately you had your teaching and music to keep you busy. You contributed to the household finances. You had known Joseph B. Jones since high school. He had gone off to Georgia to college, and after a long courtship you were married in June 1899. One year later, in June of 1900, you died in childbed.

Here was another life snuffed out just as it was beginning. You never got a chance to pass on any of those good (or bad) Neely genes. You were mourned by the whole family, leaving an empty spot in the heart of the family that could never be filled. And later that same year, on December 19, 1900, word came from Bocas del Torro, Republic of Colombia, that Robert Allen Neely, the eldest of Ben and Henrietta's children, had died of yellow fever. Two deaths within six months; the family of nine children was reduced to seven. In those days death was encountered much more frequently; although there was sorrow and mourning, it was accepted. Any large family usually lost a few children to childhood diseases such as scarlet fever or even measles. And to die in childbirth was really not unusual. Susan Henry Neely, Ben's mother, had died in childbirth, and now Elizabeth Eve Neely the First had met her Maker in childbirth.

1900—the turn of the century—Allen and Elizabeth Eve Neely Jones had seen the twentieth century arrive, but they would never witness the tremendous changes in life-style that developed from 1900 to 1985. Frank, the youngest, lived the longest, from 1884 to 1979. He witnessed the arrival of the automobile. He saw two world wars. He saw the advent of the plane, and he watched it replace the train and the steamship. He saw and was involved in nuclear research and power at Georgia School of Technology. And he saw the beginning of computers. In addition, he saw changes brought about by the demand of minorities for their civil rights. And along with all these so-called advances he witnessed the beginning of an erosion of society—first the hippies, then the drug culture, and the breakdown of the family. He saw crime come to Atlanta and the white flight to the suburbs. He did not really take in the problems of American society, for when he was ninety in 1974, his mind was beginning to slip into a state of forgetfulness and indifference. So his sister Elizabeth Eve's view of the world in 1900 and his overview of the twentieth century were startlingly different. She had no issue, and he had only one child to carry on those potentially great genes.

So Elizabeth Eve Neely II, known as Bess Neely, what about you? Whither did you go? Born March 1, 1906 in Bainbridge, Georgia, your father, John Carmichael Neely, was a physician with only one leg who made his medical rounds first on his horse throughout the area and later by automobile, one of the first ones in Bainbridge. Your mother, Anna Kate Godbee, had her first child, Mary Louise Neely, in 1902, but this child died at ten months from whooping cough. Then came John Lewis Neely, born in 1904; and then you, Bess, in 1906. You came up in Bainbridge, attended grammar school there, but when you were in the second grade your father died; your mother died when you were sixteen, and you went to live with your aunt. You must have inherited a good share of the genes of ambition as well as the tradition of schooling, for you went to college in Milledgeville, then on to Atlanta to learn interior decorating. You stuck with interior decorating the rest of your life, had the will and determination to excel in your field, and when you had sufficient funds you went to New York for a year of study; then to Paris for a year of fun and further study of interior design. You never married so those genes of determination and persistence were never passed on. Bess, your first job was teaching at a finishing school in Washington, D.C. You stayed only three months because you found that the girls there were really not interested in learning interior design. You then went back to Milledgeville and taught at Georgia College for Women for several years. In 1929 you decided to go back to New York and get a job. You weren't even aware of the depression, but you found a job in New York designing wall paper, stayed in New York for several years, and finally returned to Atlanta. There you opened an office for the private practice of interior design, did very well in your profession, and while still working in the day, you opened your own school for interior design. You started with two pupils, but within five years you had sixty pupils and needed another teacher. You loved to teach and were a great teacher. You had a great heritage of teachers behind you. World War II came along and you joined the Army as a draftsperson. After the war you worked for Davison's Department Store as an interior designer.

Now what about you, Elizabeth Eve Neely, III? You are the daughter of F. Levering Neely and Margaret Gardner Neely. You were born in 1953 in Atlanta. You are the youngest of the six children. EEN, II is your father's first cousin, and EEN I, was your father's paternal aunt. So you all are descendants of Ben and Henrietta Neely and should have some common characteristics as well as differences from infusion of other blood lines.

Does position in the family mean a lot? Perhaps. It may depend on how close your next siblings are, and the personal relationships. You came up with a disadvantage. Your sister, Sara, just two years older than you, was very bright in school, adjusted well, and you must have felt consciously or unconsciously inferior to her. She was the shining light and you were frequently in the shade. Your grades in school were never more than mediocre until you blossomed in college and graduate school. You got along well with your sister, and there was a deep affection and love between you, but you must have felt great frustration when she brought home such good grades from grammar school and later in private middle and high schools.

Although you usually were a happy child, it must have been hard on you to measure up to Sara. However, you were perhaps the gentlest of all your five siblings. You had great empathy for people and animals; their sufferings would cause you to weep. There was never a sweeter or more generous child than you. If someone was in need, you would provide help. You were tall, well featured, and really beautiful with a pretty figure and long dark hair.

When Sara left for college you were the only child at home, and you then became more of an individual. You had many friends at school and graduated in the middle of your class. Your mother and father thought you should go to a small, junior college rather than to a big university. They took you on a tour of the New England junior colleges, and you settled on Bradford, a school for girls in Massachusetts. That was a mistake. You did not like the general atmosphere, and after one year you dropped out, came home, and got a job with the Georgia Conservancy. You were beginning to feel an interest in art and botany as well as for the great outdoors—camping and skiing. You applied to the University of Colorado in Boulder and were accepted. Bradford had been too small. The University of Colorado was too big. Finally you got settled in your junior year at Colorado College in Colorado Springs. There you began to develop into a person with some idea as to what you thought life should be like. After completing your junior year, you spent a summer in Florence, Italy studying art. You began to mature after that, and during your senior year at Colorado College you did extremely well and were very well thought of by your professors.

But you were an outdoor girl now. Your degree was in botany with a minor in art. You were now hooked on the West with its open spaces and freedom from the bad features of the crowded city. Your older sister, Virginia, had been in the Peace Corps in the Senegal, West Africa, and your older brother, John, had spent two years in Brazil in the Peace Corps. You applied and were accepted for two years in the Philippines in a small town in the north of Luzon near Bontak. After a training period, off you went to the primitive town of Guinjang. It was a small rice-terraced town perched on the side of a mountain. There were no roads to the village. One had to walk four miles on rocky paths to get there.

You stayed in that village for two full years, learned the language, and taught nutrition and sanitation to the children and adults. You helped in the rice fields. You became such a trusted member of the village that when it was time for you to leave the whole town had a party for you. They killed a pig and cooked it in your honor, and there was a feast. Your parents were there at the time, and the old men of the village toasted you with their rice wine and asked your parents if you could not stay another year. Through an interpreter, your father told them that he was getting older and he needed his daughter to look after him back in America. They seemed to understand this, and after the feast there was dancing—the same dancing that there had been for hundreds of years. And your parents were given the only bed in the village which was in the Episcopal Church. It turned out that this bed was a large box or chest with a few blankets on top and some books for a pillow. They slept on it anyway, and the next day you took them back to civilization on a long jeep ride; then back by plane to Manila.

You left on a round-the-world journey to the States in a few days. And when you came back to the United States, you were a new person, mature and relaxed; still very much interested in the great outdoors and the wonderful plants. And so you went to Utah and enrolled in the botany school at Utah State University to work towards a postgraduate degree. You were still tall and beautiful and now self-assured as to what you wanted to do.

But at thirty-one (it's 1985 at this writing) you show no inclination to settle down, marry, and raise a family. So here are the three Elizabeth Eve Neely's without issue. Here are three excellent women who have as yet shown no signs of passing down those genes of quiet determination, good cheer, gentleness, and love of their fellow man to future generations. Come on, Betsy, let's get going and have some kids!

23

Ben Neely—Boston and Home

The good ship *Nerrisa* was now safely moored at a dock in the Boston Harbor in March 1848, having come from Halifax, Nova Scotia with a large load of cod fish as well as some spinning and weaving machinery from Scotland. It would take at least a week to unload the ship. The cold weather was beginning to abate a little, but it would still go down almost to the freezing point at night.

Ben had been gone from home almost six months. He was now almost fourteen and a half years old. He had grown rapidly during that time. He was now 5′ 9″ tall, his voice deepened, and although thin, he was wiry and muscular from the hard labors of ship-life. But most of all, Ben had grown in maturity and in his outlook on people and the world. He had been thrown in with a tough group of men—mostly uneducated but all knowledgeable about the ways of the world and how to survive. There also were some of the group like the captain and the first mate, Mr. Rumans, who were educated. The captain was a scholar and had a very good library which, as mentioned earlier, he encouraged Ben to use.

The ship's crew had a good morale, and in many ways was like a big family with a great variety of personalities. Thus Ben, as the mess boy, had a chance to know the seamen and the officers very well. He had learned how to get along with people and how to talk with people, invaluable abilities which would help him for the rest of his life. Being an observant person, he had a good opportunity to watch the clash of personalities of these people thrown so closely together. He had now visited Cuba and Haiti and had seen the Spanish and black French cultures. He had sailed to Liverpool and traveled to London. He had seen Scotland briefly and then had crossed the Atlantic to St. Johns, Newfoundland. From there he had gone to Halifax, Nova Scotia, and now he had just arrived in Boston, Massachusetts in the early spring of 1848.

He had had an unusual opportunity to listen to the English language as it is spoken all over the world. He had been brought up in Baltimore and Easton, Maryland. The accent in Baltimore and Easton is more northern than southern, but it was softer than that of New York with a slight Virginia and southern quality at times. Speech and accents of various parts of the country and the world fascinated Ben, and he made a hobby of mimicking various accents. He could take off on the Swedish accent of Bjorn and amuse the crew with his antics. He was careful, of course, to avoid making fun of Bjorn. This hobby which he carefully developed stood him in good stead in later years when he became a spy for the Southern Army and was able to cross the battle lines in the Civil War and be taken for a Yankee.

So here he was in Boston, Massachusetts in the spring of 1848. To Ben it was another big city. He would be in Boston for about one week. Could a young man of fourteen appreciate that he was in one of the most important cities in the United States at that time? Could he get the ambiance of such a city? He had received some help from the captain. He had been given a small book on Boston, and he had been given a history book which told of the Boston Tea Party, Paul Revere, and Betsy Ross. So he did have a little understanding of this place. Also, the captain invited him to go on a sight-seeing tour with him in a horse-drawn carriage.

Boston had been settled by the Puritans and three Puritan churches dominated the religious life of the town for fifty years. But when England sent Sir Edmund Andros as its first royal governor, the Church of England came with him and the Puritan isolation was over. But Boston did not remain docile. The rebellious citizens removed Andros from office in 1689 and imprisoned him. After the Stamp Act of 1765 and the Boston Massacre of 1770, in which the British troops fired on a crowd of civilian hecklers and killed several persons, the Bostonians were aggrieved with England and colonial rule. The Boston Tea Party of 1773 was one of the key incidents leading to the Revolution. The exchange of shots at Lexington and Concord on April 19, 1775 was the beginning of the Revolutionary War. It was Washington's siege of Boston that drove the British out. John Hancock was elected the first governor of Massachusetts in 1780.

Boston had been a great seaport but also had become an industrial and railroad center. Its sea trade was temporarily disrupted after the revolution, but was soon flourishing again, and in 1848 it was one of the three largest ports in the United States along with New York and Philadelphia. Culturally, Boston had begun to be known for its colleges and hospitals. Harvard had been founded in Cambridge across the Charles River in 1636, but by 1848 it was probably the best known college in the country. Many other colleges would soon spring up. Boston became a great center of medicine with the Massachusetts General Hospital, the Peter Bent Brigham Hospital, Beth Israel, Children's, and the Deaconess Hospitals, all of which Harvard used for their medical school. Thus began a great tradition of excellent medical care which persists to this day.

For a young man of fourteen to appreciate Boston fully was impossible, but he did get a flavor of the city from the books he had read plus the sight-seeing tour that the captain provided. He got to see the great harbor and seaport, the Charles River, the home of Betsy Ross, and Harvard College. His father had attended Oxford in England after finishing a degree at the University of Dublin, so Ben had some appreciation for and awe of institutions of higher learning. He later would attend the University of South Carolina as he too would later enter the profession of teaching, just as his father had done. Many years later he would take a sea trip from Savannah to Boston. That first view of Boston would stay sharp in his memory for the rest of his life. He had seen one of the great cities of the world, and that brief experience at that young age would vastly broaden his perspective.

Here it was in mid-March of 1848. The *Nerrisa* had come back to the United States safely and after about one week in the Boston Harbor was preparing to sail back to Baltimore, the home port. There would first be one more stop in New York. On arriving back in the United States Ben was beginning to have an inner conflict that would be very hard to resolve. Should he try to go back and take whatever punishment would be waiting for him, or should he stay with the ship for another of its journeys which might take another six months or longer? He was well liked on board the ship, and he knew he had the job if he wanted it, but the desire to see his family with all of his brothers and sisters was strong within him. He stayed awake nights trying to decide. He was not his jolly self, and Bjorn and other members of the crew realized that he was depressed. But he kept the secret to himself as he still was sticking to the story that he was an orphan and had no family.

The ship now took on more textiles from the mills of Boston to take to Baltimore. It was a cold, dark, and windy day when the ship finally took off from the dock, swung out into the harbor, and headed out to the open sea with its great rolling swells. The ship pitched and rolled, but the Baltimore clippers were well constructed and quite seaworthy, and off they sailed for Manhattan. This should be only a three-day journey if the wind kept up and the sea was not too strong, but March was the equinox, and the change from winter to spring caused a small gale; the journey took eight days.

So here was the *Nerrisa* heading toward the Verrazano Narrows which lead into the protected harbor around Manhattan with the Hudson River on one side and the East River on the other side. No bridges had been built yet; it would be in the 1880's before the famous Brooklyn Bridge would be built, and it would be in the 1970's that the huge Verrazano Bridge would be constructed, spanning the great entrance to the harbor.

Giovanni Verrazano had spotted the site in 1524, but the English navigator, Henry Hudson, discovered the Indian city on Manhattan on September 3, 1608. He had sailed up the river, later named for him, looking for a passage to India. Peter Minuit, a Dutchman, bought Manhattan from the Indians for sixty Guilders, worth about $24.00, in 1636 and then began a Dutch settlement. But in 1664 the fleet of the Duke of York sailed in and took over, and it became an English settlement, receiving its royal charter from England in 1686 with the name New York. It grew slowly, but after the Revolution it became, for a short time, the first capital of the country before the capital was moved to Washington, D. C. In 1789 the settlement on Manhattan was about thirty-three thousand in population, making it one of the largest cities in the country. By the time Ben arrived in 1848 it had nearly 100,000 people. Many would stay a short time and move on, but those who could find a livelihood would stay on. The Italians, Irish, Dutch, and Germans soon had huge sections of individual nationalities scattered about Manhattan. It also became a haven for Jews who came from many different countries and banded together. Of course all of these ethnic groups had their slums. Manhattan had been laid out in an orderly fashion with the numbered avenues running north and south, the streets east and west, and Broadway as a large diagonal across the other streets. The plans for the huge Central Park in the middle of the city were already included. There were docks for ships' berths on both sides of the island, making it a huge port. Jersey City was just building up as well. The railroads were becoming more important; the New York and Harlem railroad was built in 1832 and by 1848 was moving huge amounts of cargo from the ships to all parts of the country. New York was now the ship building center of the U.S., having surpassed Boston and Philadelphia. So in 1848 the city was already on its way to becoming the financial, manufacturing, and cultural center of the New World.

So Ben again arrived at a huge metropolis, now having visited Liverpool, London, and Boston. He and Bjorn the cook took long walks through the city, and he was amazed at the sights he saw. He saw the great extremes of wealth and poverty, carriages with footmen, the shops on Fifth Avenue, and the great masses of people crowded into the slums. He and Bjorn walked through the huge fish market and the vegetable and meat markets where they bought some provisions for the next leg of the trip to Baltimore.

He was still being torn by indecision—should he go home and face the consequences of his assault on his stepmother or stay with the ship? The time of decision was getting closer and he knew that he would soon have to make up his mind.

John Neely, Ben's father, was in the midst of his examination for the Ministry. He was a very busy man—still the principal and headmaster of the school in Easton; he had managed to pursue his studies for the ministry with classes in Baltimore on Saturdays. Sometimes he would go into Baltimore on Sunday and participate in the church service in one of the larger Episcopal churches there. He had some misgivings about speaking before so large a group; as a teacher he had, out of necessity, done a great deal of organized speaking. The congregation as well as the ministers who were present enjoyed his sermon, "Accepting the Lord's Will."

Easton, Maryland was a small town on the eastern shore. Its main industries were crabs, lobsters and fish; in addition, there were now some small factories and mills which made it a good-sized community. The Bellair Academy was quite popular with students from the entire eastern shore; because it had no dormitories, many of the students lived with families in town during the week and went home for the weekends—if they lived close enough.

John had done a good job at the Academy; Reverend Davis, the headmaster, had retired, and John had become the new headmaster. The school became well known in the area, and John was a respected and active member of the community. Now that he was headmaster, and in his mid-forties, it would appear that he might remain there the rest of his life. However, he was a man who was not afraid of change, as evidenced by his leaving Ireland at the age of thirty-one and coming to America. Even though he appeared settled into life at Easton, there was that adventurous or possibly ambitious drive that would cause him to move the whole family to Augusta, Georgia a few years later.

But it was now April of 1848. Spring was just coming in; the bitter cold was over; the snow had gone, and everyone was looking forward to warm weather. The *Nerrisa* finally arrived in Baltimore and was securely docked at the Baltimore and Ohio docks. This was the home port; it would stay here about two weeks to get refurbished, painted, and cleaned, and plans for the new trip would be arranged with the captain and the company. Ben stayed aboard ship the first day working and helping the cook, Bjorn. He decided that he would go to Easton and see what was going on. But he would do it toward dark so that he would not be seen or recognized. His appearance had changed considerably since he had left home and with his seaman's clothes he would not be recognized by the townspeople.

The family was just sitting down to dinner when he got to the house and peeked in the window. John was at the head of the table, Isabella at the other end, and the children on each side. His father was just saying the blessing when John looked in the window. He saw his father, brother, and sisters and almost burst into the house. Then he saw what he thought was the stern face of his stepmother, and he knew in a flash that he could not come home now. Off he went, crying again, back to his home aboard the *Nerrisa*.

24
Frank and Rae

Here was a young man with new ideas and burning ambition which showed through his exterior like a glowing bulb. He was humorous, full of fun, and even then had the ability to make people like him and enjoy being with him. Rae soon recognized that here was a person who would go far in this world. She had seen the ingrown and small world of Atlanta, and she had spent two years in the best women's college in America—Smith College in North Hampton, Massachusetts. She had been taught by some of the best minds in the country, and they had inspired her to like the scholarly life of music, art, and books. She naturally was somewhat of an intellectual snob at age twenty when she met Frank.

Frank in turn recognized her bright mind, and he also saw that flame of ambition which she had. She was pretty, dark-skinned, and vivacious. He had never met a girl or a woman before with such a bright mind. It was an entirely new experience for him; most of the women and girls he had met were nice but usually dull. So here was a mutual, strong attraction.

Rae's mother and father, being Reformed Jews, were much more liberal-minded than the old fashioned, Orthodox Jews. Thus they did not really mind the attentions of this bright, young man to their daughter. Frank, brought up an Episcopalian, had attended church and Sunday school in Rome, Georgia as a young man, but by the time he was at Georgia Tech, he was so busy that he attended church only if he went back to Rome to visit his mother. His brother, Ben, Jr., lived in Atlanta, having married in 1898. Frank would visit them frequently and took Rae Schlesinger to their house often for a meal. Ben and his wife, Nina, already had three children in 1904—Annie, age two and twins Nina and Mary.

Atlanta in the early 1900's had a small Jewish community which stayed very much to itself socially. There were some very good businessmen who dealt with the Atlanta business community in the day, but social mixing at that time was almost unheard of. The Jews had their synagogues, their Rabbis, and life among the Jews revolved around their places of worship. They had no special prejudices against the Christians, but they were very conscious that there was prejudice against them from the Gentiles.

The social structure of the Christians of Atlanta was also built around their churches—the Episcopalians, Presbyterians, and especially the Baptists. The Catholics were still a small minority, and there was even some prejudice against them. Also, there were several social clubs such as the Piedmont Driving Club and the Capital City Club. No Jews were allowed in these clubs. Even some of the better hotels had been known to turn away Jews. In public schools, there were quite a few Jews, and, being different, they were often the butt of ridicule and often tormented by young toughs. The Jews were always a little smarter in school and this made things more difficult for them. Atlanta was still a provincial town in many ways; finding a job for a Jewish person in a Christian community was often difficult. They thus stayed in the fields of manufacturing and selling; they varied from street peddlers to department store owners. Rich's clothing store had been started as a small shop soon after the Civil War. It had grown slowly; by 1900 it was the largest department store in Atlanta. It was founded by the Rich brothers and passed down through several Rich generations as time went by.

The Orthodox and the more liberal, Reformed Jews had only one synagogue at the time, and they had their services at different times. The reformed group usually was the more established and wealthy businessmen such as lawyers, doctors, and the big merchants. Names like Rich, Elsas, Schlesinger, Haas, Heyman, and Asher, who had lived in Atlanta since the Civil War, all had large families from several generations by 1904.

By the time Frank graduated from Georgia Tech, he was going out with Rae regularly, was invited to her house often, and had talked informally of marriage. Frank knew that he would have to be established in a good job before he could make a formal proposal. He had talked with his brother, Ben, who was now a promising, young businessman in the city. Ben had quite a few reservations about Frank's marrying a Jewish girl and told him of his misgivings; he thought it would be a hindrance to Frank's career. However, he liked Rae and really had no strong feelings. Frank also talked with his mother who was living in Atlanta by then. She also liked Rae, but she too expressed some misgivings. Frank listened but had a tendency to brush these warnings aside. He had just accepted a job with Westinghouse in Pittsburgh, Pennsylvania. He had passes on the Southern Railway and would be able to come back to Atlanta from time to time to see Rae.

Rae in turn discussed marriage to Frank with her mother and father. They were both very impressed with Frank, but they knew that there would be problems; they told Rae frankly that her life might be difficult; she would not be accepted well by the Christians. They suggested that the two of them talk to the Rabbi—this was a young, bright Rabbi who had been brought down from New York to be the assistant to the older and established Rabbi Kahn.

Frank and Rae went first to the young Rabbi Marx; he was very encouraging, but told Frank that he was sure that Frank would have to become a Jew before Dr. Kahn would marry them or permit Dr. Marx to marry them. The two rabbis, Frank, and Rae talked further. Rabbi Kahn said that definitely Frank would have to adopt the Jewish religion in order to marry Rae. Without hesitation Frank said, "Certainly." Everyone looked at him in astonishment because he agreed so quickly. The marriage would not take place for some time as Frank would soon be leaving for Pittsburgh. However, they talked it over with both families and became engaged. The two families got together at the Schlesinger house for an engagement party.

Rae had four brothers and sisters, and Frank now had six living siblings. They all came to the party, even Edgar came down from law school in Washington, D.C. He was about to be engaged to Emily Eckfeldt in Washington, D.C. John Neely, the doctor, came up from Bainbridge, Georgia; Annie and her husband, Augustus Young, came from Cedartown, Georgia. Hannah was married to William Plumb in 1902 and had moved to Augusta. Mary had gone to Alaska shortly after her marriage to Reginald Grinell; they were still there. Frank's mother, Henrietta, of course, was there.

The party was a very festive occasion and was held at the Schlesinger's home. They spared no expense and had many courses of food and a small orchestra for dancing. There naturally were speeches after dinner, and all of the Neelys made fun of Frank because he was the youngest, the baby, of the family. The two families got along amazingly well; both groups were outgoing and full of fun. However, when Frank later told his family that he would have to become a Jew himself in order to marry Rae, there was some general consternation; this was almost unheard of at the time. However, they did not try to make him change his mind. So Frank departed in September 1904 to begin his first job in Pittsburgh with the Westinghouse Company. Frank and Rae's plan was to be married in about a year; while in Pittsburgh Frank would find out what he would have to do to become a Jew. Having considerable free time, he would try to fulfill the requirements.

"Judaism, the religion of the Jews, is the complex expression of a religious and ethnic community, a way of life as well as a set of basic beliefs and values, which is discerned in patterns of action, social order, and culture as well as in religious statements and concepts" *(Encyclopedia*

Brittanica, 1982, Vol. 10, page 284). This is a very comprehensive statement, and when Frank agreed to become a Jew he had no idea as to what the whole religion was about except for a general idea; it was another religion which he could take part in or leave alone just as he had done with the Episcopal division of Christianity. His grandfather, John, had been an Episcopal Minister as well as a teacher, but Frank had had no real contact with him. He knew that Jews believed in the Old Testament and he thought he could follow the Jewish precepts if that was the only way he could marry Rae.

By talking with Rae and beginning to read some of the books on the history of Judaism, he soon realized that there was a great deal more to Judaism than the simple concept which he had previously thought. He found that

> Judaism is not and cannot be viewed as an abstract, intellectual system, although some of its affirmations may be couched in such terms. It affirms divine sovereignty disclosed in creation (nature) and in history, without necessarily insisting upon—but at the same time not rejecting—metaphysical speculation about the divine. It insists that the community has been confronted by the divine, not as an abstraction but as a person with whom the community and its members enter into relationship. It is—as the concept Torah (connoting "guidance" or "direction") indicates—a program of human action rooted in this personal confrontation. Further, the response of this particular people to its encounter with God is viewed as significant for all mankind. The community is called upon to express its loyalty to God and the Covenant by exhibiting solidarity within its corporate life on every level—including every aspect of human behavior, from the most public to the most private. (ibid.)

Frank became more interested in what Judaism really was, and as he continued to read, he began to think that the theories of Judaism often made more sense than the strict creeds of the Christians. The involvement of the religion into the life of the community seemed a good precept. The concept of one God seemed similar to the Christian ideas, but when the teachings seemed to indicate that God had chosen a particular people—Israel—as his favorite people, Frank had some questions. But he made an effort to read the Torah and books on the interpretation of it. Then there was the Jews' beliefs on death. They seemed to think that death was the end of a life, but that some of the power continued to exist. They seemed to believe that it was possible to have a resurrection. Frank particularly liked the strong moral code of the Jews, as applied to dealing with one's fellow man. Frank always exuded a kind of fixation with telling the truth, with complete honesty. These ideals would later be of great importance in his dealings with people. He often told people, "There are two things I can't abide—a liar and a thief.'' Scrupulous adherence to truth and honesty were part of his character for the rest of his life.

From history books he had read about the horrible pogroms of the Jews throughout the ages. He knew about discrimination, segregation, and rejection of the Jews by Christian civilizations, but he thought that this would not have any effect on him and Rae and their lives. It really wasn't until a few years later that he personally felt any discrimination toward Rae or himself. It would be many years later that he really understood the extent of the Nazi German effort to truly exterminate the huge number of Jews in Germany. Six million Jews were slaughtered in Germany during World War II. Frank was the president of Rich's during World War II. He was extremely active in the community, served on the War Production Board and was one of the most powerful men in Atlanta. He felt no discrimination toward himself or Rae at that time, but over the years he experienced several disturbing episodes especially concerning Rae. By World War II, even though he had become a Jew when he married, he had not made a practice of going to the synagogue and had more or less become an inactive member of the religion.

But in 1904, in Pittsburgh, he went to the synagogue on Saturdays and took courses with the Rabbis about becoming a bona fide Jew. He studied the religion and its precepts and felt no real hesitation about joining the religion. In the meantime, he was working extremely hard at the experimental factory—often doing overtime. His fellow workers were men with very bright engineering minds from the Massachusetts Institute of Technology, Lehigh University, and from many of the best technical universities in the country. The electric-powered locomotive was becoming a reality, and the perfection of this locomotive plus planning the type of safe, electrified tracks, were gigantic undertakings.

Frank was getting frequent raises from the company and was beginning to feel as if he were financially able to support Rae in Pittsburgh after about eighteen months of work. They had corresponded frequently, and he was still able to go back to Atlanta often. However, it was about this time that Henrietta Eve Neely, his mother, became very ill, first with a series of small strokes and then a massive stroke which took her life in early 1906. They delayed the wedding; not until early 1908 did they marry. Frank had now become a Jew in Pittsburgh, and both Rabbis at the Atlanta Synagogue agreed to marry them.

The wedding took place with a large part of the Atlanta community of Jews present as well as a large number of Frank's family. Ben, his wife, John with his wife, Hannah Neely Plumb and William Plumb, Annie and her husband, Augustus Young of Cedartown all came. Edgar had been married only a few months himself, and he brought his new bride, of course. As at the engagement party, the Schlesingers hosted the reception, but this time at the Kimball house; it was a large party: four hundred people attended the seated dinner.

So now Frank and Rae were finally married after an engagement of almost four years. They were much more mature now at twenty-four. Their mutual attraction had never waned, and they set up housekeeping in a small apartment in Pittsburgh close to the Westinghouse factory. Rae brought a large library with her, and the public library was nearby. She became one of the library's best customers. In addition, she worked in a settlement house in Pittsburgh which kept her busy while Frank was working at the factory.

The locomotive had now been tried out and was found to be better than expected—it was quiet, very powerful, and easy to handle. The company had experimented with a third rail but had found that an overhead wire was much more practical and safer. Frank had been involved in the development of the engine and in the plans for the factory which would manufacture it. At twenty-four he had already had broad experience in his field.

However, he and Rae really preferred Atlanta, and after about one year in Pittsburgh, during a return visit, he talked with some of the engineering consulting firms in Atlanta; one of the best ones offered him a splendid position as an advisor for building and operations of new factories coming into the area. So back to Atlanta the happy couple moved. They had saved their money and were looking for a small house. They found a real estate agent who showed them a house that they liked and could afford. But their excitement was short-lived when their real estate agent told them he could not possibly sell them a house in that neighborhood because they were Jewish.

25
Heredity versus Environment

A watershed is like the Great Divide where all the drainage water runs one way on one side and the other way on the other side. It is a very popular word and, figuratively speaking, means a turning point in a certain situation. In climbing a mountain such as Mount Everest there are resting points along the way or camping sites for the night where one rests, nourishes oneself, and can look up the mountain to see what lies ahead. But most important, there is the chance to look back or look down the mountain to see what has already been accomplished. Roget's Thesaurus does not really have a good word, but some of the words in his book are: restoration, rehabilitation, separation, renewal, restitution, relief, and recovery; all connote a rest period, but they do not imply the stopping point where one turns around and tries to survey or evaluate what has been done so far. So if an entirely new course is taken, perhaps we can call it a watershed.

Let's try to look back, see where we have been, and see where we would like to go. Twenty-four chapters of describing Neelys, mostly males, some of their spouses, and a little about their forbears. Eight generations of Neelys from Ben, who never left Ireland, to the latest edition—Thomas Gardner Neely, aged eighteen months in November of 1982—have been partially pictured. The time span from Benjamin Neely's birth (the one who never got to America) on February 28, 1778 to the present would be two hundred and four years. If we trace back to the Eves we find that Oswald Eve, the elder, was a sea captain in 1745 and must have been born in the early seventeen hundreds, perhaps about 1705. This would give us a time span of two hundred and seventy-seven years and perhaps nine generations.

Regarding the last few generations, we have considerable known and recorded information, and we have fairly specific information on four or five generations. As to the earlier generations, we have some facts, but they are much more fragmented; we have to take a few clues and try to fill in the empty spots. We have taken the liberty of fictionalizing what we don't know about those earlier generations.

Our purpose has been to see if we can trace the genes of ambition through the generations. We would like to see if we can possibly get some feeling as to which is more important—heredity or environment— or at least try to see how the two might work together in some individuals. Of course the two might work against each other too.

We have taken one person, Benjamin Neely and tried to see how one event—his new stepmother's switching of his sister in 1847—changed his life. Here was a strong environmental influence—a boy of thirteen years goes off to sea for almost two years and returns to take up his life again. He certainly had some strong drive to leave home and then some inner drive to take up life where he had left it—but now he was a changed person. In the course of those two years, he had seen several foreign cultures, visited some of the most famous cities in the world, learned how to get along with people, and perhaps attained some inner confidence which stayed with him the rest of his life. That event—leaving home—was surely a strong environmental influence, but it must have worked well with those genes of ambition.

Then there was Frank Neely; we have a great deal of solid information on him, but there are also large gaps in our knowledge. So we have taken the liberty again of using a few hard facts and filling in the rest with logical fiction. Frank was perhaps the most famous of all the Neelys. As we have indicated, he really was recognized as one of the great leaders in Atlanta. Whether this makes him greater than some of the others is difficult to say. What about his genes, and what about his environment? One might say that he made his own environment in many ways, but we really don't know. And how much of an influence was the Jewish marriage on Frank's life? Also, how much of a factor was Rae? Was she the one that spurred him on to do the things he did? We don't know.

Then we are trying to take all of Ben's sons and daughters (Frank's siblings) and see how the genes of ambition were spread around. It was as if Ben and his wife, Henrietta Eve Carmichael Neely, passed down some interesting genes to their children, but in Frank the chance split of the chromosomes just happened to produce one offspring with the proper combination of intelligence, ability, and burning ambition to use his talents.

Then we have looked at the world and America from the 1700's to the present, and what have we seen? We have seen several wars—first the Revolutionary War in which Oswald Eve, Sr. and Jr. were involved—Senior moved away during the war; Junior was put in jail for a while, but by whom—the English or his fellow Americans? The story is too vague, and we will probably never know. Certainly the Revolutionary War had a great effect and American patriotism must have started then.

We don't know much about the war of 1812; they say it was over very quickly. John Neely did not arrive in this country until about 1830. Then there was the Mexican War in 1844 from which our country obtained much land from Mexico, and the present boundaries of our country were defined. This war did not have much effect on the Neely family.

Of course the Civil War was undoubtedly the first event to affect greatly the whole Neely family. John now lived in Augusta, Georgia where he was superintendent of schools as well as an Episcopal minister. Ben had also been teaching, having finished college in 1860. He had been teaching in Augusta for six years when the war broke out. The family did not feel the effects of the war at first, but as the war dragged on, most families became impoverished, and John's family was no exception. The boys went off to join the army; the girls stayed home and worked or taught.

This terrible war affected all of the Neely family and their friends; the family, as well as the country, had to be reconstructed. John Neely, Sr., the minister and teacher, was over sixty during the war, so he was not a soldier in the war. In addition to his oldest son, Ben, there was his son, John, born in 1838. He was about twenty-five when the war broke out, and he too joined the Confederate forces like his older brother, Ben. We have no further information on John, except that he married. He might have had a tremendous progeny, just like Ben. He, like Ben, married a Carmichael. Of course Henrietta Eve Carmichael Neely had gone through the war while in Augusta and did not marry until right after the war. Here was the war that killed five hundred thousand people—a large segment of the population at that time. Never before or since has there been such a devastating war.

Then there were the great World Wars: I and II, and these wars definitely had their effects on all the Neelys. However, it is difficult to say whether these wars changed many lives. There was also the Korean and the tragic Vietnam War in which Americans, and especially young Americans, became disillusioned and from which the drug culture began. No Neelys were killed in these wars, and no known Neelys dropped by the wayside due to drugs.

So we see that world events can affect and change the lives of people, but seeing a cause and effect relationship between events and character traits is very difficult to pin down. Life before 1900 was so different from life in this world after 1900 that any conclusions about the effects of environment in the 18th and 19th centuries compared with the 20th century would not be valid.

Wars, however, were just as bad in the twentieth century as they were in the previous centuries. Our country has seen two major wars—World Wars I and II, and two lesser wars—the Korean and the Vietnam War. World Wars I and II were accepted by the greater majority of citizens as just and necessary wars. True, there were some pacifists, but there was no general protest. However, in the Korean War, to a slight degree, and in the Vietnam War, to a great degree, this country was torn asunder by the rising wave of protest over the folly and stupidity of our involvement in a war halfway around the world. The young people of that time soon found out about the futility of such a war. They felt they had been betrayed by their country by being forced to risk their lives for a cause which had no valid basis. And there still exists a great bitterness among the Vietnam veterans, as was demonstrated at the dedication of a war memorial to the fifty-seven thousand soldiers who lost their lives in that useless war. The families of the dead seem to have lost faith in our country.

Let us compare this twentieth century war torn society with those of the eighteenth and nineteenth centuries. There was the Revolutionary War, the War of 1812, the Mexican War, and the Civil War. The Revolutionary War changed many lives, and the United States was born. Patriotism was forged, providing the dynamo for the building of a great country. The War of 1812 seemed to have very little effect on the country. The Mexican War brought about the cementing of our boundaries and certainly had great influence on the people of our country.

But the Civil War, less than one hundred years after the Revolutionary War, was the war which had the greatest death toll on our country. It, therefore, probably affected the whole population of our country more than any war before or since. It is almost unbelievable that the continued existence of slavery could have caused our country to fight such a bitter war, so costly in lives. The aftermath of this war, particularly in the South, left only the slow and painful process of reconstruction. People forget easily; it is amazing that in 1985 this country has already forgotten its enmity with the Japanese and the Germans. We are buying Japanese and German cars to such an extent that our own automobile industries are almost bankrupt from the competition. There are still many Americans who remember how brutal the Japanese were; there are millions of Americans who know that the Germans killed six million Jews during the Holocaust. Many believe that the German citizens knew nothing about the mass homicide, and if they had known, they could not have done anything to prevent or stop it. Still others feel the German people could have risen up and rebelled against Hitler and his destruction.

Thus, the wars of these centuries had not only had tremendous effects on the life-style of the country, but concretely affected specific people as well. Ben Neely was one such person; his life was disrupted by the Civil War, although his burning ambition carried him through this very trying period. Perhaps his varied experiences at fourteen, traveling over parts of the world, had a stabilizing effect on Ben. His wide knowledge of the world, gleaned from travel and life among his fellow shipmates, must have had an effect on his ability to cope with the trials of starting life anew after the Civil War.

Medical science is now finding that many diseases, as well as a tendency toward them, may be passed down through generations. Medical genetics is now a subspecialty which is being recognized as having great value in predicting the dangers of certain inherited diseases to future generations. We are still not sure how traits of character are taken to the next generation: honesty, unselfishness, and industriousness; do the genes have a role, or is it all parental effect?

Studies of twins are fascinating and often revealing. Most twins come up in the same environment; however, there are studies of twins who have been separated early in childhood, but not enough studies to reach any solid conclusions. As a possible, superficial overview, one might say that in the direct line of kings one might be great while his son might be a lesser king; the third generation might be great again.

The study of certain races and nationalities should give us some solid information, but again separating heredity from environment is virtually impossible. Perhaps the study of the Jewish people will help. Here are a people who have remained fairly pure genetically for over two thousand years. Surely we should be able to see some of the characteristics of these people that are passed down. There is always the same problem—one can't separate the various factors and influences.

* * *

What now; whither go thou, O Neely spirit? Will you, like Ulysses, push off, and sitting well in order, smite the sounding furrows? Will your purpose be to sail beyond the sunset and all the Western stars until you die? Or are your genes so diluted and diffused that you will sputter along, produce an outstanding person like Frank Neely and then produce many ordinary people, happy to live and let live and not try to change this world?

We can look back over the last few hundred years, but what will be the fate and destiny of the Neelys of the next two hundred years, in the year 2184?

26

Ben's Second Trip

Ben got back to the ship in the Baltimore harbor rather late in the evening. He was still in a highly emotional state. The crew members who were aboard could not see that he had been crying as he stayed in the shadows. He later said that this was the hardest decision that he ever had to make, and that he never knew whether he had made the right one. The ship would remain in the harbor for two more weeks and would not be leaving until about the first of April 1848.

Several of the crew had dropped out for various reasons, but most of the crew, including the captain, first mate, boswain, and the cook, remained. The new trip would again go to Liverpool as the final destination. The plan was to go from Baltimore to Charleston and Savannah, then briefly to Miami, Puerto Rico, down to the mouth of the Amazon in Brazil, then across to Dakar in the Senegal, and from there to Brest in France. After France they would sail to Dublin and then to Liverpool; finally they would head back to the States. The journey had been mapped out with the captain, the first mate, Mr. Rumans, and the company representatives. It would take about nine or ten months in all before returning to Baltimore in the winter of 1848 to 1849.

There would be shipments of light machinery from Baltimore to Charleston and Savannah. Then there would be the trip to Miami where oranges would be loaded. These would be taken to Puerto Rico and exchanged for bananas. They would then sail to Brazil from which a large load of coffee would be taken to France and England. Tea would continue to be the national drink of England, but coffee had begun to make its way into the tea market and was drunk more and more in England as well as on the continent.

Ben had had his fourteenth birthday and had grown considerably in the six months he had been gone; he was more muscular but thin. He was much more self-confident now and could talk and joke with the crew as well as the captain and officers. He and Bjorn, the cook, were now extremely busy taking on supplies for the next journey. They would go into the city of Baltimore and buy large quantities of butter, eggs, flour, meat, fish, and chicken as well as potatoes and green vegetables. The ice storage room was quite large on the boat, and this would help to keep things fresh for at least a few months. Bjorn was a careful planner and also drove a hard bargain in the market. The captain and first mate trusted him completely and gave him a free hand in buying supplies. His systematic approach to preparing for the trip was a good lesson for Ben.

The first leg of the trip would take the *Nerrisa* to Charleston, South Carolina. Its history tells that the settlement was established by English colonists in 1670. It became the commercial center of rice trade and an extensive Indian trade. In the American Revolution, the city was held by the British from 1780 until 1782. Later, at the time of the Civil War, South Carolina's Ordinance of Secession was passed in Charleston in 1860. The capture of Fort Sumter in Charleston Harbor by the Confederates (April 12 to 14, 1861) precipitated the American Civil War. Charleston's population was very cosmopolitan in the early days, having attracted Huguenots, Swiss, German, Ulster Protestants, and the largest Jewish community in the colonies during the 18th century. It had many educational and cultural institutions, even in 1848, and Ben would be charmed by this city. Several years later he would return there to college in preparation for his future teaching career. In the early part of the 1860's he would serve in the Confederate Army and be based there throughout most of the Civil War, using it for headquarters for his forays into the North as a spy.

* * *

Did all of this world travel have a real effect on Ben's late life? Can a boy of fourteen years take in all that he was seeing and experiencing? And would it really make any difference in how he might react to certain situations? Would these travels at such a young age make him a better teacher? There is a narrow, provincial outlook, and there is a broad, cosmopolitan outlook. It may be difficult to see any tangible effects of the travel on his later attitudes, actions, and dealings with people, yet one feels that it must have been there. It had to have given him a little broader perspective when he was teaching thirty years later. We know that he favored desegregation of schools as early as 1882. One wonders if these experiences as a very young man—seeing all types of people in all types of circumstances—could have changed him later. One can't say for sure, but these experiences and strong memories had to have had some very strong influence on his life.

But we may be making cause and effect pronouncements where none exist. Any fourteen year old boy might be amazed at what he saw; perhaps that is as far as it would go. Let us not forget that Ben was human like the rest of us—that he had bad characteristics as well as good. So far as we have seen, he did one very human act—he got extremely angry and turned this rage into a somewhat violent act by pushing his stepmother down the stairs. Then he had another natural emotion—fear, and then he ran away. One quick episode and his whole life was changed drastically for almost two years.

If we hope to make any valid conclusions about heredity and environment, we should look at facts which are more readily available and are known to be true rather than conjecture. Therefore, it should be of some help to examine all of Ben's children in more detail. His son, Frank, is the one who really had the burning ambition to do something in this world. We have already looked at Frank's early life and have used some logical guessing, but we do have really firm information on him after he became an adult. We have looked briefly at Frank's oldest brother, Allen, and we have seen what John, the doctor, was like. We have had a brief story of Edgar, Sr. up to his beginning law practice in Atlanta. We have talked about Ben Jr. as a young man. We have not really considered Frank's sisters—Annie, Hannah, and Mary. We did consider Elizabeth Eve Neely Jones who died in childbirth in 1900. We will also have to review and consider the offspring of all the girls as well as the boys in order to piece together any pattern which may be helpful in trying to draw any reasonable conclusions as to whether ambition comes from within (inborn) or is nurtured and developed from without.

We may have pictured the fourteen year old Ben in a rather unrealistic fashion. Perhaps he did not really have such ideal character traits as it appears. A hard worker—yes—but he did not have much choice. Ability to adjust to various situations—again, he was almost forced to adjust to the life he had temporarily picked out. Studiousness and reading books given him by the captain—perhaps this was to fight boredom. Unselfishness—sometimes the most unselfish acts are motivated by an inner selfishness which produces satisfaction from the praise for unselfishness. And finally did he have a deep ambition to do something in this world at age fourteen? This is highly doubtful as most of his working hours were spent in daily activities and simple tasks.

The point is that we know that Ben was not perfect and that any fourteen year old boy of medium intelligence would have acted and reacted in much the same way as Ben did. Furthermore, we have a young man who had been brought up in a moral, church-going family suddenly thrown into the midst of a rather tough and amoral group. One of the first things he noticed was the cursing. It was somewhat shocking to him, but he soon became used to it and even joined in, on a small scale, to be one of the group. He soon noticed that there was some petty thievery among the sailors, but the disgrace of being caught stealing and the resultant peer pressure were great enough to take care of any temptations on Ben's part. Some of the sailors were religious. Frank, the Newfoundlander of the first voyage, was well known for his mean attitude. But the sailor that the others respected the most was the quiet Welshman, Bailey; short with a powerful torso, he was the one who had protected Ben shortly after they had begun the first part of the journey. Bailey was the hard worker; he could scramble up to the top of the mast in no time. Mr. Rumans recognized what a superior sailor he was and depended on him to carry out difficult tasks with the sails whenever and emergency arose. So Bailey set the tone of the ideal sailor, and the others looked up to him; Ben naturally was one of Bailey's greatest admirers.

Ben was very soon conscious of the effects of alcohol on people. In Easton and Baltimore he had seen occasional drunkenness on the streets, but he had never really had close contact with an alcoholic until he encountered Doffie on the first voyage. Also a Newfoundlander, Doffie (probably short for Rudolph) was drunk for practically the entire voyage. He was a friendly alcoholic and was never really obnoxious—he apparently kept a fairly steady level of alcohol in his system at all times—he had some special hiding place where he kept wine and spirits. While at sea, he kept the level of alcohol down enough so that he could still perform his simple tasks. Mr. Rumans knew he was an alcoholic, but he was not really intoxicated until they came into port; it was then that he overdid the drinking. He would go into a stupor and have to sleep it off. Ben became accustomed to the sour odor of Doffie's breath. When the ship came into port, the majority of the crew would go to town to one or several of the saloons, drink for several hours, and then come back to the ship in a rowdy, drunken state. Knowing that Mr. Rumans would dismiss them in any port if they did anything really bad kept them under a certain amount of restraint. Ben had a great opportunity to see what drunkenness is like in a person. He saw how it changed a personality temporarily. He, at first, had a revulsion to alcohol, but later he would experiment (as any young teenager will do) with beer and then whiskey. He became violently ill and then would not drink at all for some time, only to try again when the occasion arose. But he never developed any real taste or desire to drink.

Oh alcohol, how have you come to change so many lives? Or is it that lives are changed, unhappy, and generally not satisfying, and then the person turns to alcohol for surcease? Which is the cause and which is the effect? "He is an alcoholic." What a stigma is attached to the word. It denotes the exact opposite of ambition. The person who steadily drinks too much or the one who has frequent episodes of drinking is really dropping out of the stream of life. He simply does not care what happens. And we can see lives that began promisingly, slowly degenerate into alcoholism. Many die young, and it is thought to be a form of suicide. Others continue to live for many years, and they make the lives of their loved ones miserable. The help that is given to the alcoholic, from those who love and feel responsible for him, only helps him to continue in his ways of over-imbibing. Many may go on for fifty years as alcoholics. The Neely family has had its alcoholics and still does. Those people may have had the genes of ambition, but the drive or push becomes eroded, then broken down, and finally disappears.

Alcohol has been with us a long time. Almost every civilization has learned how to make some alcoholic beverage without the help of other civilizations. Any juice or liquid with sugar in it will ferment and turn the sugar into alcohol or ethanol (CH3, CH2, OH) if under the proper conditions. This drug affects the brain; it acts as a stimulant or excitant in small doses and as a depressant in larger doses. However, the stimulating effect of small doses is probably due to an early depressing effect on the inhibitions; social restraints are lost, causing exhilaration, emotional display, and change of personality and mood.

Alcohol consumption probably goes well back into prehistory. It was used in rites, feasts, and religious ceremonies very early on and has continued to be used in this fashion. The use and sale of alcohol had to be regulated because of its effects on human behavior. The Babylonians in about 1770 B.C. regulated drinking houses. Sumerian as well as Egyptian doctors prescribed beer and wine. The records of both early Greece and Rome indicate the widespread use of wine among all classes of people. The people of the Far East discovered how to manufacture alcohol in prehistoric times and have continued to use it through the ages.

During the middle ages, with the ascendancy of the church into the regulation of daily lives, alcohol consumption or its overuse became frowned upon and discouraged. Today, many of the large denominations consider alcohol consumption sinful.

But how has it become such a problem? Alcoholism is now considered a disease; some think it is an inherited trait, or at least the tendency toward it is thought to be so. Others will argue that it is due to some personality flaw—something called a psychopathic personality. This simply means that the person or personality cannot adjust to its environment because of some weakness. Still others say that alcoholism is the effect or manifestation of a depression. Once one begins to drink because of being depressed; a vicious cycle begins which keeps itself going. There is alcoholism in old age, middle age, and the young. The youth of today use alcohol excessively. It is more popular with the high school and college ages than ever before. Supposedly it is more prevalent than marijuana and drugs among the young people. Various estimates of the prevalence of alcoholism in the States have been made. In 1970 there were five million four hundred thousand confirmed alcoholics in the U.S. With increases in population and problems that figure could easily be seven million five hundred thousand or more at this time (1985).

So Ben, be careful as you travel around this wide world. You will find alcohol very easy to get even though you are under age. With your emotionalism and periods of depression, you could easily make alcohol your crutch; you could come to like it while young and turn to it later in life. Your forebears, the Irish, are famous alcoholics. On the other hand, you have seen so much drunkenness that you might see the problems early on and turn against it with a firm resolve. They say that children of alcoholics rarely become alcoholics themselves as they have had so many bad experiences within their own families. We hope it will be thus with you and your new family on the *Nerrisa*.

27
Frank and Edgar—Early Work Years

Hey, Frank—remember this book is supposed to be mainly about Ben and you, so let's not get distracted with so many side issues. You never were an alcoholic and you had no sympathy for alcoholics. True, you did like your martinis—you were somewhat of a nut about them. The ice had to be especially cold—some degrees below zero and just out of the freezer. The idea was that if the ice was extra cold, it would not melt so much, and this would not dilute your very special formula of mostly gin and the least bit of Vermouth. You seemed to make a ritual or a fetish of that martini before supper. But you never overdid it, and I don't believe anyone ever saw you the least bit intoxicated.

Now Frank, we have brought you back to Atlanta from Pittsburgh. You and Rae had lived over a year there, and you were working with Westinghouse as a research and development engineer. You were quite successful at your job; the two of you could have remained in Pittsburgh. But you both decided to come back south. The weather was much nicer; most of your friends and both families were in Atlanta; and you knew that Atlanta was beginning to burgeon. It was a railroad center, growing rapidly and becoming more industrialized; labor was cheap. There was considerable capital for investment and there was a great need for engineers with the ability to plan and operate the various types of new mills coming into the city. On one of your trips back to Atlanta you heard of a consulting, engineering firm that was just beginning.

You had an interview with the two engineers who were starting the company. Both were graduates of Georgia Institute of Technology like you; they needed someone skilled in electrical engineering, and that was your special field. After considerable talking and planning among the three of you, you decided to leave Westinghouse and return to Atlanta.

That was in 1907. America had passed the turn of the century, and changes were occurring in our country very rapidly. The age of the automobile had begun.

* * *

Nicholas-Joseph Cugnot of Lorraine, France, constructed the first true automobile. It was a huge, steam-powered tricycle, and in 1769 it ran for twenty minutes carrying four people at a speed of 2.25 miles per hour. Most authorities recognize Carl Benz and Gottlieb Daimler of Germany as the most important, early contributors to the gasoline engine automobile. Benz ran his first car in 1885, and Daimler ran his in 1886. They worked independently and never met each other. Steam powered carriages preceded them and continued after them, but steam became outmoded in the early part of the 20th century. Benz began producing his car—a gasoline powered tricycle—in 1885. His factory employed fifty workers in 1888. He began to make a four-wheeler in 1890. Daimler began construction of cars in 1886, and by 1889 he had developed an automobile with four speeds. Its wheels were driven by a belt and it was steered with a tiller.

Charles and Frank Duryea are credited with creating the first American, gasoline powered automobile in 1892-93. By 1898 there were more than fifty automobile companies in the United States. The three horsepower Oldsmobile was the first commercially successful American-made automobile: 425 were sold in 1901 and 5000 in 1904. By 1908, there were 241 automobile manufacturing firms in the U.S. One of these was the Ford Motor Company which was organized as a corporation in June 1903 and sold its first car the following June 23. The company produced 17,000 cars during its first full year of business.

So Frank, some might say that if you had really had vision you would have become involved in the automotive business. We don't know exactly when the first automobiles arrived in Atlanta, but we can imagine that by 1907, when you and Rae returned to Atlanta, cars had been seen on Peachtree Street. Apparently automobiles were not your interest; if you had been interested in their engineering, you would have had to go to a northern city as automobiles were not being manufactured in Atlanta. However, in your lifetime you would trace the progress of the automobile, as a novelty, to a time when automobiles were one of the most dominant influences in American life. The automobile and later the airplane literally changed the life of the American people as well as the people of the rest of the world.

But you had other fish to fry. You had your own interests and ideas, and when you arrived in Atlanta you were excited by what you saw and the part you would have in the changing of Atlanta. While in Pittsburgh you had met and worked with two men—Fred Taylor and Henry L. Gantt—who would have an effect on you and what you did for the rest of your life. They were both engineers at Westinghouse and they were your ideals and inspiration, particularly Gantt. He was a consulting engineer on production problems. It was from Gantt that you learned that management skills included participation in the civic life of the community. And it was this principle which became so deeply ingrained in you and which finally made you great. There were many good engineers and there were many good businessmen in your day. There were many that made a great deal more money than you ever made. However, it was that simple lesson taught you by Gantt that put you into history.

It was obvious that a better city would be good for business. But you never let your own business interests have an effect on your involvement with improving Atlanta. There naturally were politics and politicians at that time who had their own special interests in mind; you were the one who learned how to put pressure on them to help Atlanta become a better city. You learned how to work with them, and as time went on you were appointed to all kinds of boards; the politicians began to have a growing respect for you as a scrupulously honest man, a man of vision.

There were also two other principles which you learned from the now famous engineer Gantt. They really were one and the same: establish good working conditions and good employee relations. You stuck to this ideal for the rest of your life; no matter where you were, you insisted that the employees be helped so that they were happy in their jobs. You later were well known as the president who would talk and joke with the janitor. Even when Rich's employed several hundred you would wander around the store and exchange pleasantries with the clerks and other workers. You knew every one of them by name and often knew something about their problems. Years later after you left Rich's, when your name was mentioned to clerks who had known you, their faces would light up and they would smile and say: "That Mr. Neely, he was a real man." It was the ideas of another man coupled with your own ideas and ability, quick wit, and sense of humor which made you a superbly functioning man.

During the years from 1908 until 1915 you continued to work as a consulting engineer. The first few years were not very rewarding financially, but you made a living. And your wife, Rae, was working until she became pregnant in 1909. As time went on, the reputation of your firm spread; your consulting jobs with various companies became larger and more important. You could give advice on plant layout for new mills, production methods, and quality control. You had to travel frequently around Georgia and the South, and you became involved in candy factories, textile mills, food production, cotton ginning, and producing linseed oil. Thus your experience was quite varied and you were becoming known as a person who could help with almost any problem that a company might have.

During this period your personal life and settled down. In 1910 you were twenty-six years old and Rae was pregnant with Rachael, who would be your only child. You two had finally found a house that suited you. Although extremely busy, you made time for a happy home life. Rae was still a very important part of your rise in this world. She was forever entertaining people and would continue to do so for the rest of your lives; undoubtedly this had a tremendous effect on your life's work—both your occupation and your civic work. She soon learned to be the gracious hostess, and within a few years she had had almost all of the important people of Atlanta to your home for a meal. Often there was some type of entertainment. Also, any important person from out of town would be brought to your home, and a group of Atlantans would be invited to meet this person.

You two were not greatly affected by anti-semitism. Occasionally you would get snubbed by some important person, and you knew that you could not get into the two big social clubs of Atlanta, so you did not try. Rachael was born into the Jewish faith as was the tradition.

You, Frank, still thought of yourself as somewhat of a country boy. You liked to walk or drive out into the countryside. Even in those days you had a great yearning for some farmland where you could go to get away from the city and relax. In truth, not many years later you would buy a large farm near Norcross on the Chattahoochee River where you would build a beautiful home as well as run a very successful dairy farm. Little did you know that this farm in the country would be within the metropolis of Atlanta one day, and your farm would be surrounded by developments. The city expanded north and one day your daughter and grandchildren would be paying taxes on farmland as if it too were a development for shopping centers and condominiums. But you would get to fulfill your wish to go back to the country; you would retire there but keep an apartment in the city.

To get back to your early adult life in Atlanta, you continued to be highly successful in your consulting firm, and you were soon recognized throughout the city for your acumen in the planning, development, and operation of all types of factories. In 1915 you were thirty-one years old and had already made a name for yourself in the city. That was the year that one of the large textile mills in the city, The Fulton Bag and Cotton Mill, had production problems. The owners, two brothers named Elsas—Ben and Norman, found that in spite of a good market and hundreds of employees, they were losing money. Their director of operation affairs had been found to be not only incompetent but dishonest. So the Elsas dismissed him forthwith, and this left a very important position empty. They were familiar with young Frank Neely, resolved to see if he would take over the position, and agreed to give him a free hand and a salary considerably larger than his present income. You, Frank, saw the challenge and decided to accept their offer.

He dove into the job with all his energy; he changed the layout completely. He would work within the factory himself and soon had the respect and admiration of all the employees, from the managers all the way to the laborers of the factory. It was not long before he had turned the whole factory around—first going to full production and then expanding. The Elsas brothers were delighted with his work and gave him frequent raises in salary. Frank, himself, began an incentive and bonus system with the employees which greatly improved morale among them. Such an idea was very unusual at that time. Unions had not yet started in the south; employees were generally taken advantage of and downtrodden. Frank proved that happy workers could produce up to twice as much work as dissatisfied ones. He would stay with this job for about nine years, and the size of the factory would triple in those nine years.

Rachel Eve Neely was born in 1910 and was the apple of the eyes of both Frank and Rae. It was somewhat of a difficult birth, but mother and child came through it well. Frank and Rae were both twenty-six years old. Rae would never become pregnant again, possibly from some injury sustained during the delivery of Rachel.

Rachel was a beautiful, dark-skinned child who developed rapidly, and within the first year she showed that she would be extremely bright. In her early years, and also later, she changed Frank's life. He began to come home earlier; he stopped working on weekends, all so he could watch the rapid and interesting development of his daughter. Rae, of course, was in charge; she had read all the books of the time about how to bring up children, and she saw to it that Rachel got every advantage. Rachel picked up the inherited genes of the Neelys, Carmichaels, Longstreets, and Eves; she also inherited equal quantities and quality of the genes of the whole Jewish race, back to Moses. What was produced? A very intelligent girl and woman with definite genes of ambition along with genes for family life and children; bringing up her own family dominated her life for a while. Although she later had many useful jobs, having graduated from Smith College in Massachusetts like her mother, she never tried to follow in her father's footsteps. After her children were grown and gone, she returned to graduate school at the University of Georgia, got a postgraduate degree in social work, and had a job with the Family Counseling Service for many years.

When you went to work for Rich's Department store, your life changed. We can safely call this a watershed in your life because at this point you began in an entirely new direction. A well-trained engineer beginning work for a department store? Why, that was unheard of, and it certainly sounds as if this would be a cop-out. To waste all those years of good training on a department store? A waste? We shall see.

Before we look further into your life, let us look at one of your siblings, Edgar Adams Neely, and see what happened to him around the turn of the century. He was your next older brother and was twenty-two years old in 1900; you were only sixteen. He was working in Atlanta as a court reporter. In 1903 he went off to law school at Georgetown in Washington, D. C., graduated in 1906, married Emily Levering Eckfeldt in 1907, and began practicing law that same year in Atlanta with the well-known law firm of Black and McDaniel. The young couple had gone to Tallulah Falls by train on their honeymoon. This was a well-known resort at the time at a very spectacular falls which dropped down into a huge gorge. The water was later claimed by the Georgia Power Company, and they still use the dam along with a series of others in the area for electrical power. This whole area would later become very familiar to Edgar and Emily as it was on the way to Highlands, North Carolina where they would later have a house and just a little north of Alto and Cornelia where they would later have a farm.

After Edgar and Emily got settled in Atlanta she became pregnant, and their first child was born and named Henrietta Eve Neely for Edgar's mother, who had recently died. But infant Henrietta lived only a few days, apparently the victim of a congenital heart disease. They were both emotionally devastated, but in 1909 Emily became pregnant again and delivered a fine, healthy boy on May 8, 1910. This child was said to have been born in a hack in which Edgar, Sr. had engaged to take Emily to the hospital. Emily had a nurse with her in the hack who was able to help with the delivery. Edgar, Sr. was sitting with the driver urging the horses to go faster. The baby, Edgar, Jr., arrived just as they came to the hospital, and the doctor and nurses came rushing out to help. Edgar, Sr. vowed to have the rest of his children born at home; in truth, for the birthing of Levering and later Randolph, the doctor did come to the house.

The child was named Edgar Adams Neely, Jr., and he was destined to follow in the footsteps of his father and become a lawyer. Thus Edgar, Jr. and Rachel Eve Neely were born in the same year. Edgar, Jr. seemed quite healthy, but as he grew, his mother, Emily, noticed that he did not use his right arm as much as his left. She, at first, thought that he was just left-handed but took him to the pediatrician, Dr. Funkhouser, who examined him carefully and said that he had probably had a slight birth injury. Edgar, Jr. would be left-handed the rest of his life; he became very self-conscious about his slightly weak and uncoordinated right arm. The birth injury did not effect his intelligence. In later life he thought that this slight physical defect had a rather strong influence on his life as he sought to excel in mental skills rather than the physical efforts of sports.

But what about Edgar, Sr.? He had come back to Atlanta after graduating from law school in Washington. He worked for a short time for a law firm in New York; apparently he was not happy there and came back to Atlanta in about 1906 to begin the practice of law at age twenty-eight. He returned to Washington in 1907 to marry Emily and brought her back to Atlanta.

The practice of law for a young man was not very lucrative in those days. He joined the well-known firm in Atlanta of McDaniel, Alston, and Black. Sanders McDaniel was the senior partner. He was the son of one of the governors of Georgia in about 1885 and had a good reputation as a lawyer. Robert Alston later pulled out of the firm to form a new firm with his brother, Philip Alston. Eugene Black also later resigned from the firm to work for the Federal Reserve System. Edgar, at first, did a great deal of the research on cases which the older lawyers would try, but one day, while still a young man, he had to try a case in court for Mr. Alston. He did such a good job in court that he immediately made a name for himself as a trial lawyer. Trying cases in court before the judge and jury became his forte. He had listened to many trials as a court reporter and heard the best lawyers in town in the courtroom. He had a flair for theatrics, and a good sense of humor helped a lot; he seemed able to sway the jury with his down-to-earth manner. His skills as a trial lawyer continued to improve, and he later became recognized as one of the finest lawyers in his field, in the city and state. Along with the law, he engaged in many outside activities such as church, the Shriners, music appreciation, and he even entered into civic life to a certain extent.

How did he differ from his younger brother, Frank? He was equally intelligent and proficient at his profession, but he never had the genes of ambition to enter into the life and direction of the community as did Frank. So here we have two brothers with the same genes and environment who were very much alike and yet quite different.

28
Emily and Joshua

Some people seem to do everything right. Most of us do have hang-ups; we have worries; we have phobias; we have too much emotion; we have depressed periods; and finally we just have problems that we don't know how to cope with. Or we postpone, we procrastinate, and we often say we don't really feel like doing what we should be doing. And then the problems seem to multiply, and finally, they don't get solved at all. Most of us seem able to solve enough of the daily problems to get along in this world. We finally face what we have to do, and it gets done—perhaps a little late, but our lives continue. We call all of these little foibles being human.

Of course, there are some people who can't seem to cope with life, and they are what we call dropouts. Life just is a little too much for them; this often occurs sometime during the change from childhood to adulthood. There is a small group for whom this transition is too much. This group, usually teenagers, turn to fantasy. They turn inward and begin to see visions and hear voices. We call them schizophrenic; they used to be called ill with dementia praecox. It is now believed that there is a very strong genetic predisposition. By tracing back from the forebears of a young person with schizophrenia, we usually will find that there have been others so afflicted. Medical sciences knows that mental illness is inherited or at least has a strong hereditary predisposition. We also know that bad mental characteristics are inherited. What, then, about good characteristics? These young people have a very poor prognosis; they may remain mentally ill for the rest of their lives and often are the permanent residents of the mental institutions.

Then there is the group that somehow adjusts to life and the world, but not well. They usually have one major difficulty that keeps them from adjusting. This may be a major phobia, or they may be so emotional that trivial things upset them; they might have an uncontrollable urge to steal, or a gambling problem. They are often called psychopaths or are said to have psychopathic personalities. Most of them can get along in this world, although they may get into trouble at times. The hypochondriacs and those with chronic anxiety fall into this class, along with the episodic alcoholics. They simply cannot adjust well.

Then there is that unusual and relatively rare group of individuals who seem to be able to do everything right. They have some God-given facility to cope easily and quickly with any problems that come up. They don't get ruffled. They don't get mad or depressed. They just seem to adjust to life in a very simple fashion. They get along extremely well with other people. They may have an inner tension, but it certainly does not show. They must have a deep-seated ambition, but it does not seem to be evident to most of us. They always seem to be so cheerful. They seem to smile a lot, and they all seem to have a wonderful sense of humor. They make the rest of us smile and laugh. What is their secret? We don't know for sure. But this ability to cope easily with life seems to begin very early in life and it seems to continue throughout life unless some major influence comes along which changes a life completely.

Most of us get depressed at times. Our outlook on life is usually bright and happy, but we often become depressed—perhaps only momentarily, perhaps a good bit longer. We don't know why we get depressed. Of course there is the depression that is a real mental illness. It may be brought on by some minor, or at times major, problem. Most of us can snap out of this type of depression, but there are others who go into a state of depression and can't seem to come out of it. This is also a true mental illness, and it seems to run in families. It has a better prognosis than that of the young schizophrenic, and it may respond to psychotherapy, to certain drugs, or just to the passage of a certain amount of time. Again, we are not too sure whether it is heredity or environment or both.

Frank Neely was one of those who had a tremendous natural ability to adjust to any situation in life. He began adjusting well as an infant. In spite of all that attention and unusually good care from his mother and sisters, he did not become spoiled. Rather, he thrived on the attention; by the time he got to the first grade, he was doing everything right in school; he could do no wrong. His grades were excellent, he was already reading, and he was way ahead of his class. This adjusting ability would remain with him the rest of his life—all ninety five years of it.

Did he get these traits from his father? His father died when Frank was eight years old; he really did not have a chance to mold Frank's life very much. However, if it is a question of genes, it is possible that his father, Ben, could have influenced his life to a great extent. Certainly Ben was very successful at what he did, but what we don't know is how easy it was for him to cope with all his problems. For Frank this adjustment just seemed to come naturally. It is of course possible that there were many inner conflicts throughout his life and career, but they certainly were not obvious. Perhaps he got his drive and ambition from his father, Ben, and possibly his calm and peace of mind from his mother, Henrietta. We are not too certain about her personality, but any mother who could have raised that many unusual children had to have some special ability, probably an inner serenity, and very likely a sense of humor that rubbed off on most of her children.

Then to go back one more generation to John Neely. He undoubtedly had that strong ambition as well as the ability to cope with situations. He came to America, adjusted to the new life here, and had the strength of purpose to study for the ministry at the age of forty. How these endeavors were formed we do not know. His first wife, Susan, who died in childbirth, is an unknown personality.

Let us then look at some of the later generations.

Okay, okay, let us take an overview of one from the very latest generations—this one was born on the Christmas eve of 1968. There have been several born since then of the same generation. This would be Ben Neely, Sr.'s great, great grandchild. This particular one would be four generations down from Ben. Ben's son, Edgar, had a son named Levering who had a son named Fred who, along with Shelby Mills Neely, had a daughter named Emily Clements Neely. Where is the Clements from? This is a new name added to the rest. Clements was the family name of Jean Clements Mills Schwab, Shelby's mother. Emily was the name of the child's great-grandmother on her father's side, namely Emily Levering Eckfeldt Neely, wife of Edgar Neely, Sr. She was born on March 23, 1878 and died in the early 1960's when over eighty years of age. She never got to see this great-grandchild—her namesake—who was born in December 1968.

This was the first-born of Fred and Shelby, and there was destined to be only one more issue of this union, namely Frederick Levering Neely, the third, who would be born two years later.

But what about this Emily Clements Neely? What manner of a child is she, thirteen years later in January 1983? She had had her thirteenth birthday on December 24, 1982. This is another who can do no wrong. Let us look at her now; then let's try to go back in time and see how this amazing grace developed. She is one of those who seems to adjust to any situation in life. She is one who seems to cope with any problem, no matter what it is. She is an angel. Pretty and petite, she is always smiling, always jolly, always ready to laugh at the situation. She does well in school—now junior high school—earns good grades, has plenty of friends, and is attractive to the boys. She played the angel that came down from the sky in her high school play, and no part could have been more appropriate.

She has a great capacity to listen and take in what is going on in an adult conversation. She will speak nicely when spoken to by adults, but generally she is quiet and listens. She does well with her peers, has lots of friends in school, and she is frequently invited to spend the night with these friends. The families know she won't be any trouble, and they encourage their children to invite Emily for the night or for three weeks in the Bahamas on a boat. And the teachers know what a nice child she is, and they try to favor her, but it is not necessary. She will do well whether there is help from them or not. She has that indefinable quality to get along with the world and her fellow man.

Where did you get this facility? Frank was your grandfather's uncle. So you were not descended directly from Frank, but you both got genes from Ben, Frank's father. Like Frank, you picked up this great ability to get along with people, and it seems to come naturally. As we look back over your thirteen years, Emily, we see that you have always been the same. Everyone seems to like you, and you seem to roll along through life coping in an effortless way. However, you are not perfect; no one is. You have tantrums when trying to decide which dress to wear when going out to a party for boys and girls.

Can we pinpoint any factors which could possibly have been important? There are a few things that might have played a part. One is the great and abiding love of both your father and mother. One can immediately sense it when you are around either of them. One can tell that this mutual love, affection, and admiration is a great security for you. It gives you confidence. And it has always been there over all these years. Your parents don't have a burning ambition for themselves or for you. They want to be happy, and they want you to be happy in this world. And you will undoubtedly attain this no matter where you go.

Yes, Emily, you are the little angel. You are the one with the bright little eyes; you listen carefully to the conversations of your elders. You don't always understand what they are talking about, but you have a general idea, and you store up that information in that bright little computer mind of yours, and some day you will be a little wiser than the others and know how to solve problems a little better and a little more easily. You are the one with the smile on your face. You are the one with the little disarming chuckle and laugh. You can see that things are funny when others would be anxious or even frightened.

So here's to you—cheers and wassail—long may your spirit wave and never flag. What a nice companion, what a great wife and mother you will make. You will be thirty years old in December 1999, and you will see the new millennium come in. You will take it in your stride. You probably will be a mother by then, and you will provide an abundance of love for your children, and they too will grow up with that serenity which you have.

But Emily, you don't help much in deciding whether it is genes or environment that plays the most important role. You started out with good genes—somehow the roll of the genetic dice gave you a very good combination of genes—then your parents provided you with a great environment of love. And you will pass on this special group of genes and this very special environment.

* * *

Emily, what are you facing in the next fifty years? Will there be a nuclear war in your lifetime? Please Lord, let there never be another nuclear bomb explosion. That's all that we are hearing now. After Hiroshima and Nagasaki, there was a great revulsion to nuclear bombs, which lasted a long time. Now people seem to have forgotten that two whole cities were wiped out. The survivors, not many, were badly burned or even blinded; many suffered later effects: cancer, leukemia. The young of that time have tried to forget the tragedy. The underground shelters built in the fifties are gradually falling into ruin. However, memories still exist.

And now, almost forty years later, as Russia and the U. S. become more unhappy and hostile toward each other, the newspapers, television, and the rallies, here and abroad, tell us there is impending danger. The scientists who developed these bombs are becoming extremely concerned about the situation. We pray—O God—that Emily, her children, her grandchildren, and in fact all future generations will never have to experience the explosion of nuclear bombs in their world.

* * *

Let us take a look at another one of the new generation, like Emily. This one is Joshua Hoffman, age twelve, just a little younger than Emily. Joshua is a direct descendent of Frank H. Neely. In fact, he is one of his great-grandsons. Rae and Frank had only one child, Rachel. She married Ben Parker and had three children—Daniel, Eve, and Nathan. Eve married Howard Hoffman, and Joshua is the oldest of their three children.

Joshua's mother, Eve, her mother Rachel, and Rachel's mother, Rae, all attended Smith College in North Hampton, Massachusetts, which is probably the best college for women in the U. S. Rae went to Smith in about 1898. Her mother took her there on the train and they were met at the railroad station by the president of Smith College. This was because it was unusual for a southern girl to come to a northern college, and the president, having corresponded with Rae's mother, Mrs. Schlesinger, thought it would be a good idea to give them a special welcome as he hoped more southern girls would later come to Smith. So, after Rae was carefully settled into the home of one of the teachers, her mother, being satisfied, took the train back to Atlanta. And Rae did very well at Smith; she brought back to the south a fund of Yankee scholastic and literary knowledge as well as a little intellectual snobbism.

Rachel also went to Smith in 1928 and graduated in 1932. And Rachel's daughter, Eve Parker, went to Smith in 1962 and graduated in 1966. So on his maternal side Joshua had three generations of women who had been to one of the best colleges for women in the country.

Joshua's father, Howard Hoffman, is also a very bright, if unusual, man. His family was from New York, but he moved to San Francisco where he met Eve. Following the marriage they settled in Atlanta on the Frank Neely farm. They first lived in a house which was already there; later they built a huge rambling home in the woods close to the Chattahoochee River. Frank left the farm, originally 320 acres, in parcels to Rachel and her three children. He donated sixty acres to Georgia Tech and divided up the rest. As Atlanta grew, the city surrounded the farm, but the property had still not been sold in early 1983. The farm, still a beautiful piece of land, is bordered by the Chattahoochee and has large fields, pastures, wooded areas, several houses, and large barns for the former modern dairy. The dairy cows have all been sold, and the fields lie fallow. It is a beautiful, peaceful area, and it would be sad to see the farm cut up into developments. It would make a wonderful park. And it is also a great place to raise a child. When Joshua was quite young, the farm was still active with animals, tractors, and Mr. Ramey, Frank's chief farmer. Joshua had a chance to see all of this activity, it has only been in the past two or three years that the farming has stopped.

Howard and Eve are very good parents; they soon found that their son, Joshua, was unusually bright and intelligent. Eve and Howard both began teaching him to read at a very young age. And Joshua responded to this early teaching enthusiastically. We are not sure exactly how old he was when he first began to read, but we do know that by the time he arrived at the first grade, in the country school, he was reading simple books, and when he was in the second grade, he was reading Melville's *Moby Dick*.

He was quite a problem for the teachers at school. The family did not want to push him up to higher grades (although he could read better than the fourth graders) because he would be too small. The teachers would give him extra advanced work to do. It was about this time that Howard found out about a program for advanced pupils which was starting up at Emory University. Joshua would go to regular school in the day, and then twice a week, in the afternoons, he would be taken to Emory for the special classes. Here he was given a chance to use his full intellectual abilities. It was an experiment that Emory was putting on, and for Joshua it was very successful.

So here we are, Joshua Hoffman and Emily Neely. You two are truly the new generation. You are both bright and intelligent in slightly different ways. But you both show great promise for the future. You are both of the computer age. The age of the automobile and airplane is now old hat, taken for granted. Television is here to stay, and now everything is computers. In the home, in the office, everywhere, these instruments are taking over. You two young people will have to cope; we think that if anyone can adjust to the new age, it will be you. But take a lesson from your forebear, Frank Neely. No matter what your occupations may be, enter into the community of people and nations. Do what you can to prevent a nuclear holocaust.

29

Ben, the Confederate Spy

As Ben travelled by boat in 1848, at age fourteen, from Charleston South Carolina to Savannah, Georgia, little did he know how closely he would be involved in these two states within the course of a few years. If he had a crystal ball that would tell the future, he would not have believed what he would have seen. Exactly fourteen years later when he was twenty-eight years old, he would be making a wild ride on a horse from a little town called Lavonia in North Georgia into the western part of North Carolina around King's Mountain and then behind the Federal lines across the northern part of South Carolina and the southwest part of North Carolina. His ability to talk like a Yankee would save his life several times but would not prevent him from getting severely wounded and be *hors de combat* for over two months.

Ben's father, John, remained in Easton, Maryland for several years as a teacher and minister, but then he had the opportunity to organize the schools in Augusta, Georgia in 1852. He moved the entire family there—John was now married to his third wife; Isabella, his second wife, had developed tuberculosis and died. Two of the girls had died; four girls were left, and only the two boys—Ben and his younger brother, John. Ben would resume school in Easton, do very well when interested, and finish high school at about nineteen years of age in Augusta. He would then go to college in Charleston, South Carolina at the College of Charleston. He would stay there four years, often travelling on the Savannah River by boat and then taking the train from Savannah to Charleston. So he got to know the country very well.

Ben's father, John, had moved from the relatively neutral Maryland to Augusta, Georgia and really felt himself to be a true Southerner by the time the war started. He had never been very politically minded, but while in Easton and travelling to Baltimore, he had naturally come in contact with the abolitionists who were gathering strength, often through the churches. Maryland was a border state. Philadelphia with its Quakers and the rest of the state with the famous Pennsylvania Dutch were powerful voices and militant in their feelings about getting rid of slavery everywhere. There had never been any slavery in Pennsylvania. Virginia on the other hand, was a true and traditional Southern state, and their economy was based to a great extent on slavery. Richmond, the capital, was beginning to get a little industry, but the state was mainly agrarian with vegetable farms, fruit orchards, dairies, and horse farms all across the state. Slave labor was essential to all these activities, and there simply was not enough white labor to do the work.

Maryland was thus caught between two completely different ideologies. Virginia was beginning to feel the pressure from the North to get rid of slavery and was beginning to organize its forces. They would have town meetings, county gatherings, and the state legislature was in an uproar as the North was not buying Virginia produce with the excuse that it came from slave labor. The general feeling was that the North was acting "holier than thou" about slavery while at the same time permitting terrible conditions to exist around their factories where there was child labor, long hours, and extremely low pay for their workers. The people of Virginia felt that they were being told what they could do by the people of the North who really did not know what the true situation was. The feeling was the same throughout the South.

The northern states were becoming adamant in their contention that if America was the home of the free, slavery should be abolished. Here was an issue that was being constantly discussed in both the South and the North, and from 1850 until the outbreak of the war in 1861, the feelings were becoming more intense on both sides. It was during this ten-year span that the country was becoming totally divided.

In the border states, like Maryland, some churches were for slavery, some violently opposed it, but rarely were there any neutral churches. Neighborhoods were often divided in their attitudes, and there were even strongly divergent views within families. When John became a minister in 1848, he was beginning to hear the early arguments on both sides, but by the time he moved to Augusta, Georgia in 1852, the voices of division were not very strong. After settling in Georgia, he became extremely busy with the organization of the school system; he was also the minister of a small church. Almost all of the Augusta people were strongly pro South in their views, and anyone with different ideas was thought to be a little out of his or her mind. So John went through this period from 1850 to 1860 with increasing sympathy for the Southern cause. As a minister and a thinking man, he knew that slavery was not really compatible with Christianity, but the people he dealt with in school and church were decent and apparently Christian folks, and nearly all of them felt that the North was trying to tell them what to do. If John had come out strongly for the North by opposing slavery, he, most likely, would have lost his job. It is interesting that his son, Ben, would resign from his teaching position in Augusta thirty years later in 1882 because he thought black children should be allowed to go to school with the white children.

Now Ben had returned to Easton, Maryland in 1850, at age fifteen, after having travelled over a large part of the world. He had gone back to school in Easton and had quickly caught up with his class and graduated from Easton High School in 1852, the year the family moved to Augusta. His education during those two years on the boat had not suffered. In fact, it had been greatly enhanced. Not only had he seen a great part of the world, but the captain had encouraged him to read books throughout the voyage, and this was one of his greatest pleasures. On his trips he had seen countries which permitted slavery, and he had seen others where slavery had been abolished for many years, such as in England. He had seen the northern cities, and he had seen the southern cities with their slavery. He had heard the arguments on both sides, and he had seen slave labor at the docks in Charleston and Savannah. He had seen the cruelty to the slaves as well as their terrible living conditions. He knew he was against slavery, and he remained against it.

When the family moved to Augusta to take over the Richmond Academy, the only high school there, his father suggested that Ben take further courses in preparation for college. He was somewhat of a maverick, being much more mature than his classmates, and he did not hesitate to express his thoughts on any subject, including slavery. He got into some very contentious discussions with some of the other students at the Academy which sometimes ended in fist fights. Ben talked over his problems with the other students and his father; John calmed Ben down to a certain extent. This was in 1853 when tensions were just beginning to build between the South and the North. The lines were being drawn.

Ben had already decided that he wanted to be a teacher like his father and grandfather in Ireland. He had some rather strong ideas about the quality of colleges. John would have liked Ben to go to Harvard or Yale, but the long distances and the cost involved made it impossible. The University of Georgia was a strong consideration. It had been established in the early 1800s in Athens, but by 1853 it was still in its infancy. There was a small private college in Oxford, Georgia called Emory where a famous educator of the time, Augustus Longstreet, had been president and possibly still was president. Oddly enough, Ben would later marry Henrietta Eve Carmichael whose mother was a niece of Augustus Longstreet. Ben and his father considered Emory as a choice, but it was suffering from lack of students and funds at the time. Ben and his father finally settled on the College of Charleston. It also had been started in the early 1800s, but being in a large and thriving city, it had been well supported, was well known, and had very good courses for someone who wanted to make teaching a career.

Ben had no interest in being a minister. Whether his voyage had anything to do with this or not, we do not know. Certainly Ben had seen a great deal of the world and had seen the churches of many cities. You might call him somewhat of a cynic. However, he continued to go to church every Sunday with the family, and he thought of himself as a religious person. He simply never had the calling to the ministry.

So they finally settled on the College of Charleston, and Ben set off for his freshman year in September 1853. Ten years later he would find himself at the opposite end of the state and involved in a dangerous mission.

Georgia, North Carolina, and South Carolina have common boundaries. The northern border of Georgia has a one hundred mile boundary with the western tip of North Carolina, so it is possible to go from Georgia to North Carolina without going through South Carolina. However, the entire eastern border of Georgia abuts South Carolina all the way to the sea. In this area where the three states meet the countryside is quite mountainous, with some of the highest mountains east of the Rockies. As one travels east the mountains disappear and the rolling hills and the plains begin. However, there is one large mountain in southwest North Carolina which rises high out of the plain. It is called a monadnock, an unexpected rise in the land. Stone Mountain in Georgia is a monadnock. Unlike Stone Mountain, King's Mountain has a long ridge, or a low and narrow mountain range, extending westward almost to the large Appalachian mountains of the western part of the state.

In the 1860s there were only small settlements in the mountains. It is a rocky area and farming was difficult. There were many little settlements here and there, but no good-sized towns and not many villages. Mountainous areas just made it a little harder to maintain a livelihood. It was in the more easterly, small hills and plains that the towns and small cities were developing. Here it was easy to cultivate the land, and the weather was much more favorable at these lower altitudes. Greenville, Spartanburg, and Gaffney are close to the northern border of South Carolina. They were the largest towns in northern South Carolina. Columbia, the capital, is right in the middle of the state. There were very few important towns in western North Carolina with the exception of Asheville which was in a low, mountainous area. The larger city of Charlotte, North Carolina sits right above the North and South Carolina border in the south-central part of the state. King's Mountain is about ten miles from Charlotte, and its ridge extends westward to the mountainous area.

Mountains are not favorable for agriculture and settlements. Mountains are also a very difficult area to carry on warfare. They are a great hindrance to both sides in waging war. Hannibal found this out in the Alps. The Russians are having their problems in the mountains of Afghanistan. Both the Northern and Southern forces usually avoided the mountains during the Civil War. So towards the end of the Civil War, but before Sherman's march through Georgia, the Northern forces were pressing down from Virginia into North Carolina toward Charlotte, a major city even at that time. The next objective would be Columbia, South Carolina. The Southern forces were still fighting manfully, although their losses had been great, and toward the end of the war equipment and material were considerably depleted. Still, they were now defending their own soil; this gave them more incentive than the Northern forces that were now quite a distance from home. Charlotte had fallen to the Yankees after costly fighting in the outskirts. The Southern forces moved further south and began to re-group, hoping to turn the Yankees back and stall their drive to Columbia. King's Mountain and its ridge suddenly became very important to both sides. The mountains acted as a protection for Greenville, Spartanburg, and Gaffney, where the strongest Southern forces were. The Northern generals knew that they would have to take these towns before they could move on to take Columbia as they would be flanked if they moved too far south too fast. To know what the enemy's plan was and where their strength was could mean the winning or losing a battle.

Ben had joined the Confederate Army three years previously when hopes were high and when, in the first flush of the war, a victory was anticipated. He was at peace with his own conscience by fighting for the South. Although he was not in favor of slavery, he had lived in the South long enough to have pride in the region; the thought of the abolitionists forcing their ideas on the South was abhorrent to him. He had finished college at the College of Charleston and had been teaching in the high school in Augusta for several years as well as beginning to help his father in the administration of the school system of Augusta. He was twenty-six years old and single when the war began.

In the Confederate Army he had fought and been injured twice in battles in Virginia. Both were gunshot wounds—one in the thigh and one in the arm. Both were moderate injuries, disabling him for only a few weeks. After his second injury, as he reported back to duty, there was a call for volunteers to spy behind the Northern lines. As he was able to mimic the accents of the Northern states, he obviously was well-suited for spying. He joined a special group around Charlotte, North Carolina and had special training in spying on the North. Each man was an excellent horseman. They were trained in marksmanship, and they were instructed carefully about the area in which they would be spying. They would travel by horse through mountainous, sparsely settled country and come out behind the Northern lines. After that they were operating strictly on their own ingenuity. They often put on Yankee uniforms; at times they joined the Yankee forces and pretended to fight against the South. When they had learned some useful information, they would depart from the Yankee forces and return to the mountains and then back to the South. As the war dragged on it was becoming more dangerous for the spies: the Yankee forces were more suspicious; several of Ben's original group never returned.

Ben had been lucky so far. He had been behind the lines several times and had several places in the mountains where he could stay with friendly mountaineers and leave his horse if necessary. Ben's superior officer, Colonel Bell, called him into headquarters in Spartanburg, South Carolina to discuss how to find out which way the main thrust of the Yankees into South Carolina would be. They had already taken Charlotte, North Carolina after a bloody battle, and the question was whether the army would go east toward Charleston or south toward Columbia. Or perhaps they would try to turn westward, attempting to neutralize the three cities of western South Carolina—Greenville, Spartanburg, and Gaffney. These were railway centers and therefore especially important as far as moving men and war materials. These towns were protected in the north by the mountains, so any attack would have to come from the east or possibly in a circling movement from the southeast.

So Ben's mission in 1864 was to travel behind the Northern lines, try to see where the buildup of strength was, and try to find out some useful information from the enlisted men or officers as to which way the next offensive would be directed. Ben decided to travel by horseback across the back of the Yankee line as he would be able to go much faster, learn more, and escape more quickly if he had to. He rode from Spartanburg up through the mountains to the cabin of a friendly mountaineer near Gramling, South Carolina, changed into his Yankee uniform, then travelled north, coming around on the rear lines of the Yankees who were regrouping around Charlotte. Ben knew this area fairly well. He had some papers taken from the body of a Yankee killed in battle. The papers said he was from the town of Johnstown, Pennsylvania and that he was in the medical corps. Having travelled over a distance well behind the lines, Ben would stop from time to time, make a few notes, and then travel further. On his way back, he decided that he would take a chance and try to talk with some of the Yankee common soldiers and possibly officers on the pretext that he was gathering information about the number and condition of the wounded. With his good Pennsylvania accent he rode confidently into a fairly large encampment and asked to see the commander. The commander was a Colonel who asked to see his papers as they were aware of spies in the area.

The colonel looked at the papers and said, "Well, this is a coincidence, my aide happens to be from Johnstown, Pennsylvania." Ben knew this was bad news.

The colonel called to his aide, "Here is a man from your hometown, Fred Jamison."

Ben knew his ruse was discovered. He ran for his horse, mounted quickly, and started off. The colonel had a rifle close-by and fired two shots at Ben. One struck him in the left shoulder and one in the flank. It almost knocked him off his horse, but off they galloped at top speed into the woods. Ben escaped but was bleeding badly and getting weaker. By sheer determination he reached the house of the friendly mountaineer, and collapsed. He was nursed by the mountaineer's wife for several weeks and finally made it back to his headquarters in Spartanburg.

Soon thereafter Sherman finished his march to the sea through Georgia, and the war ended. Ben slowly got his health back and began the process of reconstructing his life in a devastated land.

30
Sara

We have briefly discussed the role of women in the passing down of genes over the last two hundred years within the Neely family. We have emphasized that, in the past, women were extremely busy raising children, and there was very little opportunity or time to accomplish anything outside of the home. Also, we don't have much information about their personalities or characteristics. Still, the women are equally important in providing genes for the next generations, but it is more difficult to evaluate them. Since we have so little information about the women, we tend to neglect them.

There have been, however, several notable exceptions either within the family or closely connected with it. These women have done certain things themselves or have had a very strong influence on the men, especially their husbands. Rae Schlesinger Neely was one of these. She was the pioneer in scholarship at the turn of the century. She went to Smith College in about 1900 and was probably the first woman from the South to take such an unusual step. She brought culture and intellectualism back to the South with her. She continued to foster the humanities in the South for the remainder of her ninety-six years of life. Literature, art, and music were all improved in the Atlanta area because of her participation in projects dealing with these fields. She had to have had some strong genes of determination to travel off to Northhampton, Massachusetts in 1900. The long list of activities and charities which she supported with her time and money made a name for her in her own right.

However, one of her major contributions to the world was the constant encouragement which she provided for Frank Neely, her husband. His accomplishments were so numerous and so important to the city of Atlanta and even the country that just listing them is a monumental task. Exactly

Benjamin Neely married to Henrietta Eve Carmichael (1834–1892).

Henrietta Eve Carmichael Neely (1845–1906).

Annie Wyatt Neely Young (1869–1947).

Robert Allen Neely (center figure).

Jordan Margaret Neely (1978–   ) and Thomas Gardner Neely (1981–   ) are both the offspring of John and Carolyn Neely.

Sarah Neely Newman, proud Mom, and daughter Rachel Anne, born July 18, 1986, the newest addition to the Neely family.

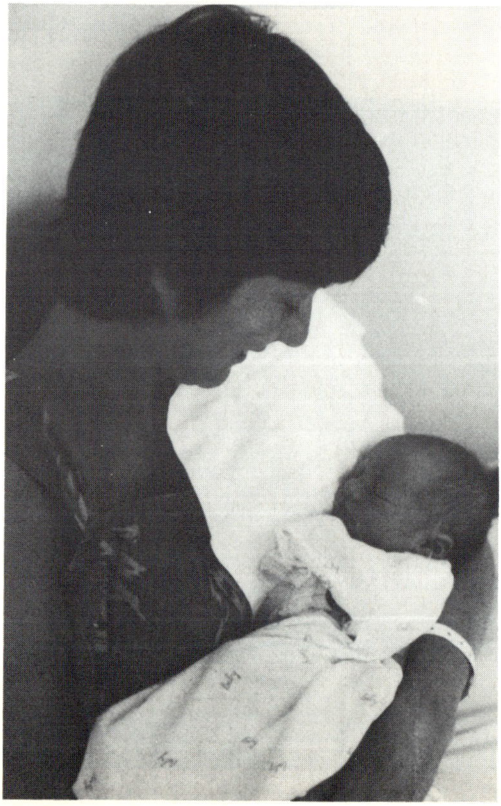

how important her help was can never be truly ascertained. It is well known that she was always assisting him in his projects in whatever way she could, but we really don't know how much she was spurring him on. Clearly Frank and Rae were equally ambitious, and they worked unusually well together. So we have at least some true information about one woman who did much more than simply raise children. They had only one child, Rachel, who certainly gained many important genes from both parents, but circumstances did not permit her to apply these genes to any single objective. She too was an intellectual, a patron of the arts, and after her three children were grown and gone, she went back to the University of Georgia to do postgraduate studies in social work; she then began a new career with the Family Counseling Service of Atlanta.

So here are two women in the family who definitely had ambition and many opportunities to satisfy these drives and aims. We have also mentioned Bess Neely, the interior decorator who never married, but who, as a young woman, went by herself to Paris and studied interior decorating for several years, returned to Atlanta, and carried on a good business of decorating for many years. We have also mentioned Virginia Neely Jackson and Betsy Neely, both of whom are accomplishing much more than just raising families. So the women of the family are beginning to be equal to the men in their pursuits and accomplishments.

But let us examine another female Neely who might someday be well known in her own right.

Sara Margaret Neely, you were born in August 1951. Sara is spelled without an "h." This is important, but I have forgotten why. You were the fifth of six children. The first four came very fast—about eighteen months apart—then there was a hiatus from John Longstreet in 1947 to Sara in 1951. Betsy came in August 1953, two years later. We have stressed that position in the family may be important, but we are not certain in this case. However, the two youngest children were always considered together; they were the "little girls." This grouping together may not have been a good thing for Betsy as we shall see later.

1952—who was president of the United States? It must have been Eisenhower. Truman took over when Roosevelt died, was then elected himself, and then Eisenhower came in for two terms. 1952—the real war, World War II, was over in 1945, and the baby boom began. America was still recovering from the war in 1952, and it had come a long way. The Korean War had not begun. America was the most powerful and richest nation of the world. It was well on the way to recovery, although Europe and Russia were a long way from recovery. So Sara, you were born at a good time, and when you found you had a younger sister to play with, and possibly to boss around a little, you were delighted.

Both of you girls were rather shy when small. You, Sara, had a tendency to hide behind your mother's skirt when strangers came into the room. You would peek around and if the stranger happened to be looking you would quickly hide your face again. Would you ever be so shy again? I don't think so. The four older children were taken out west by car in about 1955, but you and Betsy were too small for such a trip, and you were left at home with a baby-sitter and a maid. After everyone returned from the trip, your mother and father thought it was only fair to take you two on a trip, and off the four of you drove to Rock City. You were four-and-a-half and Betsy was two-and-a-half. It was a splendid trip and you may still be able to remember it. You two were brought into the main dining room—you, Sara, had a regular chair, but Betsy had a high chair. There were quite a few sights to be seen, including the rocky paths and trails of Rock City itself. You took it all in and enjoyed it greatly. You were found to be a great companion for your mother and father then, and you have continued to be the ideal person for any kind of trip, be it to the grocery store or to Rome, Italy. Also, it was about this time that we discovered how bright and interested in this great and wonderful world you were.

Sara Mockingbird Neely, Sadie, Sadikins, Sary, or even Birdie—these were some of your nicknames, but none actually stuck too well, and everyone finally settled for just plain Sara. You started out at public school at E. Rivers, and it did not take your family long to find that you were doing extremely well in school. You had that wonderful secret of adjusting to the environment; you also developed that great sense of humor which opens all gates. This was very much like your great-uncle Frank who went to the first grade in Rome, Georgia in 1890. He also did extremely well almost seventy years before you, Sara.

After two years at E. Rivers Elementary School, your family decided to build a new house. The four bedroom house was no longer really big enough, but the real motive for moving was to get the boys away from a family of children in the neighborhood who exerted a bad influence on our little darlings. In any event, the new house built for the six children was a great joy to all. Sara and Betsy had a room together, which suited them fine. But the move meant that they would be going to a new school only a long block from home. There was some slight problem at first about getting into the school, but this was finally settled and Sara and Betsy finished the six grades of grammar school there.

It was some time in the mid-fifties, possibly about 1953 or 1954, when Fernandina Beach, Florida was discovered by the Neely family. Frank Neely had bought a small beach house there in about 1950; they invited the Levering Neelys there in the summer, usually for about two weeks each year. It was the beginning of a long love affair between the Levering Neelys and Fernandina Beach, and the attraction is still just as strong as it was originally. Frank's house, called Endden for his grandchildren, was a simple cottage right on the beach. Its charm was its access to the beautiful, wide beach with the large rolling waves and smooth, clean sand. It was perfect for children; they could amuse themselves for hours building sand castles, walking in the slough if small, and jumping in the waves. The trip down by car was not so easy in the fifties; there were no expressways; driving over country roads, it took ten or twelve hours to get there. Sometimes there would be an overnight stop on the way at one of the old-fashioned motels at four or five dollars a night.

But there was a charm to the town also; it is a town built around two paper mills, but it was a seashore island with tide rivers flowing into the mainland and into the island. The town itself was small but pleasant, with the main part of town on the inland waterway and a welcome station for the boats that tied up to the town docks. There were always eight or ten shrimp boats along the docks, and inspection of these made for a very interesting walk. Momma usually came down with the six children and a friend who also had six children; since Endden had only one bathroom, it was always kept busy. Then the guests would leave and Pappa would come down and spend a week.

Sara came to love this place very early in her life, and even twenty-five years later, she comes whenever she can. Later, the family would buy a beach house themselves next door to Endden. No matter how far away Sara was living, she always made a real effort to get to the beach; she was generally the most enthusiastic of all the children about coming back when she could. There was always something so satisfying about just sitting on the "roost" high on a sand dune, looking down the usually deserted beach, and watching the waves pounding in.

During these grammar school years, Sara was maturing rapidly, still making good grades, and usually at the head of her class. At home there was always a dog and she seemed to take charge of its care. Charlie Brown was a fine dog, a good solid brown dog of several mixtures of forbears. He was usually quite brave and a good watch dog, but thunder and lightening were his nemesis, and he became a quivering amoeba when it happened. Sara often took him into a closet, sang, and played the guitar for him until the storm was over. We have found over the years that there are two classes of people, dog lovers and non-dog lovers. Sara is definitely of the former group.

She went to Fernandina first when in utero or possibly when one year old when Betsy was in utero. In any event, she has been there at least once a year for thirty-one years. So Fernandina is a part of her or she is a part of Fernandina. It's in her blood. She still takes the *Fernandina Beach News Leader,* the weekly newspaper, partly to see what is going on there, but mainly to have a few good laughs reading the society page from a small town.

So Sara finished grammar school at the public school, did very well, and then went to the Westminster School, the private high school in Atlanta. At the time, 1964, the public schools in Atlanta had deteriorated but Westminster had a headmaster who was very scholastically minded; he ran a very good school which prepared Atlanta children for the good colleges in the North. She finished Westminster quite high in her class, and with a history of Smith College women in the family along with an old friend of the family, she applied and was accepted. The family had driven her around New England looking at colleges, including Vassar; but Smith seemed to suit her and so in September 1969 she took the plane from Atlanta to Hartford-Springfield and on to enroll as a freshman just as her great-aunt had done in 1900.

She bloomed at Smith, made many friends, and did well in her studies. She majored in English, but after two years she somehow got the idea that it might be fun to be a doctor and she began to take science courses during the last two years of college, fulfilling the premedical requirements. During her senior year she definitely decided that she would apply to medical school. Her grades had been good at Smith and her premedical tests were good but not outstanding.

She applied to several medical schools including Johns Hopkins, Dartmouth, Emory, and the University of Georgia. She was turned down by Johns Hopkins where her father had gone. At Emory Medical School she probably would have been accepted—one of her father's Greek friends who taught at Emory said he would "put the squeeze on if there was any problem." One of the teachers at the medical school at Dartmouth had been in Afghanistan with Sara's parents, and she gave a little added personal touch which helped, although Sara undoubtedly got in on her own merits.

So here she was accepted to medical school in 1973 at the age of twenty-two with four years of medical school to look forward to and then probably four years of postgraduate training before she could practice. To make the decision to commit eight years of her life to this took a considerable amount of determination and ambition. Sara found out that it might be fun to be a doctor, but there was also considerable hard work those first few years. Dartmouth was a two-year school and she had to apply again to another medical school to finish up. She applied and was accepted by Columbia Medical School.

She had now been in the North for six years and was becoming Yankee oriented. Living in New York was also an entirely new experience, and the various amenities of New York were a whole new world. The art museums, the great music, the theaters were there to choose from, and these two years broadened her view of life and its wonders. Here was an experience she would remember for the rest of her life.

Sometime during these years she had opportunities to travel to Europe—once with a medical school girlfriend and once with her sister. On both trips she traveled part of the time with her parents in a rented car. She joined her parents in London, saw London, and then they traveled north to Edinburgh by car. Later she met her parents in Italy, took off to Greece with her sister, and then rejoined the family in Venice. They all traveled about by car and she and Betsy drove into Austria by themselves, stopping whenever they could find an inexpensive room. So all of this travel was another broadening experience for Sara and she would go again in 1982. She finally graduated from medical school in 1977 and then applied and was accepted for a three year residency program at Billings Hospital at the University of Chicago.

Chicago was an entirely new place for her and was also unknown to her parents. They had known New York, having lived there for several years, and they loved it. But Chicago; this was a new idea; Sara's mother and father looked on Chicago as sort of a second-rate New York, and they were not very happy over the move. The weather of Chicago was famous for its wind and cold; on a visit they almost got snowed in. But on a second trip they began to have a great deal more respect for the windy city when they found what Chicago had to offer.

So Sara finished her residency in internal medicine in Chicago and took a fellowship in hematology at the University Medical School in her same hospital, Billings. She had a very inspiring professor in her department and got into the sub-speciality of blood clotting. She became an expert on this and was doing research, treating patients in the clinic, and teaching medical students. It was an ideal situation. She liked her work and she liked her boss, and she loved the town. But romance was to change the whole situation.

Sara had always been attractive to the males in high school and college and had had many boyfriends but no very serious attachments. In Dartmouth she had a few more serious attractions, and at Columbia she had a living, or live-in, arrangement as was the custom, but when she went to Chicago, it went by the board and was terminated.

So here you are Sara, arriving in Chicago a free soul, age twenty-six. You had done exceedingly well in grammar school; you had received the Partridge award at Westminster for being the most well-rounded student. You had gone to Smith College, done well; you went on to Dartmouth, graduated from medical school at Columbia, and received an award, not for best student, but for being the most humane student. You knew New England and you knew New York. Your credentials were impeccable, your accomplishments already great, but what kind of a person were you with all of this? Were you still a little shy? Were you confident or overconfident? Were you even a little arrogant like your brother had been accused of being? By no means. You were still modest; you were bright; you were pretty, and best of all you had a great sense of humor. You thus could get along with almost anyone in work or play.

Somewhere along, during those years of residency, you made another male attachment and this one was serious. This one was the real thing. Plans were developing for a real wedding which your mother had already been through with one of her girls.

Then your mother became ill and within three weeks had left this world. Wedding plans were delayed, but in a few months it was decided to go ahead. A large contingent of family and friends traveled to Chicago to attend, and a memorable wedding took place. The groom, Steven Newman, also a great physician, had done a fellowship in California and now found an opportunity to begin practice in Los Angeles. So, off you two went and you, Sara, began a career of teaching hematology at the University of Southern California. The teaching genes of your forebears had scored again.

Here's to you, Sara. Here is to your whole new world in California. Hurry up and pass on those genes. Your one mistake has been that you did not begin to pass on those great genes about ten years ago. You could have had six children by now.

31
Frank—Later Years

The passage of time is very subtle. We scarcely notice it. The scientists say that we really begin aging at about thirty years. The speed of certain mathematical calculations slows after this age. But many other physical changes begin about this time. The most obvious one is the skin which begins to show the slightest wrinkling. As the decades go by, this becomes more and more pronounced. Apparently the internal organs also slowly change and the function of the heart, kidneys, lungs, and liver all are sightly reduced. The reserve abilities of all these organs is so great that usually no problem develops. Most of us who are sixty and over notice that our memory is changing for the worse, and in some the deterioration is much faster.

The course of Alzheimer's Disease is still not really understood. Sometimes people in their sixties or early seventies begin losing their memory very rapidly and if this is followed by a decrease in all mental functions, it is thought that this is a separate disease from the dementia of people in their eighties or nineties. However, it is now thought that the so-called premature dementia or senility is really the same process as that of the really elderly. It apparently is not just arteriosclerosis but some changes in the brain cells with complex chemical functions lost.

We are always amazed when we see an older person continuing to carry on a job or profession for a number of years longer than is usual. We all have seen physicians in their eighties still practicing medicine and apparently doing an adequate and useful job. We know of famous musicians, such as pianists, who can still play music extremely well even in their nineties. The custom has been to retire from one's occupation at sixty-five years, and this formerly was mandatory with the government and many large companies. Recently the rules and regulations have changed, and now it is often possible to continue working several years after age sixty-five.

Frank Neely was one of those to whom time was kind. His mind and body were unusually well-preserved as the years passed. And his company, Rich's, realizing his usefulness to them and the community, allowed him to continue to be very active in the board room. After being president of the company for many years, he was made chairman of the board and then honorary chairman of the board. Finally, when he was in his late seventies, he became chairman of the Rich Foundation. He continued to have an office and secretary at Rich's for another ten years or more and did not stop going to the office until it became too physically difficult or until his mind finally began to break down; this was in his nineties. He had a long and remarkable career and finally died at the age of ninety-five.

His wife, Rae, was equally fortunate in keeping a keen mind right up into her nineties, and she too had a long life to ninety-six. In fact, her mind remained considerably better than Frank's when they were in their nineties. She had more physical problems than he did and required much more attention and treatment from her doctors. They decided to move back to the farm when they were ninety and remained there the rest of their lives.

Frank owned the two hundred sixty acre farm in Gwinett County since the early 1930s. It was about thirty miles from Atlanta and was really in the country at that time. The farm was just a hobby at first, but as time went on, it became a real dairy farm. He employed several men to work the farm. One of them, Royce Ramey, began work as a young man in his twenties. He stayed on for about fifty years and finally died after Frank died. He was quite a character himself, always jolly and pleasant. He had an infectious smile and twinkling eyes with a wrinkling face, when amused. He thought that Frank Neely was just about the greatest person in the world. They got along extremely well together. He would drive Frank in a jeep over a large part of the farm every week to show him what was going on and how things were progressing. This custom lasted throughout Frank's life, continuing through into his nineties when he had to be helped into the jeep.

Royce tells the story that when Frank first hired him, he was told by Frank as they were parting, "Remember one thing Royce; there are two things that I can't abide—one is a liar and one is a thief."

Royce remained honest and truthful throughout all those years, and Frank knew it. The two men had a great respect for each other, and even on his death bed Royce was telling how great a man Mr. Neely was.

Then there was Lizzie Mae Hall who worked in the kitchen as cook and housekeeper. She was black and also very young when she started work for Mrs. Neely. For some forty-five years she was faithful to the Neelys, outliving them both. She was always pleasant to everyone and was extremely happy to be working for them all those years. She too thought a great deal of both Frank and Rae and did not hesitate to tell everyone what wonderful people the Neelys were. There were several others who worked on the farm, but these two souls literally devoted their entire working lives to Frank and Rae Neely and loved it.

In the thirties Frank had had dealings with a famous Atlanta architect, Mr. Henry Toombs, and he had Mr. Toombs out to see the farm. Mr. Toombs came up with some unusual ideas for a house on the farm, and after considerable planning and sketches Frank and Rae came to an agreement with Mr. Toombs to build a house for them. The farm had a high area with large trees in the center of the land, and this overlooked a flood plain of the Chattahoochee River as well as the curving river itself. This high spot was to be the location of the house facing the river. The house, when built, fitted into the landscape very well. It had a huge living and dining room area in the center with a number of rooms on each side, making an H formation. The living room had a very large fireplace located toward the back of the house. A large flower garden area extended out from one side of the house. It was a beautiful house in a beautiful setting, and this became the second home for Frank and Rae. They always had a house or apartment in town as well. However, they would go to the country for weekends and sometimes for weeks in the summer. Through the years they entertained many important people, including Eleanor Roosevelt. They kept a large book and all visitors would write in it. The farm grew in importance to them, although it took a good bit of Frank's time. Also, he developed a very modern dairy farm which was run by Royce Ramey and his helpers. They sold the milk to a local distributor, and it was known as being the best and richest milk in the county. Lizzie Hall kept the big house and cooked and cleaned as needed.

As Frank got to be about ninety years old, there began to be the early mental signs of old age; first it was just the usual forgetfulness, then there were the fixed ideas, and finally even some paranoid ideas—he asked for his pistol several times. This was unusual; although everyone knew that there was a pistol in the house, he had never brought it up in the past. He had the ideas that someone might be breaking into the house at night. It took a great deal of reassurance from Rae and from Royce Ramey to convince him that there was no problem. He still wanted his pistol handy just in case of trouble. Rae and Royce, realizing that he was a little out of his head, finally convinced him that there was nothing to be concerned about. In the meantime they hid the pistol where he could not find it.

These mental aberrations were usually transient. Although he went through one period of several months in which he had confused ideas and some possible hallucinations, which he got over, he never became truly demented nor did he show unusual signs of deterioration of his facilities until toward the end of his life. However, he was extremely forgetful and at times confused.

During the last seven or eight years of their lives, Frank and Rae had registered nurses with them around the clock. Rae, being the weaker one physically, needed constant help, and Frank required a great deal of assistance and care at times; he was especially happy to have young, pretty women in the house. He always joked with them and enjoyed the ones who could answer back or even tease him. Frank had several falls and had broken both hips. The first one happened when he was about eighty-six and the second at about ninety-one. Both required hip surgery; the first required a pin and plate and the second a prosthesis, a new hip joint. He went through both operations very well, in spite of his age, and really enjoyed his stay in the hospital and all of the attention that he received. After the second hip operation Frank said his doctor had told him, "If you fall again, don't call me; call the undertaker!" Though he attributed this to his doctor, he had made it up himself. He retained, to a great extent, the ability to get along well with any type of person, and his sense of humor persisted to the end of his life.

They had a large apartment in an expensive high-rise building for the last few years of their lives, and they would also go to the farm from time to time. Frank preferred the country, but Rae, feeling that they needed more and more medical attention, believed that they would be safer staying in the city. So they would go back and forth, frequently taking their three nurses with them. The nurses were all well trained, bright, young women who often became attached to Frank and Rae; many would stay on for years. One of their nurses remained with them for about seven years.

The care of the aging and aged has become more and more of a problem in this great country. In pioneer days the family stuck together, and there was usually a fairly large group of children who could look after the older people. In the early part of the twentieth century the family was still a well-knit unit, with the children looking after their parents when they were unable to care for themselves. But somewhere in the middle of the twentieth century there was a slight, and then even more severe, breakdown of the family unit. There were more divorces; families were smaller, and people began to live much longer with the advent of many improvements in medical care. Nursing homes began to develop—first as small units with four to eight people cared for by a few nurses; later, larger nursing homes sprung up. The government became involved and would lend money at reduced interest rates to anyone who wished to run a nursing home. Nursing homes were run for a profit, unlike most hospitals. It became very lucrative to maintain a large nursing home, and soon there were large groups of nursing homes in many areas of the country run by new corporations and financed by stockholders. The return on the investment was usually quite good.

However, the fact that these nursing homes were run for profit meant that the companies tried to save money wherever possible in order to provide better dividends to their stockholders. The two areas in which they could save were food and help. The food was generally poor and at times insufficient. The help was very poorly paid, resulting in unskilled and inadequate labor performance; they did not care about the patients. They did the minimum required of them, and obviously the patients suffered.

In the early fifties Congress passed the Medicare Bill. This provided health insurance for everyone over sixty-five; it gave them hospitalization, fees for doctors, and provided a considerable amount of time in nursing homes. This gave additional impetus to the building of more nursing homes, and now the government was picking up the bill and paying the nursing home the full amount they charged. There were many changes in the Medicare bill as time went on, and very soon the government did not provide much coverage for nursing homes under Medicare. However, about this time, Medicaid came along; this provided hospitalization for the indigent, and in addition it provided care in a nursing home if the patient was proved to be indigent and unable to pay for nursing home care. Very soon it was found that people from fairly well-to-do and even wealthy families were having the parents give all of their financial resources to the children. They were then declared indigent and the government would pay their bill for the nursing home. This practice went on for years, and only recently have there been any good measures to stop it. The government did finally stop it and made some fairly strict rules on the structure of the nursing homes and how they were operated. They could take the license away if their regulations were not followed. They had rules regarding safety, the number of registered nurses, and the general care administered in the nursing home. There are rules that the patient requiring skilled care be visited once a month by his doctor; should the doctor fail to show on time, funds would be withheld. The system was working out fairly well until inflation hit in the 1970s. Food and help were much more expensive as well as heating and maintenance. The prices in the nursing homes went up very rapidly, so fast that many families could not afford the good nursing homes and had to seek less expensive ones usually out from the city in outlying towns.

There are only two or three nonprofit nursing homes in the Atlanta area. One is Wesley Woods Health Center and the other is the Rhodes Home. Wesley Woods is sponsored by the Methodists, is affiliated with Emory University, and is extremely well run. It is divided into three parts; the Wesley Towers are apartments costing about $300 per month; Budd

Terrace was for intermediate care at $1,000 per month; finally the Health Center or nursing home was $2,000 per month in 1983. The latter is beyond the means of many families. Rhodes Home is partially endowed, has a board of directors of many people in Atlanta, and is also very well operated. It cost about $1,000 a month in 1983, and it accepted Medicaid patients paid for by the Federal program. Many of the nurses and aids have been there many years, and there is a good morale among them which results in better care for the patients. If our hospitals are generally non-profit, why should nursing homes be run for profit?

Frank and Rae had the ideal situation; they remained in their own home, they continued to enjoy the amenities of home, and they had three registered nurses around the clock to care for them. True, registered nurses were somewhat expensive: at least $50 per shift or $150 per day. This would come to about $54,750 per year; fortunately, they could afford it, and the good care they received undoubtedly prolonged their lives.

They both became weaker when the early nineties came along. They stayed at the farm a good bit, but as they became weaker, they remained in town in the apartment for most of the time. Frank's mind was beginning to wander a little more. He experienced indifference along with big memory lapses and periods of disorientation. Rae's mind remained clear, but she continued to be afflicted with a number of physical problems which interfered with her locomotion and kept her in bed a lot.

Finally, Frank began to have episodes of shortness of breath from congestive failure of his heart. He responded to treatment at first, but after a severe episode, he was hospitalized, and in spite of conventional treatment for a failing heart, he lapsed into a coma and died after only three days in the hospital.

There has recently been a movement for allowing people to die at home rather than in the hospital. A large number of groups or organizations have sprung up across the country encouraging this. These groups feel that a person is happier dying at home, and the family is also happier to have the dying person at home rather than in a cold, impersonal hospital atmosphere. On the surface this seems like a good idea, but there are a number of practical points on the other side.

First, dying is not pleasant to watch; there is often suffering up to the last, or more often, one lapses into a coma for several days so it does not matter to them. Also, dying requires a great deal of very good, constant nursing care, and this is very difficult to get at home. Often injections are required to relieve suffering or pain, and this should be done by a trained person rather than a family member. There are many other things available in a hospital for relief or discomfort which are not easily accessible at home. Monitoring the heart and blood pressure are especially valuable in the prognosis of how long the patient will live. Finally, a doctor's care at home is not practical.

Frank died quietly, in a coma, with no pain or suffering. Rae was heartbroken and began to fail rapidly following his death. Their only daughter, Rachel, was a partial invalid herself, having had a severe stroke following a hip operation. In addition to Rachel, Frank's three grandchildren, Daniel, Eve, and Nathan, came to the funeral with their spouses. The funeral for Frank was held in the chapel of the funeral home. Frank had outlived most of his contemporaries, but there was a large group of people there, including many nieces and nephews. He had kept up with his family ties over the years; in recent times he had not seen many of them for a long period of time, and many of them had died. Here were the remaining grandchildren of Ben Neely coming to the funeral of the last survivor of that large family of nine children.

The funeral was officiated by a Rabbi and a retired minister of the Christian Church of Atlanta. He had been a friend of Frank and Rae's for many years. Frank had become inactive in the Jewish faith and had not joined the Christian religion. The final music at the funeral was "I'm a Rambling Wreck from Georgia Tech and a Hell of an Engineer."

So here we are, Frank, at the end of an era; you were born in 1884 and died in 1979 at age ninety-five. You had seen the vast changes in the world, and you had participated in these changes. You had done a great job for your company and had expanded it through the South with wisdom, judgment and that great sense of humor. You were beloved by all who knew you, whether they were great or little people. But more than that, you waded out into the community locally and nationally and made it a better place for all of us to live. You were recognized as one of the great men of Atlanta, Georgia. And you left a great heritage.

You had your zenith about twenty years before your death, and naturally your fame had begun to fade. But when the *Atlanta Magazine,* a new publication, described the ten great men of Atlanta and left you out, they showed their abysmal ignorance.

32
Family Man Par Excellence—Ben, Jr.

Benjamin Neely, Jr., son of Benjamin Neely, grandson of John Neely who came over from Ireland, you were born in Augusta, Georgia in the "Cottage" on the Carmichael Plantation on July 18, 1873. You were the fifth child and the third son of Benjamin and Henrietta Eve Neely.

You were a pleasant, sweet person from your earliest childhood, and you would remain a saint in the eyes of many up until the time of your death in 1938 at age sixty-five. You had the same chromosomes as Frank Neely and the rest of the children, but your ambition in life was to be a good family man. And that you were, par excellence. Your greatest pleasure was to help others and you kept doing that your whole life.

You were somehow caught in a series of circumstances where you had to be the titular head of the family after your father died. Allen, the eldest, did not quite live up to the expectations of his father, who hoped he would follow in his footsteps and become a teacher. But Allen was not a great student and had no interest in teaching. John, who lost his leg early in life and later became a physician, was a good student, but he did not stay at home much. So when Ben, Sr. died in 1892 young Ben was just nineteen years old. He had stopped school in the eighth grade and gone to work to help support the family. He placed all of his salary checks in the family bank account and never kept anything for himself.

He studied hard, and after he stopped attending school, he took night classes in accounting and worked in Rome for several years doing clerical work and accounting. He was very good with figures and this would be a great asset for the rest of his life. He was working in Rome for the L&N Railroad as an accountant but was offered a better job in Atlanta also with the L&N Railroad. The salary of the Atlanta job was much higher than that of the job in Rome, so he took it. He would travel back to Rome on the weekends; his mother, Henrietta, and several of the girls still lived in the old house, and he had become interested in a young lady in Rome, Miss Nina Hume, whom he had known in school. He met her again in church in 1896 and the courtship began. Ben was getting frequent raises with the L&N Railroad, and with Edgar and John working the finances of the family were better. Ben, the accountant, looked after the family's money. By 1898 Ben, now twenty-five, felt that he could safely marry and support a wife as well as help the family. So he married Nina Cornelia Hume on June 1, 1898 at a small wedding in Rome. He took his new bride back to Atlanta and settled in College Park, a quiet, little, country town on the outskirts of Atlanta; he could easily commute to downtown by trolley. He began his married life in an apartment in a part of an old house and immediately began saving to buy a house.

In the meantime, Frank had finished high school in 1900 at age sixteen, and decided he wanted to go to Georgia Tech and study engineering. Ben, being the only moderately prosperous one in the family, told Frank that he would help put him through college even though it was somewhat of a hardship. So Ben would help Frank through four years at Georgia Tech, going into his savings at times, and delaying his ability to buy a house for himself. Edgar was working in Atlanta as a court reporter and making a fairly good salary, but in 1902, at age twenty-four, he decided to go to law school in Washington, D.C. He had saved a fairly good nest egg but still needed help with tuition, transportation, and living expenses. Ben volunteered to help financially.

Ben, who had been only to the eighth grade in school, was sending his two brothers to college. Meanwhile, his wife, Nina, became pregnant and their first child, Annie Eve Neely, was born in 1902. The twins Mary and Nina came very soon thereafter, in 1903. Ben was able to purchase his house in College Park in 1903, and within a few years Benjamin, the third and then Laura were born.

Ben had been working for the L&N Railroad as an auditor and was highly regarded. He was excellent with figures. Southern Bell Telephone Company was just beginning in Atlanta, and it was growing very rapidly. One of the vice presidents knew Ben at church and told him about the audit department at Southern Bell, which was badly in need of a good auditor. Ben was offered the job and took it as it paid a better salary.

Ben had to travel a good bit for the telephone company to audit the books in major cities. In 1908 he went to New Orleans by train and had to stay several weeks. The sanitary conditions in New Orleans were not very good, and with the many floods of the Mississippi, the drinking water became polluted. Ben felt a little sick while there, but he finished his job and came home. The day after returning home his fever went up to 102 degrees. He thought he had the summer grippe and stayed home in bed. He then developed severe diarrhea and the high fever persisted. He called the doctor to the house. After examining him, the doctor made the tentative diagnosis of typhoid fever as it was known to be prevalent in New Orleans.

The fever persisted for weeks. Extremely ill, Ben lost weight. Typhoid fever in those days had a high mortality rate; the doctor told his wife, Nina, that he might not survive. Nina nursed him day and night and just about wore herself out, but she would let no one else do the nursing. Finally, in the third week, the fever broke and the diarrhea subsided, but he was left extremely weak for several months. The doctor advised him to get a job where he would be outside more. Ben looked around and found a job with Equitable Insurance Company, selling life insurance. Instead of being confined to a desk, he was making calls most of the day. Life insurance was coming into prominence in those days, and Ben had many friends and acquaintances. His gentle and likable personality along with his honesty and dependability paid off and he soon was doing quite well.

He now had his whole family of five children, and one of his greatest pleasures in life was to come home in the late afternoon and be with his wife and children. He loved them all very much, and this was reciprocated by the children. His home had a large garden for flowers and vegetables and often the whole family would be out in the garden pulling weeds, fertilizing, and picking the fruit and vegetables for their meals. They also had a barn with a cow and chickens and a pony for the children. The family was very close and Nina, his wife, was an excellent mother, but tragedy was due to sadden the family very soon.

The week before Christmas everyone was preparing for the holiday. School was out, the children were excited. Nina and Mary had gone to a neighbor's house, Mrs. Bell, who had girls almost the same age as the twins. They had come to see the Christmas tree.

In those days they had candles on the tree which were lit with matches from time to time. This was December 18, 1910, and the twins were seven-and-a-half years old. As it was getting dark, they began to light the candles on the tree. Nina and Mary had on long, pretty, cotton dresses. One of the lit candles fell and rolled under the edge of Mary's skirt. The dress seemed to flare up in a flash. A loud cry went up from everyone as well as screams from Nina and Mary. Little Laura, age two-and-a-half, was there, screaming. Mrs. Bell was so flustered and excited that she did not know what to do; instead of trying to put out the fire with a blanket or water, she told Mary to run home as fast as she could. Mary obeyed, but she was half a block away from home. She was now a flaming torch running down the street. By the time she got home, she had severe burns over most of her body. The family rushed her to the hospital in College Park, and then she was sent to Crawford Long Hospital in Atlanta. In spite of all treatment available at the time, she passed away on December 19, 1910.

The family was devastated. Here was a jolly family preparing for Christmas, and suddenly tragedy descended upon them. The girls would cry for days and even Ben and Nina could not hold back the tears. Ben and Nina were strong members of the Episcopal church, and though heartbroken, they took it philosophically as death of a child in those days was not unusual. Christmas was indeed sad, but slowly the family recuperated and were able to carry on.

Ben and Nina had always been quite active in the St. John's Episcopal Church in College Park. The whole family attended church and Sunday school regularly. Ben first taught in the Sunday school and later became superintendent of the Sunday school, a job he held for over twenty years. He also served on the vestry of St. John's for many years and was junior warden and finally senior warden of the church. The entire family took it for granted that they would go to Sunday school every Sunday and also to church as they grew older. The church and its activities was thus a major part of the life of the entire family. The family was instilled with the principles of the Christian religion; as time passed, the children continued to be active in their respective churches throughout life. In addition to church work, Ben served many years on the College Park Board of Education and was its president, a volunteer position which he held for twenty years. Later, one of the new elementary schools was called the Ben Neely School in his honor.

Thus Ben's life revolved around his family, his work, and the community as it related to his family. He had no ambition to be a politician, to make a lot of money, or to get a lot of power. He was now general agent for the State of Georgia for the State Mutual Life Assurance Company of Worcester, Massachusetts and was quite successful in this position. However, at the height of his career, he was driving to Atlanta in a new Buick when a drunk driver, coming out of a side street, hit Ben's car on the driver's side. Ben was thrown to the other side of the front seat, struck his head on the dash board, fractured his upper right leg, and his upper right arm. He was unconscious for several days in the hospital; he was in casts over most of his right leg as well as his right arm. He was hospitalized for several weeks and out of work for about three months. Fortunately he had good sickness and unemployment insurance. However, the accident left him very weak, and he decided to give up the General Agency and just sell life insurance for the company.

The remaining children, Annie, Nina, Ben, and Laura, went through the public schools with good grades; they all went on to college. Annie, the eldest, became a dietician and worked for several large hospitals. She did not marry until late in life and continued working in a hospital in South Carolina where she retired with her elderly husband. Nina, the second child, also went to college but married soon thereafter and had two children. Ben, Jr., who had been born with a heart murmur, was rather frail most of his life; he married but had no children. He died relatively early in life at age thirty-six. Laura went to college and became a teacher at the high school in College Park. She married an army officer and moved around until his retirement. She says that, after teaching for twenty-five years in the high school there, when she walks down the street people whom she has taught through the years come up and embrace her; she often has forgotten what they looked like as children.

Ben loved the country atmosphere of College Park. He was able to keep cows that grazed in a pasture behind the house. Young Ben would milk the cows and little Nina would churn the milk, reading most of the time. Papa Ben put emphasis on school work while the children were small, and he insisted they finish their homework in the afternoon before they could go out to play. The children obeyed even though other children in the neighborhood were out playing right after school. Apparently the old Neely tradition of excelling in school work was ingrained in them.

Nina, Ben's wife, was a great mother who looked after her children very carefully but did not spoil them. She and Ben showered love on their children and this was returned by the children. Nina helped them with homework, stayed busy most of the time, and looked after her husband in every way, in sickness and in health. Thus the foundation of love which these parents provided to their children was the incentive for the children to emulate their parents and to do something useful in this world. And indeed that was the case. They did not have grandiose ambitions, but they were all useful citizens.

After the automobile accident in 1926, Ben continued to work with the insurance company selling life insurance. He remained quite active, however, going into Atlanta every day and coming home at night to enjoy the company of his wife and children. In 1926 Annie was twenty-four, had graduated from college, and was working as a dietician in Spartanburg, South Carolina. Nina was twenty-three and had married Frank E. Hankinson two years before. Ben, Jr. was twenty, and Laura was eighteen. They were both in college. Ben, Sr. had made it possible for all of his children to go to college. The children began to drift away as their careers evolved, but they would all come home frequently.

Ben's greatest ambition had been to make a reasonably good living and look after his family, and he achieved both of these aims. In his last year of life, Ben began to fail in health and became weaker. Apparently due to a heart condition, he would get chest pains and shortness of breath on exertion. Finally, he had to give up his business completely as walking half a block would cause discomfort. He had sufficient pension funds accumulated over the years, and giving into the suggestions of his doctors, he retired.

The family would come back home as often as possible, but no one realized how sick he was. On the morning of August 8, 1938 Nina found him apparently asleep in bed. She did not disturb him at first, but in a little while she became concerned, tried to wake him, but soon found that he was not breathing. He had died in his sleep. The funeral at St. John's Church was very well attended by his many friends and admirers.

Here was a true family man, a man of great integrity, a man who was beloved in the town and local community. He had no ambitions to change this world nor did he try. He was unselfish to the core. Nothing gave him more pleasure than doing something for another person even if it meant doing without himself. His relationship of love with his wife, Nina, was on a firm foundation and never wavered. Nina lived twelve years after his death and kept the homestead as long as she could. Laura and Ben had both married, and Annie would marry later.

So here's to Ben—cheers and wassail. You lived a good, full life. True, you had disappointments, sickness, and even tragedy—the loss of your daughter, Mary, at a very tender age. But you weathered the storm; you kept the faith; your spirits never flagged for long. You had a good rebound. You were the one who helped put Frank through college. You never went to college yourself, but you became very skilled with figures and learned to be an auditor and later an insurance man.

Were you not as great in the eyes of the Lord as your younger brother, Frank, who became famous for his achievements in the city as well as the nation? Who can judge? I believe you were a happier man than Frank, who seemed driven and whose problems spurred him on to accomplish more in this world. You both got along well with people and you both had a good sense of humor. Frank did have an unusual magnetism or perhaps outgoing spirit which endeared him to people very soon after he met them. Ben was slower to make friends, but when he did, it was a friend for life.

What is this unselfishness with which you were so endowed, Ben? Where did you get it? Was it living in the middle of a large family where there was much sharing of whatever was on hand? Or did you get it from the genes of your mother and father? I would say that generally all of the children were unselfish, but you seemed to carry it to an extreme.

The divorce rate is mounting. Perhaps it is better to break up a marriage than to continue a farce, but this world would be a much better place if there was more of the elixir of unselfishness, which you had, than seems around now.

So, Ben Adam, may your tribe increase! You were another one who loved your fellow man.

33
Margot—Early Years

Life in Saranac Lake, New York in 1920 was a constant battle against the cold. High in the Adirondacks, close to the Canadian border, in the midst of a string of natural lakes, it was indeed beautiful, especially in the winter. But winter lasted from six to nine months. The old joke that Saranac has two seasons, winter and August, was almost true. There was snow on the ground from October into May. The ice did not disappear from the lakes until middle May, and then suddenly the ice was gone overnight. Twenty to forty degrees below zero in the midwinter was not unusual.

Lake Placid was only ten miles away; it was a summer resort and almost deserted in the winter. But Saranac carried on winter and summer. A great majority of the residents had come to "cure"—to get well from their tuberculosis. Then there were the natives—people who had lived there for several generations. The word native had a very bad connotation. They were looked down on for some reason.

The train came from New York on its way to Montreal and stopped at Saranac; often the snow drifts would prevent the train from going through for several days or at least delay it for several hours. But unless there was a terrible ice storm, the trains, though delayed, would get through. So the train from New York would bring new patients to Saranac almost every day.

Trudeau Sanatorium was the best known institution in the U.S. for the treatment of tuberculosis. It was internationally famous and was always full. So the patients spilled over into the town and rented houses or rooms. There were quite a few well-known tuberculosis specialists who treated these patients in town on a private practice basis. Other sanatoria were springing up in the town. The Veterans Administration had a home for the tuberculous veterans. Money was given by the Will Rogers Foundation for a new sanatorium to treat retired actors. So it was a small town with the chief industry tuberculosis, its care, its treatment, and the services that it would generate.

So Roy Gardner, you had taken that train ride from Grand Central Station in New York City in the fall of 1917, arrived at Saranac Lake, and became a patient at Trudeau Sanatorium along with several other physicians as doctors had a high rate of tuberculosis because of frequent exposure. You were fortunate, Roy; many people still died in spite of rest, fresh air, and the cold weather. But in six months' time, you were much better, up and around, and had been offered a job in the laboratory at Trudeau. You took it, and now you could send for Carabel and Margaret as you had rented a small apartment in town. So on a cold day in March 1918 the family was reunited. The snow was still deep.

Here you were, Margaret (later to be called Margot because you had gone to a French camp, L'Ecole Champlain), beginning your life high in the Adirondacks, fighting the cold weather like the rest with heavy coat, mittens, and hat when you got out of the house. And there was always an open fire made by Roy at night. It was a great comfort to the family to sit around the fire at night. Years later you, Margot, would have open fires at your home in Georgia through winter and into the spring. Some people said you wanted an open fire even if you had to turn the air-conditioning on.

You, Margot, were obviously a bright child at thirteen months when you arrived in Saranac. You learned early how to get what you wanted from your father and your mother. But Carabel could not be manipulated, whereas your father was putty in your hands. He would do anything you wanted. This relationship continued through the years, and Carabel and you, Margot, being strong characters, developed a mild but sometimes fierce opposition to each other. Roy and you, Margot, on the other hand, got along fine, had a deep respect for each other, and a long and abiding love. Both Carabel and you had the Irish temperament with big mood swings. Roy was always even-tempered and calm no matter what the circumstances were.

Carabel soon became pregnant again, and Dorothy was born in 1919; she was two-and-a-half years younger than Margot. Dorothy got her father's easy going outlook on life. So here were two girls with the same parents and parental chromosomes, yet the temperaments were entirely different. Margot naturally was a little jealous of the attention Dorothy was getting as a new addition to the family, but very soon she realized that here was a new, living doll and later a great companion and fellow conspirator to plot getting what they wanted from Carabel. If this failed, they could go to Roy; they knew he would help them win their wishes.

Margot was the leader of the two girls; she did not hesitate to make known her desires. Being older, she was naturally dominant in those early years, but by the time they were both in school, Dorothy had developed her own personality, and in school they became more involved with their own classmates. Two years difference in school meant somewhat different interests. But they would remain the best of friends for the rest of their lives; although during the last forty years they were geographically separated by the width of our country.

School was private at first for the girls. Public school was full of "natives," and these people were to be avoided. However, they would later go to the public high school as there was no other high school in town. Margot, who obviously was very bright before beginning school, showed her quick mind to the teacher in the first grade and was one of the first to learn to read. However, it was soon found that she was holding the books very close to her face, and she was taken to the optometrist in Saranac who found her extremely nearsighted. Roy was concerned about her eyes and took her to a famous eye specialist in Montreal, driving her over in the family car. It was only one hundred miles, but it was over curving and sometimes rough roads. By starting early, they got to Montreal in about three hours, saw the ophthalmologist, and came back that night. It was thought by the doctor that she might have progressive myopia, but this did not prove to be the case. However, she would have to wear glasses for the rest of her life. She would put them on when she got up in the morning and take them off when she turned out the lights at night.

Momma Carabel and daughter, Margot, began quite early to develop a competitive spirit as to who would get her way. Margot did not like oatmeal, considered the mainstay of a child's breakfast as it would "stick to the ribs," and Carabel felt this was extremely important. Margot had all sorts of excuses for not eating it; Carabel was quite firm that she should. When Margot was six years old a crisis developed in the family. Carabel kept insisting that Margot finish her oatmeal, and the problem had gone on for weeks. Finally one morning when they were all there at breakfast, the usual controversy began. Suddenly Margot got up from the table and shouted that she would never eat oatmeal again because it had "prickers" in it that hurt her throat. She ran upstairs, lay on her bed, and sobbed loudly. Carabel was for bringing her downstairs and giving her a spanking, but Roy tried to calm the waters, suggesting that oatmeal was not that important and Margot could eat eggs or cream of wheat if she wanted to. It really was an important time in the mother-daughter relationship; Margot had proved that she could get her way and it was the beginning of a long series of minor confrontations which Margot somehow seemed to win over the years.

She continued to do well in the early school years but developed an antipathy for mathematics which would stay with her the rest of her life. English, foreign languages, history, and spelling were her *forte* throughout school. She became the leader of the children in the neighborhood, both boys and girls. She would organize games, the parties, and the activities of the group. She was a good ice skater and later a good skier, but the latter had not become popular when she was a child. She could run and climb as well as the boys and was a great favorite with them. One of her early schoolmates, Thomas Cantwell, was also a leader and organizer and they got along very well. Later the two of them would begin the first ski-tow in Saranac. It was made out of an old hearse pulled to the top of one of the low mountains. They figured out how to attach a cable around one of the wheels of the jacked up hearse. It was quite a novelty in the town and also quite a success. Downhill skiing became a very popular sport for the whole town.

There were a fairly large number of children in the town, and some would remain friends throughout their lives although scattered over the country. Tommy Cantwell was one of these. He later went to Williams College and then to law school. After that he went into the army and was stationed in Atlanta, Georgia. He came to see the Neelys in Atlanta, and Margot said he was the same old cocky Tommy as he had been as a young boy. He was one of the few who returned to Saranac Lake after the war, married, set up law practice, and proceeded to have a large family. His first wife was from Texas. He had met her during the war and brought her to Saranac Lake after the war. They had two children, but she did not really like this terribly cold country, and she took the children and went back to Texas. Tommy married again and this time had eight children; unfortunately he had a heart attack and died at about age fifty-seven. Another life snuffed out before its time.

* * *

Heart attack—what a dreaded two words. John Neely, the doctor, died suddenly at fifty-eight without warning. LeRoy died at fifty-eight with no history of heart trouble; Emily Eckfeld Neely had a heart attack in 1938 at age sixty but survived it and lived on into her eighties. Others in the Neely family must have had heart attacks, but apparently it was not prevalent throughout the generations as far as is known.

Coronary thrombosis with myocardial infarction means that a blood clot or thrombus forms within one of the branches of the arteries supplying the heart muscle, cuts off the blood to part of the heart, and the muscle supplied by that artery dies. This is called a myocardial infarction. But the true cause of the blood clot forming is a small, damaged area in the lining or endothelium of the artery. This is called a plaque which is from arteriosclerosis or hardening of the artery. The plaque is made up of cholesterol and may ulcerate in the center. Platelets come to cover the ulcer; these in turn attract the protein molecules fibrin and form a mesh or net. This in turn traps red bloodcells and the process repeats itself forming a growing clot which blocks the artery. If the clot happens to be in an artery which supplies a major part of the muscle, there can be heart failure, an arrhythmia, or a precipitous drop in blood pressure leading to shock and possible death. So these lives have been cut off by a tiny ulceration no bigger than the head of a pin which, for some unknown reason, begins the death-dealing chain of events.

Why does this happen? We don't know for sure, but we do know that certain things seem to predispose to arteriosclerosis or hardening of the arteries. Diabetes is well known to be a big offender. Even relatively young diabetics in their forties often get severe arteriosclerosis. Smoking is a known hazard; heart attacks definitely are more frequent in heavy smokers. High blood pressure predisposes to increased hardening of the arteries, and other apparent factors are lack of exercise and nervous tension. The exact roles of the latter two are not well proved but seem to be important. Finally, heredity is a well known factor. High cholesterol runs in families and seems to parallel the incidence of heart attack. So one must be very careful in their choice of parents. Being overweight and high fat diets are also bad. The young executive who is a heavy smoker, overweight, under much tension, and does not exercise is much more likely to block his coronary arteries than his easygoing counterpart who jogs, does not smoke, and keeps his weight down.

How has this evolved through the ages? For some reason the female of the species before age forty-five is not subject to heart attack nearly as much as the male. After menopause the female will have more heart attacks but still does not approach the dangerous status of the male who can drop over anytime and die of a heart attack without warning. 500,000 to 600,000 people die of heart attacks in the U. S. each year; seventy-five to eighty percent are men.

There is a theory that 70,000 to 100,000 years ago man was the hunter while the women looked after the home. The male, after running and hunting for several days or weeks, would bring home the results of his hunt, cook it and eat it, and go out to hunt again. Perhaps the exercise was protective for him. Women, on the other hand, became used to staying home and not exercising as much. In the 20th and somewhat in the 19th century the male went into the office, sat all day, smoked heavily, over ate at night, and did not run or do any form of strenuous exercise. He therefore became much more likely to clog up his delicate coronary arteries and die young of a heart attack.

Ah coronary arteries, what delicate and important structures you are.

Coronary means crown, and you do in truth circle the top of the heart like a crown, but you send down branches to supply the whole heart. So one tiny plaque in one of the two main coronary arteries may ulcerate and then the clot is formed which blocks the artery completely. It is nature gone wrong. The platelets and the clotting are a supposed protective mechanism, yet in these little but important arteries the clot is the killer and life is cut off in thousands of people, never knowing what struck them.

What can we do to prevent these terrible tragedies? Try to avoid the risk factors, get exercise, and keep the weight down as well as try to live healthy lives. Sometimes this seems to work; still, if one had bad heredity, there is little that can be done in a preventive way. We are told to take one aspirin every day. This coats the platelets and they will not cling to the ulcerated plaque like normal platelets. The coronary care units have helped; the mortality rate has been cut from about thirty-five percent to fifteen percent with careful care and attention to every problem that may arise. The latest treatment is to try to dissolve the clot as quickly as possible after it forms. This involves a treatment with a strong enzyme. This treatment, just beginning, shows great promise especially if administered early, a few hours after the onset.

Coronary artery bypass surgery is a highly successful operation in the majority of suitable patients. A vein is taken from the leg, attached to the main artery of the body, the aorta, and then attached below any partially blocked coronary artery. Blood flow thus becomes much improved to the heart muscle and the angina disappears. This is said to be the most common operation there is now.

Angina, pronounced an-gí-na, simply means pain in the chest. However, it is a fairly specific symptom and the doctor can almost always make the diagnosis of narrowing of the coronary arteries simply by a careful questioning of the patient.

The patient says, "Doctor, I get this tightness in the middle of my chest when I walk up a hill, or if I am in a hurry, or if I get very upset or nervous, but it goes away in a few minutes if I rest."

Those words are enough to make almost surely a positive diagnosis of coronary disease. Stress testing will usually confirm this by showing changes in the electrocardiogram; a coronary angiogram will show the exact degree and position of the narrowing in the coronary system. This is done by injecting a radio-opaque dye directly into the artery.

The pain of a true heart attack may be equally typical. It is a severe, often crushing pain beneath the sternum (chest bone) which may radiate into the left arm and sometimes the right arm, possibly into the neck, and occasionally into the back. It is often accompanied by extreme weakness, sometimes nausea, vomiting, and shortness of breath. It does not go away with rest, and will persist for hours if not treated. This type of pain usually means that the clot has formed and is blocking the artery.

Treatment for an angina attack is nitroglycerine under the tongue. This dilates the coronary arteries and quickly improves the circulation to the heart muscles and the pain goes away. There are now good drugs to prevent angina such as a long acting nitroglycerine, drugs that block the nerve response in the arteries called beta blockers, and now a new set of drugs called calcium blockers which are quite effective. But for the cure for angina, a coronary bypass must be performed to alter and improve the blood supply.

The treatment for a real heart attack includes oxygen, strong pain medication, and specific medications of any disturbance of heart rhythm that may arise. But the latest and most exciting treatment is the new enzyme streptokinase; this is injected into the veins to dissolve the clot as soon as the diagnosis is made. So great strides have been made; still thousands and thousands of lives are suddenly coming to a tragic halt because one tiny plaque in the heart's coronary arteries becomes ulcerated and the clot forms. This problem is being researched by many great minds in medicine; perhaps someday we will have an answer.

* * *

Margot could recite all of the poems in A Child's Garden of Verse at age two and-a-half. Her father taught her the Greek alphabet before she learned her ABC's by drawing the letters in the sand at Lake Clear. She was always the leader of the children in the neighborhood. She would always accept a dare even if it meant falling down and hurting herself. These early traits of intelligence, courage, and leadership persisted throughout her life, and her list of involvement in this world was never fully documented. Someone called her the flame, and her sister says she was the flame by which we lesser mortals warmed ourselves. She died at her zenith, going out of this world like a shooting star. She never could have tolerated being really old.

But the one amazing trait that she had was a warmth of spirit, a charming personality that seemed to radiate and infect others. She could talk easily to anyone, a prince or a pauper, and they would be able to converse easily with her. It is extremely difficult to characterize this ability to communicate quickly and easily with your fellow man, but she had that ability.

Was it from a gene or was it an acquired characteristic? It was something entirely individual. As we look at her early childhood we see the beginnings of it; it was there in her high school days, and it seemed to flower in college; that was when everyone began to recognize that warmth of spirit. This *joie de vivre* was not just within her, but came bubbling out and somehow made others happy just to be with her. So we hope that this trait is in the genes and chromosomes and that it will go down through the generations.

34
Frank, Edgar and Piedmont Hospital

Let us look briefly at the lives of the two brothers, Frank and Edgar Neely, as they began their own lives as young married men soon after the turn of the century.

Here we have two brothers with the same genes and chromosomes beginning their careers at the same time. Both had the same background in early life. Frank did not know his father well because he was only eight when Ben died. Edgar was fourteen when his father died. Both had to make their own way in early adulthood, although both were helped financially at times by their older brother, Benjamin, Jr.

Both were bright, and they both felt that learning was one of the most important things in life. They both married well, although Edgar had married a Yankee and Frank had married a Jewess. Physically both were tall, handsome, young men. Their personalities were, however, quite different. Edgar was easy going, relaxed, enjoyed life, but was quite emotional. Frank was hard-driving, restless, and active all the time. Both had a good sense of humor and both got along well with people. Edgar made more close friends than Frank, who had a wider range of acquaintances.

The wives of both were quite an asset to them. Emily became accepted in Atlanta society and had many friends herself. She helped Edgar but did not push him. Rae was always a driving force behind Frank, although it was not obvious. It is very difficult to say how much influence she had on Frank and how much his own drive and ambition shaped his career. Certainly Rae helped in his rise to greatness.

Edgar's ability to talk and his love of theatrics very soon were found to be valuable in the courtroom. He showed the makings of a great trial lawyer soon after returning to Atlanta. He was with a very good law firm, and they soon began to depend on him to conduct the cases that came to trial. The firm represented many large companies and especially insurance companies who were frequently being sued. They also represented several of the major railroads which were also being sued a lot. Edgar became a trial lawyer for the defense. He had been a court reporter for several years before becoming a lawyer and had learned the ways of a good trial lawyer then. He prepared his cases well, and he knew how to cross examine witnesses effectively, stressing their testimony at times and deriding it at other times. He knew how to present a case to the jury. He would get their sympathy early on, and his good humor kept them amused. In accident cases, he brought maps, charts, and photographs. He did not try to over-influence the jury, but his clear presentations often found the flaw in the prosecution's case.

As time passed, he matured, continued to try the difficult cases, and very soon had the reputation of being one of the best defense lawyers in the city. Big corporations from out of the city would employ his firm because of his abilities, and he would be called out of Atlanta for consultations. He was extremely busy at his profession, often working Saturdays and some Sundays. He was too busy to become active in the community and he had no special desire to do so. He did do considerable work for charitable organizations, was very active in the Episcopal Church, and was a member of the Piedmont Driving Club where he was well known and liked. He liked to write poetry at night. He also loved music and would compose songs with one of his friends, George Lindner, who was a musician.

He liked reading and was a Shakespeare lover. He and Emily would go to the theater often; they were also among the music sponsors of Atlanta. Edgar was able to get passes on the railroad because of his legal involvement, and he often took the whole family on long train trips—once out West and to Seattle then back through the Canadian Rockies. He and Emily travelled to Europe many times.

Their first child, Henrietta, died shortly after birth in 1908. Edgar, Jr. was born in 1910, Levering in 1912, and Randolph in 1915. Edgar, Sr. loved his family and enjoyed being with them. The family would go to Highlands in the summer and later would go to the farm in Alto which Edgar bought when the children were young. He would come up frequently on weekends, but the family would stay all summer.

He was doing reasonably well financially as a young lawyer, but he was not good with money and had a tendency to spend it. He did save a fair amount once, invested it in real estate in East Point, but the city went north instead of south and he lost it all. He continued to live well, and spent most of his money as soon as it came in. He was never rich, nor was he ever in real financial trouble. He and the family lived through the crash of 1929 and the depression that followed, and they never really felt it. The railroads and insurance companies continued to pay his fees.

Frank had much more financial acumen, saved and invested his money wisely, and by the time he was sixty-five, he had bought the large farm, built the big house there, and was probably a millionaire in the days before inflation had eroded the value of money.

So it is difficult to compare the two—one was in business and very active in the community; the other was in a profession in which he excelled and did not care to try to change his environment. Frank continued his drive or purpose right into old age, and once he had showed his ability in the community, one volunteer job led to another and kept him constantly occupied with very little free time except at the farm. Edgar also was very busy with his legal practice, but he enjoyed life more. He used alcohol much more freely than Frank, and although he could not be called an alcoholic by any means, he often stayed up at night writing poetry with a tumbler of whiskey at his side. Also, Edgar had more of an eye for the ladies than Frank, although we have no certain information on this about either of them. Edgar played golf; Frank took off to his farm for the weekends where he stayed busy improving the operation of it by careful overseeing. Edgar became ill in his mid-seventies and died at seventy-three of cirrhosis of the liver. Frank lived to the ripe old age of ninety-five.

During their lifetimes great strides were made in medicine. Dr. Lewis Thomas calls medicine the youngest science—this because it really was not a true science in the 19th century and early part of the 20th century. It was not until the mid-1930s that sulfa drugs were discovered, and people were miraculously cured of diseases which had once been fatal. Sulfa was followed by penicillin and a long string of antibiotics. Tuberculosis, one of the greatest killers, began to be controlled in the 1940s with the antituberculous drugs, and is now almost eradicated. Vaccines of all kinds prevented disease. Heart disease, the greatest killer of all, began to have a much lower mortality with surgery and new drugs in the 1970s.

Edgar died in 1951 at age seventy-three of a disease that might have been helped more now but still could not be cured. One wonders if the long life of Frank and Rae was really influenced by the advances in medicine or if it was just chance that they lived so long.

Rae was a perfectionist—she wanted the best medical care that was available for her family. Very early in her young married life she heard of a young doctor in town, Dr. James Edgar Paullin. He was an internist, had gone to medical school at Johns Hopkins, was very active in the Emory Medical School at Grady, and was on the staff at the small but elegant Piedmont Hospital, located in a large house on Capital Avenue. Rae was referred to him for a minor problem with her kidneys. They immediately developed an empathetic relationship which continued for the remaining thirty years of Dr. Paullin's life.

Dr. Paullin brought many new ideas to Atlanta from Johns Hopkins. He trained in pathology and laboratory science and knew what a great help these could be in practicing internal medicine. He organized and developed the laboratories of Piedmont Hospital and Grady Hospital, areas which had very little precedent in Atlanta. He was interested in new ideas, but he took a scientific approach to any new treatment. Medical care from 1910 to 1920 was in its infancy, and although advances had been made concerning the cause of illness, treatment of most diseases was sadly lacking and good treatment of the sick consisted of good nursing care and reassurance. Dr. Paullin was probably the single most important person in bringing the best and latest in internal medicine to Atlanta. He also was a great teacher and began working actively at the medical school, spending considerable time there every day though his own private practice might be suffering. He later became a full professor of internal medicine at Emory and director on a volunteer basis.

Piedmont Hospital had been begun in about 1880 by a Dr. Amster, who was interested in physical therapy and soon thereafter by Dr. Floyd McRae, Sr. who was a surgeon and who soon became well known and respected throughout the city. These two doctors attracted other good doctors. Small additions were made to the hospital, and it was beginning to get a good reputation for good care. Dr. Paullin had begun practice in 1912 and he soon took in a younger partner, Dr. Cliff Sauls, who had gone to the old Atlanta Medical College which later became Emory Medical School.

Soon the hospital added young specialists, some of whom later became famous in their fields. Dr. Mike Hoke, the orthopedist who also worked at the Crippled Children's Hospital, developed new operations for bones and did amazing reconstruction on club feet and deformities from poliomyelitits, which was prevalent at the time. He also needed an associate very soon, and Dr. Lawson Thornton joined him to make an outstanding team.

Dr. William Shallenberger, who had been trained at Johns Hopkins in obstetrics and gynecology, joined the staff and later became one of the outstanding doctors in his field. Next came Dr. Montague Boyd, also trained at Johns Hopkins, but in the field of urology. Dr. Charles Dowman joined the staff, the first neurosurgeon in Atlanta.

Most of these doctors were trained at Johns Hopkins which was considered the best medical school in the country. They had joined forces to make a hospital dedicated to the best medical care. Very soon almost an entire block of houses on Capital Avenue was converted into part of the hospital. Finally, right after World War I, they realized that they needed a more modern building, and after several meetings, they decided to build a two-hundred bed hospital behind the old houses. The doctors financed this themselves and partially owned the whole hospital until the 1930s, when it became the nonprofit institution it still is today.

The building was completed in the 1920s. When it opened up more doctors were added to the staff, and it was now a complete hospital, providing every specialty in medicine then available. Almost all of the doctors were very active on the teaching staff at the Emory Medical School, which in those days used volunteer physicians from the city to provide all of the clinical teaching of the medical students in their last two years.

As the doctors realized how important good nurses were to good medical care, they launched a nursing school early in the 1900s. It also was small at first, but gradually expanded and became one of the best nursing schools in the city. It continued to be an excellent school until its final closure in 1983. This was a sad day, for Piedmont nurse graduates had a great admiration for their school. The decision to close the school was partially financial, but also due to the fact that nursing schools were becoming four year institutions offering a college degree.

The director of the nursing school, after it had been established for about fifty years, was Miss Genevieve Garren. She ran the nursing school with an iron hand. Any little student nurse who stayed out after 10 p.m. against the rules was promptly dismissed from the school, sent back to her south Georgia home in tears. She conducted a very good nursing school with high standards; those who got through the program were proud. The school continued to train nurses well. Most of the student nurses were from the country. They had to be graduates of high school. They were very dedicated young ladies. The doctors were quite active in teaching in the nursing school, and this was good for the general morale of the students.

So here was a small but excellent hospital with a good group of physicians who were doing more than just caring for their patients; they were teaching nurses and interns.

So it was an elite, small hospital in which everyone knew each other well; they were almost like a large family with everyone working together with a common purpose. Dr. Paullin had developed the laboratory so that almost any blood test done in the country could be done there. The x-ray department was enlarged and strengthened and the physical therapy department, now under the direction of Dr. Mike Hoke, was helping to rehabilitate all types of orthopedic cases. Dr. Paullin would encourage medical students to go to the northeast at Harvard or Hopkins for interning and residency training. One of these was young Dr. Carter Smith who went to Peter Bent Brigham Hospital. He came back to Atlanta to join the other well known internist at Piedmont—Dr. H. Cliff Sauls. He brought the first cardiogram machine back to Atlanta. At that time it was a huge cumbersome machine, but it was the basis for good cardiology and was putting the diagnosis of heart attacks on a scientific basis. Young, well-trained doctors were coming to Piedmont in large numbers as the years went by, and it was becoming more difficult to get a bed.

Another of the young doctors who devoted his life to Piedmont Hospital is Dr. Charles F. Stone. He also had been trained at the Brigham Hospital in Boston in one of Harvard's best residency programs. Stone had that great combination of being a fine doctor and having a superb sense of humor. These qualities helped him make Piedmont Hospital a great institution.

Dr. H. Cliff Sauls was the one at Piedmont with financial acumen. He realized that Atlanta was growing north instead of south and that Piedmont was going to have to move some day. He had the hospital buy a large tract of land on Peachtree Street at Collier Road in the depression years of the 1930s for a very small price. This was later to be the site for the new Piedmont Hospital built and finished in 1956 and financed mostly by contributions from doctors and their patients.

Dr. Floyd McRae, Sr. was the chief surgeon for many years, but he died relatively young, and his son, Dr. Floyd McRae, Jr., trained at Johns Hopkins in general surgery, soon became the dominant figure in the hospital, making a great combination with the internist Dr. Paullin. Not only an excellent surgeon, he also had a most unusual and charming personality. Almost all of his patients loved him and many thought he was God.

These two men, Dr. McRae and Dr. Paullin, became the doctors for Frank and Rae Neely; also, they all became very close friends. The two doctors were becoming famous in their fields, and Frank Neely was getting very well known in business and in Atlanta community affairs. Rae was the one who saw to it that the ties with these men and her family continued and were strengthened. They often came out to the farm for meals and other social events. Rae continued to have minor health problems and would go to see Dr. Paullin from time to time; Frank would go for check-ups. Their health remained good over the years, and the good medical care they received must have contributed to their long lives.

Edgar Neely was not so fortunate, although he did live to the respectable age of seventy-three. He and his family also used the Piedmont doctors for their medical care.

Edgar, Sr. was in Piedmont Hospital during his last illness and was mentally lucid up until the last few days. His son, Levering, had become a doctor and was practicing internal medicine at Piedmont. Dr. Carter Smith was Edgar's physician and was well known for coming early to the hospital.

When Levering came into his father's room at 9 o'clock one morning, Edgar said: "Son, why don't you get up earlier? Carter Smith makes $100 before you get out of bed!"

Two days later he died of a massive hemorrhage.

35
The Flavor of Fernandina Beach

A beach, a beach wide, long, flat, and white. Large sand dunes, the one toward the sea larger than the height of two men, then a trough and the main sand dune, perhaps the height of three or even four men, but nice solid sand dunes. It would take a tremendous storm to wash into the outer one, but it would have to be a hurricane which went on for days to hurt the inner mammoth dune. Sea oats grace the outer dune in a lacy line. The inner dune has grass and vines and cacti. And that sea: green-blue with large rolling waves coming in with that beautiful symphony of sound. Let the eye take off up or down the beach—three miles, five perhaps. Very little sign of life except on the weekend when a few people walking and an occasional car (unfortunately) mar the beauty of it. Then let the eye wander out to sea—slowly drinking in the expanse of it. Go on up to the horizon and then look left and slowly to the right. Take in the edge of the water and the sky. Usually there will be a few clouds—cumulus towering up. Or perhaps just a haze. Whichever, it rests the soul. It doesn't take away pain, but it makes one better able to stand the pain—to cope. It does what Bach or Beethoven or Mahler can do when the music booms in from both sides in stereo—just as if you were sitting in the front row of a great orchestra. So drink in that view of the sea—take it home with you, you might need it later.

And little town—wherein is your charm? You are a group of about six thousand people within the city and perhaps three thousand more just outside the city. Most of you work at the mills—two large paper mills, I.T.T. Rayonier and the Container Corporation—or you provide services for those who work at the mill or for the summer people. For a long time the town was not developed like other resorts because of the smell of the paper mills. That has long since been corrected, but it slowed things down for a long time and only now are the developments burgeoning.

The town itself has great charm; it is old and the main street goes right to the docks where shrimp boats come in with their catches. When you drive down the main street, you turn around at the open dock, where a visitor or a summer person can park the car and get out and inspect the shrimp boats which are often cleaning or unloading. Big yachts coming up or down the inland waterway are often docked here for gas or food or a short rest. Fernandina has a welcome station for boats—a small, round, modern building with free orange juice and information for those travelers by boat. Wander a little further and you will see the gas station on the wharf which services boats. And the local private boats are berthed here; it's fun to look them over and comment on them.

Then wander back to your car and across the railroad tracks which take logs to the mill, and there you will see a quaint, little, red station which was the main railroad station for Fernandina when people traveled by train. Now it is restored and is the office of the local Chamber of Commerce. Drive slowly up the main street and you will pass Ladies' Street. Yes, you are right, this was a very old port, but very busy in the old days and the sailors did need a Ladies' Street. On a little further and you see the Palace Saloon, well over one hundred years old with a beautiful long bar and wooden busts of ladies with expanded bosoms looking down at you. Have a draft beer—the ladies waiting on you are pleasant, smiling, and easy to talk to.

The street has been restored with old fashioned fronts on the stores; flower and bush beds have been built out into the street in a tasteful manner. Hardware store, clothing store, drug stores, gift shops, restaurants, you name it, it's there. There's the old red brick city hall on the right and the large post office on your left. You can see a couple of small, pretty churches and then the beautiful funeral parlor. There are the two new modern-architecture banks and then the school on your left, and the playground and tennis courts are on your right. Oh yes, you can see the small stadium over to the right where the high school plays its football and baseball games. Much of the town gets out to see these events, and the school spirit is fierce. Then we begin to see the pretty houses lining the broad street. The street goes straight to the beach about one or two miles from the dock. Then the beach road goes both ways, north and south, for about four or five miles. Oh, I forgot, just before you get to the end of the road which would take you right into the sea were you to go on, are two institutions—on the left, the entrance to Fort Clinch, a pentagonal bastion fortress as the sign says. Built in 1850, it was used during the Civil War, but is a museum now. Ships come up the river from the sea, either to St. Mary's or Fernandina. It is now a state park and a very interesting one covering over a thousand acres; the fort is about five miles from the entrance on Atlantic Avenue—fifty cents for adults and twenty-five cents for children.

The fort itself is quite interesting, pentagonal as it says, with a moat and a drawbridge, then a big courtyard, and old cannons mounted upon the parapet. There is a museum with Civil War articles, bygone buggies, and Minieballs. The guards, dressed in Civil War uniforms, are glad to answer any questions and have been known to smell of beer. But, of course, it must be a boring job. So take a good look, get up on the high parapet overlooking the river. To the right is the open sea; straight ahead is the tip end of Cumberland Island, the famous Yankee refuge of the rich such as the Carnegies and later the Atlanta Coca Cola Candlers. Great decaying mansions, a few of which have been restored, still exist and one, Greyfields, is a hotel run by the Fergusons. Mrs. Lucy Ferguson, born a Carnegie, age ninety, will talk to you about old times if you would like. But the island is now a national park, a wildlife refuge with broad deserted beaches, fresh water lakes close to the beach, and wild horses and wild hogs seen as one drives around in a jeep. The boat from St. Mary's goes twice a day, and the conducted tours are usually sold out. Campers may come and stay several days if they wish, but most tourists go back at night. The Federal Government bought the land, but the property doesn't have to be given up until the living children of the present families die.

Look to the left. There is the paper mill of St. Mary's, Georgia and the town in the distance. Just around the bend to the right is Kings Bay. Yes, that is the huge atomic submarine base which is already operating. A small town has sprung up five miles up river from St. Mary's. Said to be about ten thousand people now, it will grow to thirty thousand. Real estate prices in St. Mary's have already skyrocketed. The channel in the river has been dredged and huge quantities of sand have been transported by pipes to restore the beach on the upper part of Fernandina. Submarines go up and down the channel, but one never sees them. Now look to your right and going several miles out to sea are the jetties, great blocks of stone placed years ago to keep the river open for ships. Now next to the jetties is a beautiful new fishing pier that goes out almost a mile and very close (fifty feet) from the jetties. People used to walk out on the jetties to fish, but it was dangerous on those big, irregular boulders.

Okay, come back to your car at the fort's parking lot and drive back about one mile; turn left; go another mile and you come to the entrance of the fishing pier as well as to a large beach area which has been built up by the dredging. It's well worth a walk out to the end of the long pier. Watch the fishermen and drop a line yourself if you have time. Then back and see the campground and on out the main entrance again. Here is a state park, very well cared for, and across the river a huge twenty-mile-long island which is a national park and also beautifully cared for. Environmentalists, you have done two great jobs here.

Okay, at Atlantic Avenue take a left and there on the right is the other institution—the Fernandina Beach Care Center which is of course a nursing home which most communities need now. There is also a new hospital in town, and more doctors have come to Fernandina because of improved facilities. Of course, if you become very sick, you will go to one of the more complete hospitals in Jacksonville. Now, proceed straight to the traffic light. Straight ahead is the sea with a blocked road to the beach. Across the street on your right is the beach market which has recently changed hands, is modernized, and has a new name. On the left is miniature golf and close by the great slide for kids to go down; it's a huge and tortuous water slide into a small pool. It's very popular in the summertime. They say that Holiday Inn is going to build a huge hotel here, but we haven't seen any signs of it beginning. Fernandina has always been a summer resort; the big hotels have been wary of opening up here as they might not have customers in the winter. With the large influx of retirees that has occurred, a good hotel might be full in winter as well as summer, especially if they can get a few conventions.

Okay, let's go south on Fletcher Boulevard. This is the beach road that will go along the beach for at least five miles before it turns inland below the city limits. There are houses on each side of the road, those on your left on the water front and those on the right perhaps with a small view of the sea but not really adequate. As you drive along, you are struck with the modest appearance of the houses on both sides, but as you go a little further, you begin to see larger, more expensive houses mingled in with the cottages. A typical beach area which sprang up with no restrictive zoning rules. Further down the beach the expensive houses become more frequent.

You pass the lighthouse on the beach which actually is a house which looks like a lighthouse, and there is a sign on it "For Rent." Further down you come to the Seaside Inn on your left. This had been there for many years; it was very popular in the old days but now has declined and is deserted. There is a traffic light here and Sadler Road turns off to the right. This will take you to a five points, and by turning left there you would come to the great bridge spanning the inland waterway; this would take you off the island.

But let us proceed down the beach on S. Fletcher Avenue a little further. About one mile down from Sadler Road you see three large structures on the beach. Two of them are eight story condominiums on the water front and beyond them is a four story condominium. These are the first and only relatively high-rise structures on the beach. There was considerable controversy when the zoning finally allowed them to be built, and they haven't really made much impact on the single-family home owners on the beach. Then we travel south for about one mile—the houses are a little further apart, several are modern, and the one on the street-side is said to be completely solar. Then comes a cluster of fairly nice houses on both sides, although none are luxurious by any means. This is the area where some retirees, some local workers from the mill, and a number of people who have summer houses live. The Neelys are summer people, although they come winter and summer. Other people's houses are more vacation dwellings, as they live in Atlanta or New York.

Endden was named for Eve, Nathan, and Daniel when they were little. Rachel Eve Parker had three children by Ben Parker, but they later got a divorce. Rachel raised the children. Rachel is the daughter of Frank Neely, the son of Ben from Rome, Georgia.

Frank Neely came to Fernandina Beach in 1947 when he was at a meeting for the Federal Reserve Bank of which he was chairman that year. Some of his friends took him up to Fernandina Beach and he became charmed with its simple beauty. He bought the small house on the beach with two lots next to it. The price of the house and lots has risen dramatically over the last thirty-five years. Frank and Rae would come down often; Rachel would bring the children frequently as all children seem to know that a beach is an ideal playground.

Rachel would invite the Levering Neelys to come and stay a couple of weeks. The Levering Neelys had six children and, although the house was small, the beach was large and could accommodate them. Margot would frequently bring a friend who had five children, and they would stay a week until Levering came down, and he would stay with the family another week. This went on year after year, and the children thought it was their second home. But often the Levering Neelys could not come when the house was offered so they bought the beach house next door and could come whenever they wanted. Two doors down, Edgar Neely, Jr. bought another beach house, so it really became a small enclave of Neelys and Parkers, all related and all living in Atlanta. The children of the Levering Neelys became increasingly attached to their house and would come back as frequently as possible even though they might live a long way off. Margot and Levering would go summer and winter, and to them it definitely was a second home. Nathan bought Endden from the other children and has improved it considerably. He comes down from New York where he works, brings wife and children, and stays about a month.

All right, go south a little further and the houses seem to get bigger and more expensive looking. Mills B. Lane, formerly of Citizens and Southern Bank in Atlanta and now retired in Savannah, has a large colonial tabby house on the beach with three hundred large imported palm trees around it. Then there is a dentist's house. There is the huge house of a judge from

South Georgia and a tremendous house of a rich Yankee. A small hiatus and we have the Sandpiper, a group of closely packed, low-rise condominiums priced for the middle-class. There are about thirty in a solid row. The road turns inland here and crosses the large road which proceeds south. But let us explore the road straight ahead; it goes into the woods with overhanging tree branches shading it. Straight ahead is a huge tree growing in the middle of the road. It was such a nice tree that they didn't want to tear it down, so the road splits and goes around it on each side. Then you come to the inland waterway again; turn left and again left and we are at the famous Sandbar Restaurant, an old, low, converted house. The seafood is still superb and has been for fifty years.

Okay, back to the caution light and then go south on the main Highway A1A—first a large nursery for plants is on the left and then the Honeydripper Bar for blacks. This is at the upper end of American Beach, one of the few black beaches in America, which is fast becoming deserted and the houses are falling to pieces.

Then, and here bring out the trumpets for the rich, The Amelia Island Plantation, the mammoth real estate development begun by the then famous Fraser who had done so well with Hilton Head two hundred miles north in South Carolina. The Amelia Island Plantation, like its older brother Hilton Head, is built around the shrines of golf courses, tennis courts, shops or shoppes, and swimming pools with plenty of outside tables and bars. The entrance is to the left off of the highway, but a bridge has been built over the highway which leads to several golf courses and eight hundred golf condominiums on the marsh side. On the beach side there are about twelve hundred condominiums of every shape and description. Oh, but they are tastefully done, they fit into the landscape and the shrubs and trees are the careful work of a landscape architect. There are very few houses on the beach, but they generally have a good view of the beach. The whole area is over a thousand acres and covers at least one hundred square miles. Plans were too big at first. Fraser got out when he saw it was not going to succeed. The area changed hands several times, but now, after ten years, it is financially solid. The majority of the condominiums have been sold and they are building more.

So here on the south end of Amelia Island is a gigantic development with armed security guards at the gate. You cannot come in unless you have been invited by a home owner. Oh, if you want to talk with a real estate man, they will be glad to show you around and give you the full sales pitch.

So you have a choice if you are coming into the area and thinking of settling here. In the south you have a planned community with rules and regulations that cannot be broken or you will have to leave. On the north end of the island you may elect to be part of a quaint, small town with a great history and all the advantages and disadvantages of a natural community with all kinds of people. And how can we characterize the people of the town as compared with the people of the Plantation? The people of the town are a cross section of America with all types of people from every walk of life. The people of the Plantation are similar in one way—they all can afford or think they can afford a condominium which sells for $125,000 to $250,000. So they may be from different professions or business, but they have a stamp on their backs which says, "I am a monetary success, and how are you?"

Both beaches—the north and the south—are broad and white. both have beautiful expanses of sea out there. Shrimp boats dot the horizon of both. Sea gulls fly up and down the beach. The graceful pelicans come gliding through the air in that symmetrical V formation. You may be lucky enough to see the black skimmers flying low on the beach in the late afternoon or early evening. Consider yourself most fortunate if you should see a group of dolphins following the wake of a shrimp boat.

So the two areas—each a part of the same island—are very much alike yet very different. There is no accounting for taste. So make up your mind carefully and choose which you want.

36
Margot—Until Marriage

Margot moved from grammar school to high school very easily. It was from a private elementary school to the public high school in Saranac Lake. It was the only high school in the vicinity. The town had grown but was still only a village of about five thousand people. At that time, 1933, most of the families there were involved with tuberculosis in some way. There was usually a family member who had come to cure; often the family just remained in Saranac, took up roots, sent the children to school, and settled down for the rest of their lives. It was a small melting pot. People came from all over the eastern United States. There also were quite a few Southerners from Virginia and Georgia. There were a few from the West, but a resort center for the cure of tuberculosis had begun in Denver, Colorado, and, later, Arizona.

Tuberculosis strikes both rich and poor. However, those arriving in Saranac Lake to cure were usually fairly rich but also well-educated; the majority had been to college, and there was a large number of good doctors from all over the country. Many of the doctors later became famous in the field of tuberculosis. There were Drs. Lawreson Brown, Baldwin, Fred Heise, Leroy Gardner, and Edward Trudeau, the founder, to name a few. It thus was somewhat of an intellectual community for such a small village high in the mountains. And this intellectual heritage spilled over into the children, most of whom went on to college. As the times changed there were very few who came back from college to settle; rather, they moved away and scattered all over the United States.

Margot thus began school and high school in an intellectual atmosphere where much emphasis was put on learning, music, art, and medicine. Her father, a Yale graduate, knew Greek, Latin, and the classics. With Margot's great respect and love for her father, it is easy to see that she would be encouraged in her studies. And indeed she did very well in grammar and high school. She made excellent grades throughout except in mathematics: algebra, and geometry. She had some kind of block on these two subjects. She finally finished algebra with some coaching, but geometry was a real bugaboo, and she had to take it several times before passing it and graduating in 1933.

High school was a fun time for her. Her joy of living, her bubbly personality were obvious in grammar school but became nurtured and really developed in high school. She was a very pretty girl as a teenager; she was a little taller than many of the boys but became quite popular with the male gender and always was asked to the dances. She was also an outdoor girl, loved the snow and ice, and began skiing before the sport became popular. Her friend, Tommy Cantwell, was a frequent companion as they had the same interests, but were not romantically involved. Margot had many different boy friends in high school and Tommy had his girl friends.

How did this outgoing, unique personality develop? We really don't know how much were genes and how much environment. Her sister, Dorothy, had the same parents and a similar environment and yet her personality was different. Margot was much more mercurial with emotional swings; Dorothy was the calmer of the two. They both had a good sense of humor which would stay with them the rest of their lives, and they always enjoyed each other's company throughout the years. Dorothy kept her restrained, dry sense of humor which was always present in her letters. Margot's humor was more hearty, more raucous. She made people feel good and they laughed at her humorous view of life. She could somehow express herself in such a way that it was funny in a self-deprecating way.

College was a whole new way of life for her. She had graduated from Saranac Lake High School at age fifteen and the family thought she was too young to go to college. Northfield Seminary, a girl's boarding school in Massachusetts, seemed like a good solution, and so off she went in the fall of 1933. Margot thought that the school was a disaster. Here was a fancy New England boarding school for girls whose pupils consisted of children of the moderately wealthy of New England. Margot would never discuss that year except to say that she hated it. So we will probably never know the details. We do know, however, that the first year at Middlebury College in Vermont was a heavenly contrast to the boarding school. Margot had made good grades in Saranac Lake High School and at Northfield, in spite of her dislike for the latter. She could have gone to Smith, Vassar, or any of the good colleges for women. She spurned these places, however, for the coeducational, small, but excellent college in the central part of Vermont known then for its foreign language department. Also, it was well known for its ski team and the good skiing in the area.

So she arrived at Middlebury College in the fall of 1934. Roosevelt had been elected in 1932, and took office in 1933. He would be reelected in 1936 and begin his second term in 1937. The depths of the depression were from 1929 to 1933. In 1934 Roosevelt had just begun the New Deal. Roosevelt had tried to do something to help the economy with his creation of jobs and it was beginning to pay off. More jobs produced more money to be spent on goods which in turn produced more jobs, and the economy, although creaking and needing oil, gained momentum. There still was a depression, but Roosevelt had the image of leading the country out of the depths of the depression. The working man had confidence in him and reelected him. However, there was a growing group of conservatives who thought he was ruining the country by his deficit spending. This group was getting more and more paranoid about what was going on in this country.

Margot was almost apolitical at this time. She was usually a person of very strong opinions, but in her freshman year she really did not know that there was still a severe depression going on. As time went on she would follow the Republican banner for a few years, but later, as she learned more and more about the world, she would be a strong liberal and have a tendency to espouse the causes of the underdog.

After the confinement of Northfield school, the freedom in college was such a pleasant contrast that she was like a bird in a cage suddenly given its freedom. She seemed to love every bit of college. Here was a huge smorgasbord of courses—she wanted to take them all, and the choice of only five courses was very diffcult to make. Then she had her skiing. It was mostly cross-country skiing and she loved it. Then there was the male interest. There she was in a class with all these attractive young men and she was a very attractive young woman.

She made many friends at Middlebury and many remained friends for the rest of their lives. Even though they might see each other only every few years, the two would seem to pick up the pieces of the old friendship very quickly and have a fine time with each other. When she moved south one would think that the contacts with the old friends would be lost, but no—the old friends would come to visit on their way to Florida, and Margot was often in New York and New England and would look them up. Then there were the reunions. She didn't really have much enthusiasm for the reunions but did go to the 25th and 40th, enjoying them fully. Then there were all those with whom she kept up Christmas cards through the years.

College is a place to make strong friends. One is a free and independent spirit. One begins forming his or her own philosophy and if one finds a kindred spirit, a bond of friendship grown. Then there are always shared experiences and after a time, the ties are begun which grow imperceptibly stronger. There was a group of about four or five very close friends made in college—several were roommates at one time or another—Florence Hume, Mary Heckman, Emily, Dottie, and several others. She would see her closest friend, Florence Hume, every few years, and as their birthdays were very close in February, the two families began meeting in Fernandina Beach once a year over the last five or six years.

Florence had married Jimmy Minor, also a Middlebury graduate, who had dated Margot at times during their early years at Middlebury.

Margot soon found her special interest in American and English literature. She fortunately did not have to take mathematics in college as this was her *bete noir*. However, she was an avid reader of early English literature and was charmed by Chaucer, whom she could quote many years later. Poetry was her special love and she knew all of the English poets well and later the American poets. Frost was one of her favorites. She actually did a little poetry writing herself but did not let many people know about it.

The first two years were a little wild in that she soon found herself a nonconformist. She joined a sorority but soon got out of it. She just did not fit in with a group of girls who thought along the same pattern. However, the sororities were not so important there at the time, and Margot made friends quickly throughout her class. She was already a good skier when she arrived at Middlebury and she was able to ski with the boys on equal terms. So weekends in the winter were full of outdoor activities—skating, skiing, and then the evenings by the fire drinking beer and often dancing.

Tommy Cantwell and Tommy Clement, also from Saranac, were both at Williams and one or the other would ask her frequently for weekends of skiing and parties. It soon became the joke that Margot really went to Williams to college. She continued to do well in her studies and was a favorite of several of the teachers. Her early charm and ability to get along with almost anyone was greatly enhanced by all of her new interests. She would still go home for weekends at times, and during the summer she would work for her father as she had become very adept at the typewriter. She was a stickler for the proper spelling and punctuation and would continue to be a good speller and to insist on good sentence structure for the rest of her life. She had a great disdain for bad spellers and was very fond of finding misspelled words in the local and Fernandina papers.

In her last two years of college, one might say that she matured and settled down more but was always ebullient in her outlook on life. She smoked in college and she continued to be a heavy smoker most of her life. She was well-rounded in her knowledge of the arts—knew and recognized the great painters and sculptors and had an abiding interest in classical music and would love Mozart, Bach, and many others for the rest of her days.

Upon graduation from college in 1938 she went with her parents to Europe by boat. Dr. Gardner was attending several international meetings on dust diseases of the lungs and presenting research papers himself. In the summer of 1938 Hitler was coming into power and there was a growing controversy everywhere as to what he was doing. Margot had learned enough in college to know that she had an adamantine opposition to Hitler and his Nazi and Gestapo principles. She was quite outspoken on the subject while on the boat, a German ship. Her father thought she was a little too outspoken, but did not try to restrain her. She did form an alliance with a young man on the boat who was travelling over Europe on his own, and he would follow them in order to court Margot.

Carabel and Margot took off on the train by themselves and went to Florence where Margot had a superb time with the arts. She had already been to London and Paris on the way over and ended the trip in Geneva where Dr. Gardner was finishing up his meeting. They then took the train to Cherbourg and came home on the boat from there. The trip was a wonderful experience, enabling her to see the places that she had read about in college.

When they arrived in the States they went back to Saranac Lake and Margot resolved to go to New York City to enroll in Barnard College of Columbia for postgraduate work and to find a job. This was still in the time of the depression and jobs were not easy to come by in New York City. However, Margot found a cheap room near Columbia University and Barnard College, got out the want ads from the papers, and began to write down the best prospects. It was only after several days of disappointment that she realized that in spite of being a college graduate she would have to take almost any job she could get.

One of the first jobs she got was in a refrigerator factory in Brooklyn. This job consisted almost entirely of typing reports and letters. She kept this job several months and proved that she could support herself, although she lived in a small room near Columbia which had a kitchenette. She was making twenty dollars a week. She managed to survive and to enjoy the Big Apple. In the fall, she enrolled in Barnard in the night school, taking more literature.

Finally, the job became quite boring and she began looking around and reading the want ads. She found a job in the import business which she thought she would like very much. She found that this job was really a great deal of long distance telephoning, usually overseas. Often she could not understand the people calling as they would be Japanese. If she could not understand what they were saying, she would simply hang up. Finally the boss found this out and fired her.

After a series of short-lived jobs, she found a job in Manhattan in a doctor's office. He was an ear, nose, and throat specialist. Margot was one of two general flunkies and had to do about everything including typing, assisting the doctor, running errands, and the like. She did not like this job nor did she like the doctor, but she kept the job longer than any of the others because it paid well and was convenient to get to. She knew New York very well by this time and was an expert on the subway and elevated trains.

Somewhere along the line she met Danny, who was from St. Louis and was in a brokerage firm. She had been rather lonesome in the big city, although there were a few college friends scattered about in Manhattan. Danny began taking her out fairly regularly. Their relationship was that of two lonesome souls in the big city and they did not have much in common. However, the alliance lasted several months, but finally exploded and Margot stopped her job and went back to Saranac Lake to get over the breakup.

* * *

So who could have predicted her future at this point? She was twenty three years old in 1940 when she returned to Saranac. She had become fed-up with her job, with the stupid doctor in New York, and fed-up with her misalliance with Danny. So she made a clean break with both, went home, and got a job with her father and also on the newspaper in Saranac Lake. And that breaking away would change her future completely and give it a totally new direction. It would ultimately be to the South. She herself would never have dreamed that most of her life was to be spent in Atlanta, Georgia, that she would have six children, and that she would become deeply involved in community affairs in that city.

But what we want to know is how that remarkable personality developed an entity that was somehow different from the run of the mill—the ability to get along with people almost immediately after she met them. She would have them smiling or laughing with only a few words. She could talk with almost anyone whether rich or poor, bright or dull. Whence came this sparkle? Was it back a generation or two? Was it several generations back, or did it somehow develop, grow, and flower from cultivation?

Her father was brilliant but steady; unflappable, he stood as a man with a purpose. He attained greatness by his efforts, but he was extremely modest and his ambition, if present, was never obvious. Her mother, Carabel, was a home body. Her ambition in life was to make a good home for Roy, as he was more important to her than even the children.

So Margot arrived in this world with high intelligence manifested before grammar school. She had the drive to excel in school which she did. Her personality was unusual as early as grammar school and continued to develop in high school. College nurtured her unique personality with contact with her many friends who will attest that she had the charm throughout college. New York during her first stay there was a period of trying to find an identity, failing, and returning home to start again.

Life began again at age twenty-five when she married. This most unusual personality would be the key that would unlock doors, and she would begin a life of service to her community and would long be remembered by her children and friends as a bright flame that was snuffed out earlier than it should have been.

37
Vignettes of Frank Neely

That Frank Henry Neely was an unusual man is not to be questioned. That he was a great man will have to be decided by history. But that there was a large group of people who came in contact with him and thought he was great is a certainty. There still are many people who had close contact with him when he was president of Rich's, and who, when his name is mentioned, give a slight smile, shake their heads slightly, saying: "That was a man." Meaning of course: "That was a great man." From the lowest clerk at Rich's to the highest executive, they all are in agreement. The reasons may vary a little, but the general feeling is that he did great things for Rich's, for Atlanta, for Georgia, and for the country.

Rich's was his vehicle to greatness. He could have used many other vehicles but this one seemed to suit him the best. It first gave him contact with a large number of people, people of all types from executives to porters in this huge institution. As president of the corporation, it gave him an open door into the community, namely the city of Atlanta. He knew the mayors, the city council, and for some reason had an excellent rapport with the head of the board of education in Atlanta, Miss Ira Jarrell. Then he knew the state people, all the way from governor on down through the Department of Agriculture. He had that ability to make friends immediately—a smile, a few words, a little kidding, and a little cussing and he had a close friend.

From the state politics he moved on to the national political scene communicating, transforming the problems of Atlanta into those of the country. Having been president of the Federal Reserve Bank while still president of Rich's, he had many contacts in Washington, and when World War II broke out, he was put on the War Production Board and as such came to know President and Mrs. Roosevelt. Later Mrs. Roosevelt would visit the farm in Norcross.

How does one achieve greatness, and specifically how did Frank arrive at greatness? What were some of the ways he got there? He achieved great power, but he did not misuse his power. Many say he could have been governor easily, but that he did not have to be governor because at his height he had more power than the governor. With his thousands of friends, he could make one phone call which might change a situation completely.

He was somehow able to build up a loyalty among his employees which was one of his strengths. He could talk to the janitors and sweepers and knew them by name. He demanded and got the loyalty of the executives of the company. He had been employed in 1920 by Mr. Walter Rich, who recognized him as a person who might turn a fine company into a great firm. He began as a vice-president in charge of running the store and he very soon demonstrated that the engineering methods he learned at Georgia Tech could be applied to a department store in such a way that each department meshed with the others to make a huge, highly efficient machine.

He had most people afraid of him, but he also had their deep respect for his ability. His favorite people were those who would talk up to him, at least to a point, and not be afraid of him. Those who simply agreed with him all the time did not interest him. He used a mild form of cursing to emphasize a point and it was highly effective.

"Get that damn thing out of here" meant he wanted it out quickly.

But if he said, "Get that goddamn thing out of here" he meant to remove it immediately and "I don't want any questions!"

However, he never used foul language, and the hell, damn, and goddamn were as far as he went. He would frequently get very angry over some situation that displeased him and would not hesitate to give the one responsible for it a real dressing down. But they all said that his bark was worse than his bite, especially those that knew him well.

Then there was his sense of humor. He was quick on repartee. He came up with a sharp answer before a question was fully formulated. He was able to make you laugh with a very simple word or statement. That ability might be called one of the dominant aspects of his character. This led to his ability to get along with people and to lead people. When the government put in the forty hour week, he was very much opposed to it. He thought people should work until their job was done, no matter how long it took. He believed in the work ethic, and that play and relaxation should come only after work was done. He stayed until any hour at night and he made his several secretaries in the office do the same. The rest of the employees followed the law, but not those in Frank's office. Nor did they get overtime. This went on for quite some time. The secretaries brought this to his attention. He would have none of this. Only after many months and logical pleading by Vice President Dick Rich, that Frank was breaking the law, did he finally give in.

Characteristically, he always insisted to the employee that if something needed to be done, it should be done now, not later. He was so adamant about this that the entire work force knew that they could not postpone or procrastinate about anything. When a meeting had been planned to consider opening a new store in Nashville, he had casually asked for some aerial photographs of the land and countryside where the store might be. The meeting time arrived and a large group had assembled to discuss the situation. After a very short discussion, he quietly asked to see the aerial photographs. No one said a thing. The room was suddenly so quiet that one could have heard a pin fall. No one had taken the responsibility to obtain the photos.

"Well," he said, "where in the hell are they?"

No one answered. Finally the Vice President said, "They haven't been done, sir."

"Well," said Frank, "We will just adjourn until they are done."

This was at 10 a.m. By 2 p.m. aerial photographs of the Knoxville area (120 miles away) had been taken, developed, and brought to Atlanta and the meeting was continued through the afternoon.

The farm was not just a place for him to rest on the weekend, but it was a place to pick up a new occupation. Accordingly, he dove into the job with just as much vigor as he did at the office. Royce Ramey was his caretaker, director of farm operations, and confidant. Royce had come to the farm as a young man of nineteen and stayed on the farm until his death at sixty-nine. He also was completely loyal to Mr. Neely, and he had a profound respect for him. Royce ran the dairy farm of about three hundred fifty cows. The farm had the most modern equipment with the latest and best milking and bottling equipment. Frank had some of the first automatic milking and bottling machinery. He had consultants from the Georgia Department of Agriculture as well as professors in farming from the University of Georgia. All of the records of the cows with the amount of milk they produced along with their names were kept at the office. Mrs. Dorothy Lavine, who had just been given the job as one of Mr. Neely's four secretaries, was given the job of keeping the farm records.

After about three months at this job, she went to Mr. Neely and said, "I think you are going to have to get rid of some of those cows."

"Why? asked Frank.

"Because some of them are not giving any milk at all."

"Let me see the record, Dorothy."

She handed him the record with the names of the cows. He looked at it briefly, gave it back and said gently, "We call those bulls, the ones who give no milk, Dorothy."

The farm, for Frank and Rae, became a great place of entertaining. It is true that almost all of the important people of Atlanta visited there at one time or another. There were all types of people, from U. S. Generals, such as Lucius Clay, to politicians (if honest), to musicians, and all types of artists. Mrs. Roosevelt came out while in town during World War II. The food was always superb, the help was most gracious, and there was the flavor of the old South. However, there was something different about these gatherings. These were the interesting people of the city—architects, the best artists, and musicians as well as lawyers, doctors, and journalists. Rae had a big hand in the entertaining; she knew who to invite. She knew who to sit next to whom, and she knew how to keep the conversations going and would often go from group to group and add a little infusion of interest if the conversation lagged. Then she would go to the next group. No one really realized what she was doing. So the great parties at the farm became famous.

The question has often arisen—how important was Rae to the meteoric rise of Frank to greatness and power? There is no question but that Frank was a personality with self-propelling qualities. He had to have his own ambition and drive, and it would be foolish to say that Rae was the power behind the man and that she was really the one with ambition. However, there is also no question but that she was a great help to him, that she worked with him and for him. They made a great team. Rae was not the self-effacing, subservient wife by any means. She was one of the few people who could contradict him and get away with it.

Once there was a large group of businessmen who came to the farm for a meeting in the late afternoon. Rae served them tea and cookies.

Frank stormed in and said, "Rae, this is not a ladies tea party; these men need something stronger to drink."

Rae spoke right up: "It is too early for them to drink liquor. I am serving tea and cookies now."

Frank never said another word about it and had tea and cookies himself.

Frank had a way of springing surprises on his employees at the store. He enjoyed making unusual requests and watching the reaction to these requests. He once told his secretary to get him an almanac.

She asked what kind of almanac, and he said, "I want an almanac with the hours, the minutes, and the seconds of every day."

"What will it be used for?" asked his young secretary.

"I want it for the milking record at the farm—I want to know exactly how much and when those cows give milk."

His secretary dutifully found a big almanac and added pages for minutes and seconds. This was a carry over from the old engineering days when everything had to be precise to the second. He soon found that cows' milk production cannot be measured in seconds.

During World War II, the word had come down from Washington to conserve energy. Frank's method in the office with his assistants and secretaries was to reduce the number of light bulbs in the fixtures. Being an inside office, it was getting difficult to see, but Frank continued to have more light bulbs taken out. The secretaries did not say anything for awhile, but finally asked him if he wanted to contribute to the candle fund.

"What the hell are you talking about?" he shouted.

"Well, we just need a little more light in the office," she said.

He turned red and she thought he would explode, but he suddenly broke out laughing and said—"Well, we will give you two more bulbs apiece."

One of his trusted employees, Mr. Travis Guest, would get a lot of requests at the storé and at the farm. One of Mr. Guest's duties was to be in charge of sign making at the store.

Frank said, "I want a large sign that says: 'rake,' and another that says, 'hoe,' and another that says, 'shovel.' "

Mr. Guest said, "I will be glad to make them, but do you mind telling me what for."

"I want them for the farm, so we can tell where these implements are supposed to stay when not in use, and when I come by or the workers we can see what is missing."

He got his request and rarely had any lost implements.

He invited Mr. Guest out to the farm and then asked him to ride around with him in his jeep. This jeep ride had become a ritual with him to look over the farm almost every day. Mr. Guest agreed and off they went down the road at quite a clip. Mr. Guest, being a city boy, was holding on for dear life.

After they got to the river, the road ended and Frank said, "I want to go over and see those pine trees that are dying," pointing to an area across rough terrain about five hundred yards away.

Off they went again at quite a rapid speed. This time Mr. Guest could scarcely hold on at all.

When they got there and stopped, Mr. Guest got out of the jeep observing, "Is there a road around here? Because if there isn't, I am walking back."

Frank laughed and let him walk to a close road and picked him up.

Frank insisted on promptness. It became known very soon that you got to his meetings on time. Once he was making a speech at a 9 a.m. meeting. One of the new executives walked in at three minutes after 9 a.m.

Frank stopped his talk and called out to the man, "What the hell are you getting here at three minutes after nine for? When I said nine o'clock,

I meant nine o'clock and not any time later. Just don't bother to come if you are going to be late."

The man learned his lesson quickly.

One of the employees became vice-president of Rich's during Frank's presidency of the company. One story on how he became employed at Rich's is as follows. Frank was driving slowly up the side street beside Davisons. He saw several people unloading a truck to take the merchandise into Davisons. There were several black men and one white man who was dressed in a business suit. Not only was he directing what they did, but he was actually doing some of the labor himself.

Frank stopped the car, got out, and asked the white man, "What is your name?"

"Jack Jones," was the reply.

"How would you like to work for Rich's at a good increase in salary?"

"Fine" was the reply, and that was the beginning of the association that led the man to the vice-presidency of the company.

Rich's began having a huge Christmas tree on the bridge between the two stores in the late forties. The tree was always over sixty feet tall, decorated, and could be seen for miles around. Mr. Neely put Mr. Guest in charge of the tree and it was a very difficult job (and expensive) to put up the tree and decorate it. As the cost rose, many people in the company suggested that they make an artificial tree and use it year after year. Getting the tree entailed trips to North Georgia with a huge truck and crew, hauling it to Atlanta, and then raising it into position with a derrick. Once in place they could secure it. This became a tradition, with Mr. Guest telling many stories about how he got the tree.

When the other executives suggested the artificial tree, Frank declared: "Listen, we have had a real tree all these years. The people expect a real tree. It's now a tradition and, by God, we'll continue to have a real tree."

Fifteen years after Frank left the store and eight years after his death, they still have a real tree.

Many people didn't know that Rich's had a famous basketball team at one time. The team grew when there were several good players working there. The team challenged other companies but they became so good that they began playing against local colleges. Coached by Mr. Travis Guest, the team drew Mr. Neely's attention. Soon they began recruiting just like a college. They would offer the graduates who were on the college team a job at Rich's—and a chance to play basketball on the team. Such recruiting continued for several years and you can still see some very tall employees around Rich's.

Frank began another tradition: the institutional ad. Once a week there was a full page advertisement in both the *Atlanta Journal* and *Atlanta Constitution,* although it did not advertise any merchandise. Rather, it was an editorial type ad, supporting such a cause as the Community Chest (United Way) and describing what the United Way was doing for the Atlanta community. They placed many other types of editorials for Atlanta, but the advertisements never got into politics, unless it urged people to get out and vote. These ads continued until the store was bought.

Frank's daughter Rachel was not only the apple of his eye; she also embodied his genes of intelligence and drive. Like her mother, Rachel went to Smith College, came back to Atlanta, and went to work for Rich's. Frank, of course, didn't think anyone was good enough for her. But he had connections in Washington, heard of a bright lawyer named Ben Parker, had him introduced to Rachel, and they were later married. Parker worked as Rich's lawyer, but soon a rift developed between Frank and Ben and later between Rachel and Ben, so, after producing three fine children—a girl and two boys—they divorced.

If we ask what was the one thing that made Frank so great at Rich's—and great in the community and indeed the country—one would probably say that it was his uncanny ability to instill loyalty to Rich's and to himself. Frank generated a morale within the firm such as we have rarely seen in any other institution. That morale, however, faded soon after he left.

38
Emancipation Proclamation Brown

Yes, Mancy Brown, you were an integral part of the Neely family. True, you were black, true you were considered a butler if one must have a name for your profession. But also true is that at least for one branch of the Neely family, namely the Edgar Neely, Sr.'s and their children and the grandchildren, you, Mancy, were as close and as loved as anyone in that particular segment of the family. So let us look at you, Mancy, as one with no Neely genes and of another race and see if any Neely characteristics rubbed off on you. Let us examine your life and try to make a judgment regarding your ambition, if any, and regarding the ambition of your children. We would like to know if ambition can be passed along simply by association.

You began knowing the Edgar Neely family when you were nineteen. You continued to have fairly close contact with them until your death at age sixty-five in 1964. So for forty-three years you either served, helped, or worked part time for Mr. Edgar Neely, Sr. So you must have been born in 1901 in the country. But in the early part of the twentieth-century, blacks were leaving the country and coming to the city in large numbers to get jobs.

We are not sure just when you arrived in Atlanta, but we think your family brought you when you were about fifteen. You went out and looked for work right away, and you found a variety of odd jobs. Your family were good church members and joined a tiny little church on Decatur Street called the Ebenezer Baptist Church. You went to Sunday School every Sunday (it was taken for granted that you would), and the minister was a young man named Martin Luther King (Senior). He would later have a profound effect on your life, as we shall see. Perhaps he is the one that inspired you, Mancy, and helped you with your own particular type of ambition.

In 1920 Edgar Neely, Sr. was forty-two years old, had been practicing law for over ten years, was with a good law firm in Atlanta, and had begun to make a name for himself as a trial lawyer, usually for the defense. His office was in the Hurt Building in the center of town, and there was a new garage for parking next to the Hurt Building. It was called the Ivy Street Garage. Edgar kept his Buick there while at work. Black men parked the car in the morning and brought it down at night. Mancy had just got a job as one of the men to park the cars. It paid very little, but the business men usually tipped the one who brought their car down.

Mancy was a tall, thin, gangly young man with a courteous manner and a constant big smile. He parked Mr. Neely's car several times and Mr. Neely frequently engaged him in conversation.

"What did you say your name was?" Mr. Neely asked Mancy one day.

"Mancy Brown, suh, dat's my name."

"That sounds like a girl's name to me," Mr. Neely said.

"No suh, no suh, that ain't no girl's name. That's my name."

"Well who gave you that name?" Edgar asked.

Mancy answered, "My daddy, he done it—he said it's a good name cause it really is a shortening of my real name, and I will spell it for you—EMANCIPATION. Emancipation Brown, and my Daddy says that is a name to be proud of. Dat's what he say."

Edgar laughed, "Emancipation Brown—your father is right, that is a name to be proud of. Say, what about coming out and working for me for a while? I will pay you better than what you are getting here."

So that was the start of a long association of mutual respect and trust. Mancy did come out to Ansley Park and began working for the Edgar Neelys. They had a cook named Addie. Mancy helped her, but mainly he worked in the yard and drove Mr. and Mrs. Neely wherever they wished to go. The pay was five dollars a week and it was taken for granted that he would get his meals free as his hours were seven in the morning until eight or nine at night. Mancy liked his job much better than the impersonal one at the garage. He felt that the gods were smiling on him to get this new job.

In the 1920s there was no integration—streetcars were segregated with whites in front and blacks in the rear. Schools were segregated completely, and black schools were of poor quality.

No one could have been more polite and pleasant to white people than Mancy. He wasn't subservient—he was just plain nice. The three Neely children were ten, eight, and six years old in 1920, and they all liked Mancy immediately. The children often ate early in the pantry next to the kitchen and Mancy would usually serve them. Three feisty little boys fussing and often quarreling at supper were a handful, but Mancy was able to handle the situation and distract them by engaging them in conversation.

There was another facet of his inner structure which was not immediately perceptible. He had some type of inner drive to do something in this world, and we shall see where this led. He was working as a faithful manservant at the Neely's, but he also became more active in the Ebeneezer Baptist Church. Martin Luther King, Sr. took him under his wing, encouraging him to study for the ministry. There were several Baptist seminaries in the South, but they also had mail courses. Mancy took the mail course, studied at night, and met often with the elders of the church. The mail courses took years to finish, but Mancy stuck to it and finally in the forties became ordained. He became an Assistant Minister to Martin Luther King, Sr. but never gave up his work as he had a wife and three children to support.

Ebenezer Baptist Church appeared to be just another black church with its long Sunday services, its good choir of black voices, and the flamboyant sermons of Martin Luther King, Sr. One would have never known at the time that there was a nest of intelligent black people in the church who were working for the betterment of the black race. Nor would anyone have predicted at that time that the whole civil rights movement was born, flowered, and almost achieved its goals right from that church.

Martin Luther King, Jr. was born in the twenties and was a young child when Mancy became involved in the church. Mancy knew him from childhood and later both men had great respect for each other. But it was really Martin Luther King, Sr. who had started the good people of his church meeting together frequently to see what could be done to improve the lot of the black man in this world. At first it was just discussing things; later it was organizing small groups to go to the good white people to get help with some injustice, but it was years later, in the 1950s and 60s, that under Martin Luther King, Jr., the Civil Rights movement mushroomed and exploded. Mancy was involved in the development of that bomb. We don't know exactly how much he did, but we do know that he was deeply involved.

In the 1930s, after Mancy had worked full time at the Neelys for over ten years, he talked to Mr. Neely about getting a second job. His children—Margaret, Christine, and Arthur—were growing and would soon be going to college as this was Mancy's avowed purpose. Mr. Neely represented General Motors, whose big assembly plant was located in Doraville. He called one of the vice-presidents, told him how reliable Mancy was, and Mancy went to work within a week. He was started as a janitor or clean-up man. The hours were from 7 a.m. to 3 p.m. and then Mancy would come to the Neelys to work the rest of the afternoon and into the evening. General Motors paid very well and Mancy's earnings were more than doubled. Later he got a job on the assembly line and this also raised his wages.

While at General Motors, he worked at his job quietly and as usual was pleasant to one and all. However, he began noticing how much the blacks were discriminated against and he began talking with other blacks. Any blatant discrimination or mistreatment of a black would be pointed out to the foreman.

When Mancy started with General Motors there were practically no unions. Several labor organizers had come down and were beginning to have increasing effects. Mancy, still always the gentleman, became active in the labor movement and helped organize the union at General Motors. At least he organized the black people and he became their spokesman. Throughout his efforts, he always seemed to be fair and reasonable, and even management recognized him as a good influence rather than one who did things behind their back. His Christianity extended into his work place.

Mancy continued to work in and with the church, although he held two jobs—his General Motors job in the day the Neely household job every afternoon and often on Saturday and some Sundays. He became Martin Luther King, Sr.'s right hand man and would often preach on Sunday and also teach in Sunday school. In addition, he was often meeting with that bright group of blacks whose avowed purpose was to better the lot of the black man. Martin Luther King, Jr. was a student in the thirties and in the army in the early forties. After the war, he came home and studied for the ministry at theology school in Washington. During this time, Mancy was Assistant Minister to Dr. King, Sr. When Dr. King, Jr. returned in the late forties, he took over many of the duties of Mancy because Mancy just did not have time with his other two jobs. So Martin, Jr. brought back some new ideas to the church, namely, peaceful resistance to the injustices which were going on against the blacks. But he was working with a small group who had been inquiring about and documenting the discrimination against blacks. Martin, Sr. and Mancy, although apparently old fashioned blacks, were nevertheless the ones who began a movement which Martin Luther King, Jr. organized, strengthened, and spread. Marches were beginning; Martin Luther King, Jr. was put in jail several times but managed to get out because he was always non-violent—despite the blatant violence against him.

Mancy was active in the movement, often marching and protesting, but he never was put in jail. Mancy had a certain dignity and was older than most of the other protesters. The police seemed to recognize this and usually picked the younger ones to arrest. Mancy's ablity to talk with anyone was a rare trait that helped him get out of any situaton. He remained active in the civil rights movement as its ranks suddenly swelled to flood tide under Martin Luther King, Jr., who had now become a national figure in the fifties.

Mancy still had two jobs. At General Motors he was now on the assembly line, making a good wage since the union came in full force. The union brought desegregation and no black discrimination along with it, and Mancy thus had achieved the dual purpose of unionization of the assembly plant and equal opportunity for blacks. Although he was not personally responsible for the changes, he had been a part of it and this was a great satisfaction to him.

Meanwhile, he was still working for the Edgar Neely, Sr. household. He was the butler at night serving the family, washing dishes, and driving home around eight o'clock. He would drive Mrs. Neely to the market on Saturdays. He would do odd jobs around the house as well as the garden. Mrs. Neely considered him her property, at least on Saturdays. Mr. Neely and he would have great conversations, always with a humorous side. He also got along unusually well with the three children and they came to know him well. He was always ready to help them in their endeavors. Thus, over the years, he became more and more a part of the family.

Mancy was married to Louise and they had three children; Margaret, Christine, and Arthur. It was in his children that Mancy really showed his ambition. He knew the value of college and wanted all of his children to go there. He had saved money over the years for this. Margaret, the oldest, did not diasappoint him, and in addition to graduating from college, she went to law school, graduated, and became a lawyer. She later married a lawyer and moved to North Carolina. Christine, the second child, also went to college and became a teacher. She married soon after college, but the first marriage was a mistake as her husband developed mental illness and could not cope with life. She became the prime baby sitter for the young married Neely's infants, and so the second generations got to know each other. Mancy's third child, Arthur, turned out to be a great disappointment to him. He did poorly in school, never could go to college, became a petty criminal, and spent much time in jail. All of Mancy's efforts couldn't change Arthur.

Mancy had suffered from high blood pressure for years, but had no symptoms which interfered with his activities. However, in the sixties, he began having troublesome nose bleeds and went to his doctor, a black physician who found his blood pressure to be extremely high. He was put on blood pressure medication but it did not seem to help much. He later began to have shortness of breath on slight exertion and during the night. He went back to his doctor who found that he had developed early congestive heart failure. He was hospitalized at Hughes Spalding Pavillion and would spend a great deal of time there until his death in 1966. He had a series of high blood pressure complications, including poor circulation to his right leg, heart attacks, and frequent small strokes. He would go home for short periods but usually had complications which would bring him back to the hospital again. Finally, he had a massive stroke in 1966 and fortunately lived only about one week after that.

By this time, Mr. Neely, Sr. had died and Mrs. Neely, Sr. was chronically ill. Levering Neely had grown up and gone to Emory College and Johns Hopkins Medical School. He had begun practicing and Mancy asked that Levering come to see him at the hospital. All of Dr. Neely's expertise did not help, and Mancy developed one complication after another and finally died of a massive stroke.

The funeral was held at the Ebenezer Baptist Church in 1966. Martin Luther King, Jr. had already been killed. There were not many of the Neelys who could come to the funeral. Mrs. Edgar Neely, Sr. was in a nursing home, Edgar, Jr. was out of town, and Randolph Neely and his family lived in Montreal. So Levering and his wife Margot were the only Neelys at the funeral. It was held one Saturday afternoon at four o'clock. Levering and Margot got there early and sat toward the back of the church. People were coming in and sitting down here and there, and it wasn't long before the church was filled to capacity. Margot suddenly tapped Levering on the arm and said "Look!" pointing straight ahead. There high on the chancel was Mancy Brown lying on top of his coffin covered with a white lace netting. He was dressed in his Sunday best suit complete with tie, shirt, and dress shoes looking as if he were asleep.

About this time, someone from the church tapped Levering on the shoulder and asked, "Wouldn't you like to say a few words during the service?"

Levering was surprised and chickened-out by saying, "I would be too upset to talk."

This seemed to be the proper answer and was readily accepted. The choir, dressed totally in white, was now coming in and the sweet sounds of good black voices filled the church with a beautiful symphony of sound. The choir slowly filed up on the podium and took their seats, after finishing the hymn. The speeches began one after another, all praising Mancy.

Finally Martin Luther King, Sr. spoke. His speech went something like this:

"There we were in the big city of New York where I was attending the Baptist Convention. I had just got back from one of the big meetings and had gone up to my hotel room to rest before the banquet which was to be held that night. I took off my shoes and shirt and was dozing in an easy chair. The phone rang and I picked it up as it was right next to me. That was when I got the terrible news that my lifelong friend and helper, Mancy Brown, had died. I could not believe it, although I knew he had been sick in the hospital. I turned to my wife and says, 'Bunch (I calls her Bunch), me and you have got to go to Atlanta right now.' When I tell her the sad news she says she is ready. We called the airport and sure enough they have a flight to Atlanta at 8 p.m. and they had room. We picks up our bags and gets a taxi to the airport and makes the flight on Delta just in time. The first part of the trip was easy, but then the light comes on and the captain says we better fasten the safety belts as there was rough weather ahead. Very soon that old plane was pitching and we had to hold tight. Then we could see the lightning coming down in big bolts, then claps of thunder and the whole plane shook. I will have to say I was scared and that lightning and thunder kept on until we got to Atlanta. When the plane finally landed in Atlanta, I said to Bunch: 'The dear Lord has brought us through just so we could get to Reverend Emancipation's funeral. Praise the Lord!' "

A thirty minute eulogy followed.

So Emancipation Brown, you were well named. You were part of the great transition from slavery to the recognition and dignity that you so richly deserved. It was your amazing grace that helped bring about a new order, not perfect it is true, but so much better than what was before it. You were loved by the Neely family and almost a part of it, but when history is written, you were the one that loved your fellow man the most.

39
Emily Levering Eckfeldt Neely

You, Emily, came down from the North to the heart of the South and you overcame. There was no question but that you were a Yankee who invaded the South in 1906. Although you were married to Edgar Neely, Sr., a true Southerner, you arrived in an environment of relative hostility. The Civil War had been over a fairly long time—thirty-eight years—but the Southerners still had a strong territorial sense. In Atlanta in 1906 Yankees were still the enemy who had come down to lay waste our beautiful country below the Mason-Dixon Line. There were still many people who remembered the Civil War well as they had lived through it. True, almost two generations had occurred, but people in their fifties, sixties, and seventies experienced the devastation of our land. So it was only natural that Emily would not be accepted by Atlanta society even though her husband was in one of the best law firms in the city.

Emily, you were born in Philadelphia on March 23, 1878. Your father, Frederick Levering Eckfeldt, was a physician. He had a long tradition of good family behind him on both sides. The Eckfeldts were from Philadelphia and lived in Ambler, a suburb. They were said to have come from the Netherlands, but there is a question as to whether the Pennsylvania Dutch are not really Germans. In any event, the Eckfeldts were a well-known family from Philadelphia to Washington, D.C., and they had a tradition of working in the Mint of the United States. One of them was the director of the U.S. Mint for many years.

The Leverings, however, are a much more famous family, both in Philadelphia and Baltimore, and both phone books have the names of numerous Leverings. They were teachers, coffee merchants, doctors, and they had many other professions. Emily's father was a physician in Philadelphia. The name Levering apparently was English as there still is a small town in southeast England named Leverington. There are no Leverings in the phone book there now, but the old church has many tombstones with the name. The story is that the Leverings left the town in the 1600s due to religious persecution. They migrated to Holland and at least one of them came to America—Josiah Levering—in the late 1600s. He is said to be the one responsible for the vast number of Leverings in Philadelphia and Baltimore now. There is a small town in Michigan named Levering, and there is a street in Los Angeles named Levering Street. But who is the cause for these namings? We don't know. There is a Levering Hall at Johns Hopkins, but which Levering was it named for?

So on her father's side Emily had great forbearers. Her mother was Kate Kugler. Certainly that is a German name. Kate married Frederick Levering Eckfeldt in about 1875. Emily was their only child. Kate had at least a sister that was a medical missionary to India, but we don't have much information on her.

So, Emily, you and Edgar came to Atlanta after the wedding and went up to Tallulah Falls on your honeymoon. Then you two returned to Atlanta and found an apartment on 8th Street not far from town, and Edgar could ride the trolley to work. So there you were, Emily, right in the middle of the old South. The first year was a little lonesome for you in the daytime but you fixed up the house, did the cooking, and the biggest time of the day was when Edgar came home in the evening.

Frank and Rae Neely lived close by and you often saw them on the weekends. Emily and Rae got along well together as they had some of the same interests such as music, art, books, and cultural events. Although they saw each other frequently throughout their lives, they did not become the closest of friends.

Edgar and Emily joined St. Lukes Church as it was in the neighborhood. Emily had been a Lutheran, but there were no Lutheran churches here then. Church is a great place for young couples to get to know each other and it was here that some lifelong friendships were made for Emily. The hostility toward Yankees broke down very soon when people got to know Emily as she was vivacious, outgoing, and bright. And she was always a lady. She dressed properly, learned the southern amenities very soon, and somehow gained the respect of the people of the church and those from the law firm.

What was her childhood like? We don't really know. She was the only child of Fred and Kate Eckfedlt. He practiced medicine in Philadelphia and was doing well until his untimely death as a relatively young man of forty-six. Kate and Emily moved to Washington, D.C. as there now were many relatives living in Washington, and Kate's sister suggested that they come there. Emily had been brought up in a household where learning and reading were held in high regard and she went to a private school through high school. Very few women went to college then, but Emily continued her own education and became quite skilled at the piano, often practicing for hours at a time. She also was widely read in the classics and had studied French with a private teacher. The sudden loss of her father was especially hard on Emily as she had no brothers or sisters. Her father had left them a good income from a moderate sized estate. Although never rich, Kate and Emily did not have any real financial problems. So she had a good literary and cultural background when she came to Atlanta.

Then came the children. First Henrietta Eve (named for Edgar's mother) was born in 1908. But she died in a few days due to a congenital heart problem. Then Edgar was born in 1910, and legend has it that he was born in a hack on the way to the hospital. He was delivered by a black nurse just as the hack pulled up at the hospital door. Next Levering was born in 1912, and finally Randolph was born in 1915. So Emily had to be satisfied with three little boys. She would have very much liked to have had a girl, but it was not in the stars.

She was a very conscientious mother; she saw to it that the children had the best medical attention. She always had them dressed properly, and she encouraged their learning and development by reading to them every night. But boys are a care and disappointment to a mother who wants to bring them up properly; they fuss, they cry, they fight so much that momma almost gets to the end of her rope, and when they go to bed there is a big sigh of relief.

Big Edgar helped with discipline and it was occasionally necessary to give them a spanking and a good whipping. They usually had a nurse who lived in—there were two sisters from Highlands who alternated in looking after the little darlings. There was just enough difference in age to cause the older to tease the younger, and then an explosion into a general free-for-all fight would occur. Miss Emily, the lady, could never understand why they should want to fight so much. But she continued to try to calm them down and soothe the waters.

Miss Emily became active in the community at a very young age and early in her married life. She was quite active in the church and this led to other projects. She became involved in the League of Women Voters, the Crippled Children's Hospital, the charitable organizations, and the Sheltering Arms (for orphans). Later she would be in the Drama League, The Musical World of Classical Music, and also the Alliance Francais, and the English Speaking Union. She employed a French governess to help with the children and teach them French. Her name was Mademoiselle Tamayo, and in the daytime she taught at Washington Seminary. Emily encouraged the children to take music lessons while they were young, but this was not too successful. There was a large Steinway grand piano in the living room which Emily played. But she could not induce the children to play, although Levering would later take lessons when he was about twelve or thirteen years old.

Edgar, Jr. was found to have a weak right arm when about six months old, and he was taken to all of the specialists in Atlanta. It was finally decided that it must have been a birth injury. It would persist the rest of his life and he would try to compensate for this disability. But Emily insisted that he get all available treatment for it.

Highlands, North Carolina came into the picture very early. The family would travel by train to Cornelia on the Southern Railroad, transfer to the Tallulah Falls Railroad, and get off at Dillard. They would then take a horse and buggy and go up into the mountains about eighteen miles on curvy dirt roads to Highlands. Edgar, Sr. would ride a horse alongside. Later they would make the trip by Model-T Ford and often got stuck in the mud and would have to be pulled out by the mules of a farmer.

But Highlands suited Emily perfectly. The high altitude kept it cool all summer. The scenery was beautiful and it was a different world from Atlanta. Also, the summer people were from Altanta and New Orleans, and many friendships made in Highlands continued for life. The people from New Orleans were quite cultured ladies and gentlemen and they brought along their little offsprings for the summer. Emily fitted right in with this group. So the family would go back in the summer for a long time and later buy a house up on Satulah Mountain in Highlands. There was a great view of the valley and other mountains from the porch. Emily really came to love Highlands and it became part of her.

Meanwhile in Atlanta, Edgar, Sr. and Emily were falling into a close circle of friends as well as a large group of acquaintances. There would be frequent dinner parties and some bridge parties with the Perrys, the Thorntons, the Reeves, the Lokeys, and the Krensons. Harold and Esther Swope were friends from Florida and the families would frequently visit each other. The Edgar Neelys had now built and moved into a house on Barksdale Drive in Ansley Park. It was bought as soon as the three children were born as they needed more room. It was a modest but comfortable house with three bedrooms and a big screened back porch which was for the boys to sleep in during the summer. The house would stay as the principal home of the Neelys for about forty years.

Addie was the cook for many years and Mancy did the heavy housework and all of the yard work. He would also chauffeur Mrs. Neely wherever she wished to go. This she greatly enjoyed, although she was able to drive around town by herself quite adequately.

Edgar and Emily's relationship was usually quite good, but at times there were problems on how to bring up the children. Emily was always the lady; Edgar was perhaps a little more raucous, although he liked music, drama, and art. In later years he probably strayed a little. Emily probably knew it, but did not admit it. In general, though, it was a happy marriage. Emily insisted on the best possible education for the children and they were all sent off to Episcopal High School at one time or another. Randolph went for three or four years, but Edgar and Levering went for only one year.

Edgar, Sr., now a very successful lawyer in Atlanta making a good salary, tended to spend it all. Emily had a moderate inheritance which Edgar managed poorly and she finally had the bank take care of it. Edgar had bought real estate in East Point in the thirties, but even then the city was moving north and Edgar lost a good bit of money as well as some of Emily's inheritance.

They now had a wide circle of friends in Atlanta, had numerous dinner parties, and would often have celebrities in music, art, or drama from out of town. Emily had become fully accepted in the city and people completely forgot she was a Yankee. Edgar had joined the Piedmont Driving Club, the elite social club of Atlanta. Barksdale Drive is only two blocks from the club, and during the summer the whole family would get up before breakfast and go take a swim. Emily joined right in.

Emily was short, about five feet two, but this never bothered her. She wore her hair up in a bun and it slowly turned grey, but never totally white. She wore pince-nez glasses, and being very nearsighted, had to wear them all of the time. She remained pretty all of her life. She never allowed herself to become too fat, although she may have been a little plump. She wore long dresses for many years but then changed with the style. She was always well-dressed and when driven around, she always wore a hat and white gloves.

Edgar and Emily made many trips to New York, and with Edgar serving as the lawyer for the Southern Railroad, he had passes on many other railways. One summer he got passes for the whole family to go out West, all the way to California by way of Denver and Pike's Peak, then to Los Angeles, up to San Francisco, and then to Seattle. From there to Banff and Lake Louise, across the Canadian Rockies to Chicago, and finally home. It certainly was a great trip, but the children were eight, ten, and twelve, and it was an exhausting trip for all. The children went out in a boat by themselves on Lake Louise; they began fighting in the boat, and the parents—on the shore—did not know what was going to happen. Finally they got in to shore and sure enough, Daddy spanked them all. Edgar and Emily also took quite a few trips to Europe, especially England and France. Of course, in those days, it was entirely by boat from New York. They also encouraged the children to go to college; Edgar went to North Carolina and then Emory Law School; Levering went to Emory and then to Hopkins Medical School, and Randolph went to the University of Georgia.

When the children were still young, Edgar bought a farm in North Georgia. It had a large farm house on fifty acres with a barn and trees. Big Edgar thought it would be nice to try a little farming again as a hobby. There was a good-sized peach orchard with large fields and cotton was grown in one area. They had a hired man to do the farming. There were a number of unusual people in the neighborhood, with several from England. Emily, at first, thought that she would be unable to adjust to a farm; later she came to like it because of the unusual group of neighbors. The boys had a pony with a little cart. There was always a large group of boys there in the summer as each of the Neely boys would invite a few friends. So here was another chapter in Emily's life made possible by the rather cultured neighbors.

Bringing up those boys was no easy task, but Emily stuck to her purpose of trying to make young gentlemen out of them. Edgar and Emily encouraged good grades and hard work at school just as big Edgar's father and grandfather had done. Edgar, Jr. was the most brilliant of the three but also the most erratic. He would later make three A's in college one quarter and three F's the next. Levering was the quiet one as a child but made good grades in school and in college, usually by hard work. Randolph was the most conventional one; he made good grades if he wanted to and later became a successful businessman. Edgar was destined to become a great trial lawyer. Levering thought he wanted to become an engineer and was entered in Georgia Tech. He says that his mother wanted him to become a doctor, so he gave up the idea of engineering and went to Emory to later become a physician.

Well, what made you tick, oh Emily Levering Eckfeldt Neely? Did you have great ambition for your family as your sister-in-law, Rae Neely, had for Frank? No, your ambition was in a different direction; you wanted Edgar, your husband, to be a gentleman and you wanted your three boys to grow up to be gentlemen—whatever that is. I guess it was to be courteous, to be helpful, to be unselfish, to get along with people well, and to be cultured. By cultured, she meant to know literature, music, and the arts. But the goal was very difficult to achieve, and although you had some satisfactions with the boys, there were also disappointments. When Edgar, Jr. became a real lawyer she was happy. She was especially delighted when Levering went to Johns Hopkins and became a doctor, and though disappointed that Randy went to the University of Georgia and left in his senior year to go into the army, one sees that Emily felt she had at least attained part of her goal. She was unhappy with Edgar's first marriage. She did not think Claire was a lady by background. They never got along too well. She thought Margot, Levering's wife, was great, and that Kay, Randy's wife, was fine.

The story about Emily's role in the meeting of Levering and Margot was one of her tales. Levering had tuberculosis and was under teatment in Trudeau Sanatorium in Saranac Lake, New York. Emily came to visit him in that cold and snowy area. She was standing on the corner waiting for the bus when Mrs. Thompson drove by, recognized that this was a lady, stopped and offered her a ride to the hotel. The two ladies talked and Mrs. Thompson invited Levering and her next door neighbor's daughter, Margot, for dessert, and that was the beginning of a forty-two year melding.

Emily had a heart attack in 1937 at age fifty-eight. She was treated at home, made a complete recovery, and never had any more trouble. After Edgar, Sr. died in 1951, Emily lived in the house on Barksdale alone with the cook, Addie, and Mancy. She began to have failing health in the sixties and went to live with Edgar, Jr. and Claire. While there she picked up her breakfast tray one morning, stumbled and fell, breaking her hip. She became worse, her mind began to fade, and she finally had to be put in a nursing home. Fortunately, she did not live long there and died at age eighty-two in 1960. The funeral was at St. Lukes Church and she was laid to rest next to Edgar, Sr. at Westview Cemetery.

One legend about her promulgated by Michael Neely her grandson: He says she was an excellent marksman with a pistol and could hit playing cards thrown up in the air at the pistol range. Levering questioned this but does have a vague recollection of her taking shooting lessons.

So Emily, you were in truth a *grand dame*. You held the banner high and would not waver. You had great satisfaction in life and you had great disappointments, but the former predominated. There was something missing in your life, and in looking over the years, I believe that if you had had a daughter to confide in, your life would have been complete.

40
Keefe

To Keefe

(Written at his birth, November 1977)

To be or not to be,
To have or not to have,
All those misgivings before,
Will we be good parents,
Will we have to change our lives so much,
Should we bring someone
into this uncertain world—
All these and more—natural
enough then, but now—
He is, he lives, we have
'Tis true—He's here—
Hallelujah.

So forget those rhetorical questions—
But remember this—
For life to be really satisfying
It must be fulfilled—
To know joy
One should know pain.
To be really happy—
One should know suffering.
A life of service,

of doing, of action is
great and almost fulfilling,
But one day it grinds to a halt,
And to have one part
of you continuing,
Is the ultimate fulfillment.

Not only that—but To watch a flower grow from a seed—
to a mighty oak
from an acorn, is
wonderment enough—
But to watch
the flow of life
into a babe,
To see that gradual process
of moving, grasping, crying, hiccuping,
feeding, excreting,
burping, looking—
Then later the quick
smile, the amazing growing And then the crawling, pulling
up in the crib and that
first early step.
These are the simple actions that are so natural,
but to watch them in your own flesh is that
part of life which you would never know,
had he not arrived upon the scene—
And had he not come—
there would have forever
been an emptiness of
which you would not
even be aware.

After those first months or
first years—
The march of events goes
all too fast—
Somewhere he becomes a
person, an individual,
the clues of which are
seen before six—
Then communication, school, reading, playing,
learning, always learning.
And you are the onlooker
now—Oh yes, you

point the way but the
genes pop out in such
an odd way—
An individual but part of
you, an extension
of both parents—

So hail to thee Keefe—
You have a name uncommon
And I predict a person
uncommon.
I predict a good mind
active and curious,
intelligent, one who
will not take from his
fellow man, but one
who serves them.
I predict a good
sense of humor, that
part of life so essential
to endure the ups and
downs as they come along.
I predict a gentleness
of spirit—not meek,
but slow to anger,
who weighs his words.

Perhaps a judge?
but no matter
What ere he does or
where ere he goes, he carries
with him a priceless treasure,
—the heritage of two great
spirits.
Hail to thee Keefe
Here's a toast to
Your life—and
may you one day
watch another Jackson grow.

* * *

Some people are different, perhaps slightly or remarkably unusual. They seem to show some characteristics that make them stand out from their fellow man. We call them characters, or we say, "They broke the mold when they made him." Sometimes they are unusually bright, but not necessarily so. Some have an especially good sense of humor. Some may be great men, but others may be reclusive and never venture far into the world of people. All great artists, musicians, and actors have some outstanding group of traits which have catapulted them to their positions. Some have unusual traits but never use them.

Whence cometh these unusual individuals? Why are they different? Is it genes or environment? Certainly environment can influence how one's life evolves, but it seems more likely that the origin centers in the genes—the DNA or RNA which makes up those characteristics come out like the roll of dice—each of the forty-eight chromosomes has thousands of genes, some dominant, some recessive. Some characteristics may have been recessive for several generations and suddenly burst forth to make an unusual person.

Ben Neely, born in 1834, you were a character as we have seen. From the time you ran away from home at age thirteen until your death in 1892, you stood out among your fellow men. You had nine children and all of them were a little different. Frank was the most exceptional; Edgar, Jr. was perhaps the most unusual one of his generation, but Edgar, Sr. was definitely a character. Levering, Edgar, Jr.'s brother, was slow to become different but finally fell into that category. Virginia, Levering's second child, is a remarkable woman, but let us take a good look at her son, Keefe Jackson, age five years and eleven months, son of James Jackson of Oklahoma.

Keefe, you were born on October 11, 1977 and will be eight years old October 11, 1985. You are four generations down from Ben Neely, Sr. He was your great-great-grandfather. How do we know that at age six you are different and undoubtedly will continue to be different from your peers? Let us see if it is possible to put you under the microscope and examine your exterior as well as your interior mechanisms, motivations, and possible aspirations. Let us see if you are beginning to get some stirrings of ambition and then let us try to predict in which direction you might go in this life.

I think your father, James Jackson, was probably the first to realize that you were a little different. Your father is a reflective, observant man. All children seem to come out with some remarks or answers to questions which, in their simplicity or purity of thought, make the parent smile or laugh. So it was with Jim. However, he finally began writing down some of the things you said. They seemed like answers that any child might make, but somewhere along the line you, Keefe, began to show an unusual interest which slowly grew, developed, and evolved into a passion. This wasn't just a passing fancy; this interest persisted over a period of several years, at least three years, and only now, at age eight, has a new interest taken over and pushed the first interest into the background although, by no means, is the first interest dead.

The psychologists say that the mind and early development of boys and girls is the same. This means that at least up to three or fours years there should be no great differences in their responses. Girls may begin to identify themselves with their mothers early on and exhibit behavior more feminine. Similarly, boys will soon realize that they are masculine and again may identify with the father. But the women's libers will say that there is no difference in the mental capabilities of a boy or girl.

But I say that a very small girl, age one to three, will become interested in dolls and that it is most unusual for a young boy to play with dolls. Furthermore, a boy begins to play with balls very early. We have seen this in the Neely family, in Thomas Longstreet Neely, who, before age two, would hold on to a tennis ball for dear life, and who, at age two-and-one-half, is able to drop the ball and hit it with a tennis racquet. He now can throw a ball, kick any type of ball, and will keep working and playing with a ball for hours. And don't try to take a ball away from him, that is, not unless you want to hear a real howl. Jordan Neely, Thomas' older sister, age five, almost six, never did play with a ball much but has always liked dolls.

The story which exemplifies Keefe's overpowering interest at about age four-and-one-half occurred at Fernandina Beach at the Levering Neely beach house, Frogmore. The Jacksons were there, along with Keefe, also the John Neelys, along with Jordan. Both families as well as the Neely grandparents were sitting out on the deck (or better known as the roost) overlooking the beach. Keefe was kneeling down pushing a small truck around. Jordan was jumping around and someone suggested that Jordan show us a little ballet as she was taking ballet at the time. She, not being shy, danced around and showed us the basic steps with some of her own interpretation mixed in.

After she finished someone asked Keefe, who was still playing with his truck, "Hey Keefe, can't you do a little ballet for us like Jordan?"

He looked up briefly and then turned back to his truck and said: "No, I only do construction."

His father, Jim, says that somewhere between age two and three, Keefe started his fascination with trucks and construction. Jim thinks that he first began imitating the sound of trucks even before he could talk much. He thinks Keefe has a phenomenal ear and memory for sounds and that this started him on trucks. Very soon he was beginning to identify and call out the basic types of trucks, from the eighteen wheelers to all the various types of construction trucks, and from those he went onto the bulldozers, graders, and grinders. He demanded and received many books on trucks.

Jim and Virginia helped him with his interest and Jim would read to him often.

After the types of trucks and road construction machines were mastered, he began to learn the parts of a truck and then the major parts of the engine such as the carburetor, spark plugs, battery, and exhaust system. He was under five years old and learning more and more details of a subject which usually might interest a boy of ten to fourteen. From trucks he learned about the parts of a bulldozer and could identify one of the special types of extra wheels that bulldozers have called a bull wheel. His father was talking to a construction man one time and the man said that Keefe identified the wheel correctly, although Jim had no idea what he was talking about.

Next the library provided a picture book on the evolution of the automobile, from the first steam driven vehicle in the early nineteenth-century up through the German gasoline car of Daimler and later Benz and finally the modern cars. Keefe insisted that this book be read to him several times and very soon he had practically memorized the book and could spout out all the details of each of the twenty-five automobiles described. Soon he was looking at the cross sections and plans of automobiles just as he had done with trucks previously. Then he got into books on road building and the history of road building going back to Roman times.

When Keefe was three years old, he was enrolled in a private school, the New School. He had been in a nursery school the previous year but this was just a place to leave children when their parents were at work; it had no real program. The New School, however, was directed by a bright young Ph.D. in psychology, and he had a dedicated group of young teachers. The school fostered any interests that might be shown in these preschool children. They were given the opportunity to do a great deal of drawing if they wanted to, or they were encouraged in the appreciation of music as well as given an opportunity to participate in singing and dancing. Another field of endeavor was to learn a language early in childhood. They did not push the children to learn to read or write, but they gave them ample opportunity to learn. Many of the children could read and write simple words; some were much more advanced than others. Thus in many ways it was an *avant-garde* school. It had many of the Arkansas University people on the Board of Directors. This included several doctorate graduates in psychology, especially those skilled in testing children and their intelligence. However, they did not make a great point of intelligence testing nor did they tell the parents the results of any testing they had done. They did have frequent conferences with the parents and would talk over how the child was progressing.

Soon after Keefe went to the New School, the principal had a conference with the Jacksons. He told them that they had an exceptional child.

He did not say he was brilliant or a genius. He did tell them that Keefe's interest in construction was a little unusual, and that this interest should be fostered until some other interest came along. The parents may have wondered what this single interest would lead to and if it really was healthy. The principal assured them that he was drawing well and a great deal, often pictures of dogs or other animals and that he had recently developed a new interest, music. Also, he was doing very well in the language courses. Thus his interests were not too out of balance. The other report was that he got along well with the other children, both boys and girls. He also had a great creative talent and was beginning to make things with paper such as musical instruments and construction machines. Thus Jim and Virginia were highly pleased with the report. They continued to get books out of the library about every week. They knew the librarian very well and she was helpful in selecting books for him.

For his fifth birthday Keefe had a construction party of about six boys and one girl from the school. They all made various machines and brought them to the party. Keefe made a cement mixer—it was a participation party and a great success. Most of the children at school had seen some of the Star Wars movie series, and the children were once asked to make some object or thing they had seen. Many of the children would make laser guns or something they had seen in the Star Wars series. Keefe was making a complicated instrument and the teacher asked him what he was making.

Keefe replied, "That is an agroma."

"What is that for?" asked the teacher.

"Oh," said Keefe, "that is an instrument that the Romans used in olden times to make roads and to be sure they were level."

The teacher thought this was a little unusual and told his parents.

It was shortly after his fifth birthday that a new interest appeared and it soon became as all-consuming as the construction interest. Jim had realized that Keefe had an unusual ear for sounds and could reproduce sounds extremely well. He could repeat a difficult Italian word with an excellent accent. He could make the sound of a horn or an organ such as the one at church. He talked a lot about his grandfather playing the recorder and the flute. His parents had recently acquired a record player and were beginning to collect records of classical music. They say that the day Keefe heard the Nutcracker Suite he became a different person. He wanted to play it over and over *ad nauseam* to Ginny and Jim. He was taken to a classical music concert in the park and was sitting up close to the orchestra. He was completely fascinated by all of the musical instruments. He had to have a description of every instrument in the orchestra. He learned the difference between the woodwinds and the brasses. He learned the instruments that had a single reed and a double reed and made a big point of it.

If you didn't know that an oboe had a double reed, you really were ignorant musically.

He sat up front in church and watched the organist play. He became especially interested in organs, had the family get a large technical book on the history and mechanics of the various types of organs. He understood the principal of having an air pump either mechanical or powered by someone pumping behind the organ with their feet. From the organ he progressed to bagpipes, found pictures of them, knew the sounds they made, and constructed a model out of a paper sack and cardboard tubes.

The school encouraged his new interest in music. While playing music on a record player, he would often take a baton and mimic the conductor whom he had seen. When he went to Atlanta to visit his grandparents, he was given a small recorder. He kept it with him, even taking it to bed. He was able to blow it and make sounds and he is anxious to take lessons on how to play it.

As his parents would get more records of various composers, he would insist on being read a short biography of the composer. The composers seemed to intrigue him and he remembered what their lives were like. He made his father bring out his clarinet and had him play at night. The 1812 Overture was one of his favorites, and he knew exactly when the cannons would be shot off. He began writing a few songs. When his grandfather came with a flute and another recorder, he insisted on a concert along with his father on the clarinet.

Well, Keefe, you are indeed unusual. First let us try to analyze what is going on, then why. You definitely have a voracious appetite for knowledge. Furthermore, you are extremely interested in detail, and when you get something on your mind, you must know all about it. The details are terribly important to you. And you have a great amount of tenacity. Your interest persists for months and years. Then you are creative—you draw, you make things that will move or work. You build toy trucks and musical instruments. The teacher says you are not a genius—you are not reading or writing way ahead of yourself. However, you do have a phenomenal memory—almost photographic at times.

Your principal, after watching you for three years, thinks that you have been encouraged in your creative ability by the fact that you do not have a television in your home. You thus have to entertain yourself and you do it in an active way. I think that the lack of a T.V. may be a factor, but I am much more inclined to think that you got an unusual set of genes and that you were very wise in your choice of parents as they are the ones who have helped you to develop some very unusual traits.

The future? It will be most interesting. Ten, twenty, thirty years from now—where will you be and what will you be doing? And Keefe, there is now a new change in your life and what effect will it have on you? The sudden change that has occurred in your life is the acquisition of a new sister—Margaret Elizabeth Eve Jackson—to be called Margot after her grandmother. You came into the delivery room right after the birth, looked at your parents, then at the baby—pointed at her and said, "That baby does not even know anything about television!"

But you, Keefe, know precious little yourself about television. So here's to you, Keefe, engineer, musician, or whatever—long may you wave. And here's to little Margot—you have a great heritage attached to your name. We will watch you carefully, little Margot. You may be even more unusual than Keefe.

41
Self Portrait of Levering

F. Levering Neely
Age Twenty-two

The Johns Hopkins University School of Medicine
Baltimore, Maryland
Second Year
Prior to Comprehensive Exams
May 1936

Dear Eddie,

I am at a dangerous corner, old man. You made the statement that the majority of my letters to you were colored by a morbid inferiority complex. I disagreed with you at the time but must frankly admit that it may be true of this one. Therefore, if you be in no mood for having your shoulder wetted with a few, nay several tears, cast this letter aside for I realize that your own troubles weigh heavily.

Here are a few facts: that is, I believe them to be facts, which I have gained by cautious objectivism:

L.N.'s mind is not retentive for long periods of time, although it is good for short.

L.N.'s mind deals better with a problem than it does with description. Medicine is a highly descriptive subject, involving frequently unrelated masses of fact.

L.N. throughout his life has had little practice in speaking, the presentation of a subject before people.

A certain degree of speaking is essential not only throughout courses but in examinations.

L.N. is weak in anatomy and neurology, although in chemistry and physiology, definitely superior to the average.

Anatomy is a vast, deep subject; neurology more alive, less vast.

L.N. has a load of present courses which can well nigh obliterate opportunity for review.

A systematic thorough review of previous work is essential.

L.N. is in the midst of a group of people who by and large have no intellectual interest (not that Neely has lovely intellectual interests), but who have an ability to drive themselves through the processes of acquiring necessary knowledge and doing necessary work as mechanically as turning one bolt in the Ford Factory.

Twelve students failed their second year exams last June.

Don't think that I am not frightened, because I am slightly. You see, it's the fact that I can't evaluate my powers; I don't know where I stand. That is the disheartening factor. Not an exam, not a grade, not a word of praise or condemnation. Nothing. He comes, he goes, and that is all. Not only that, but it is difficult to evaluate the ability of my fellow students as there are many that you see like shadows. I know the kind of suits they wear, the shoes and collar, and tie effect—the moles on their faces—all the little mannerisms of posture that they have. Thus far, I know better to speak, pass words, laugh, jest. But how the majority will bear up at judgment day, I am forced to say, I am ignorant.

I am being swirled off into a field of knowledge which is infinite. I must organize, choose, pick.

I arrived here with the idea of living a marvelous life, studying the essentials decently, but spending more time on those things which actually interest me. I pride myself in a few ideas which I should like to investigate. I thought the whole thing would be just a breeze. I thought of living a social life, of reading history, of learning German cold and in its entirety, of exercise, tennis, and golf, of writing a bit, of learning calculus (a lovely subject), of learning psychology, of Latin, of going to shows, of playing the piano, of becoming smart and witty, of having a swell girl, of making money, of sleeping enough (7 ½ hours), of eating marvelously, of making a wide circle of friends, of being second in my class, of discovering something, in short of, *living*. These things I want and others too—a hunger as wild and inchoate as Tom Wolfe's, but gradually I see that getting through (God, have I fallen this low?) must come first.

I am afraid only of being afraid (who said that first?).

Oh, I am confident, I can do it, I will do it, but I must come down to earth and gut myself on essentials—driving knowledge down in such a way that it will come back to those peoples' satisfaction. So, I am going to hermit myself and study back-exams, and do all the conventional things until I know that I have it, until I can laugh at exams, learning until my mind is an index, a perfect file. It is not pleasant, but upon making or breaking on this one hurdle hangs my life. God, give me strength.

Best to Claire,
Lev

* * *

Self Protrait

Okay, the book is almost over and you have not written anything about yourself except the earliest memories. If we are going to try to trace genes and characteristics, we are going to have to look at you—oh, Frederick Levering Neely, known as Levering or Lev. There are now three Frederick Levering Neelys—Junior is called Fred and the third is called Little Lev or just Lev. There was a dog named for Levering, but he is long gone. So we will have to look at both Neely and Eckfeldt genes as well as trace the events and factors in the environment that might have had an impact on you.

You were the middle son, with Edgar two-and-one-half years older and Randolph two-and-one-half years younger. Edgar and Randolph were in a totally different age group, but you, Levering, were fairly close to both of them. You were the peace-maker when tempers flared up at each other. You also were caught in the middle and had to tread very carefully to avoid the anger of both. They say you were a quiet child, but with Edgar talking all the time and Randolph being aggressive to survive, it is only natural that they did most of the talking. This could have been a factor in choosing another profession rather than being a lawyer like your father and older brother, Edgar.

So let us pick up the story in grammar school and take it on from there. Four years at Tenth Street Elementary School. Many friends were made there whom you would see through college and several are still around Atlanta, sixty years later. The teachers are still fairly clear in your memory, although the first grade teacher is somewhat hazy. The second grade teacher is very clear as a heavy, pleasant type. The third grade teacher was tough, and the fourth a pleasant matronly type. You made good grades, like an A and two B's, there and would do the same the rest of your school career. You did your homework fairly conscientiously. You got along well with the boys but were quite shy with the girls. The school district changed between your fourth and fifth grades, and you transferred to Spring Street School. A few of the old group came with you from Tenth Street, but it was really a whole new group of students. The north side of Atlanta sent their children there. The girls would transfer to Washington Seminary; the boys to O'Keefe Junior High and then Boys High School. There was not a good prep school in the environs, and so some would go off to Woodberry and Episcopal High in Virginia.

So, it was Spring Street School in 1922 and 1923 at ages ten and eleven for you, Levering. These were the preadolescent years and probably important. Ansley Park was a nice, middle class neighborhood, and there were many young people your age there, going to school with you. You could walk to school if weather permitted. If not, your father would take you and some of the other neighborhood children. Your mother and father insisted that you do your homework well at night, just as their parents had done to them. You did quite well again and seemed to be doing especially well in arithmetic.

Your relationships with your parents were generally good. You loved them both but probably didn't know it at the time. Your father was a very busy man but was full of fun and humor and kidded all the children in a friendly manner. He did not have time to join in recreation much with the children as is stressed now, but you saw a good bit of him on weekends and then on those trips to Highlands and the farm in Alto. He did take you swimming at the Piedmont Driving Club in the early morning in the summer. There were many Sunday trips in the family car to see friends and relatives in the general vicinity. Of course, the three boys in the back seat had a strong tendency to fight, and your dad would have to stop the car frequently to threaten and often give a spanking. You remember only one good whipping which you got as a child when it was discovered that you had been taking money from a forgotten church contribution box, in the attic, which was full of dimes, quarters, and fifty-cent pieces. You were always "finding" money under a rock on the street, and no one knew where it was coming from until your mother became suspicious and remembered the church box, called a mite box. When Dad got the news, he took you upstairs and gave you a formal whipping with his razor strap, which he used for such occasions. You remember that very well because he was crying the whole time he was doing the whipping, and you had never seen him cry before. It was very traumatic to you emotionally, but I think you stayed honest for many years to come.

Well, we are going to have to go rather fast if we are going to cover the full seventy-years in this chapter. We will try to hit the highlights and perhaps come back and do it in more detail later.

Okay, on to O'Keefe Junior High in 1924. You would be there for three years from age twelve to fourteen. These were the adolescent years, and I am sure they were important for you, but you can't remember much about them. Some of the teachers are still clear in your mind—Mr. Hawkins, the geography teacher; Miss Rae, the english teacher; Miss Stopfer, the math teacher; Miss Colvin, the history teacher—all good, inspiring teachers. You again did fairly well in school but not outstandingly. Sports included a little football and basketball, but baseball was not your favorite.

Summers were at the Athens Y Camp above Talullah Falls, and these summers probably had as much effect on you as the school in many ways. It was a way of life and you did begin to excel at speed swimming and were beginning to become good at tennis. This was good for your ego. So you got through adolescence fairly easily with a good many friends among the boys and only slight contact with the girls at school or Sunday school.

Oh, the family was quite active at St. Lukes Church and you would all go to Sunday school; as a young man you were an altar boy with a robe and carried a cross in the procession. Of course, you misbehaved while sitting in the altar stalls during the service by shooting spit balls at the other altar boys across the chancel. You did have some good Sunday school teachers. It would be at church that you discovered girls as there was a Sunday Night Service League in which you would participate. It was here that you began to have your first dates, but that was probably a little later, in high school, when you had a car, that you had any formal dates. Prior to that time, you just went around with a large group of other males to a popular girl's house on Sunday afternoons. There would often be as many as thirty or forty young bucks at a girl's house during the afternoon. Light refeshments were usually served. One of the most popular girls was attractive, but her father, a well known lawyer, was so entertaining to the young men with his wit and jokes that she had a great added attraction.

Then to Boys' High. You had just been there a few weeks when you went to Augusta for the weekend to visit your cousin, Neely Plumb, who was almost exactly your age. When you came back, you were sick with a high fever. Your parents took you to a doctor who did a blood test. He called your parents out of the room and told them that you had leukemia and that there was no treatment for it. They took you home with a heavy heart. In the next few days they decided to try another doctor as they did not like that diagnosis. This doctor was an old friend of the family. He took another blood test and found that you had malaria which you had contracted in Augusta while hunting in the Savannah River. You were given quinine and the fever promptly went away, but you were in a very weakened condition for about a month; when you recovered, the family decided to send you to Peacock School for Boys. You still tell the story that you flunked out of Boys' High.

Peacock was a high school run by three brothers who, at that time, were getting quite old. The school was a big old house on Peachtree. It was dubbed "flunker's retreat" as many students who had flunked out of school came here. It actually was a very good school. There were small classes, and the three Peacock brothers were all excellent teachers. So you stayed there three years and graduated from high school in 1928 at age sixteen. The family thought you were too young to go to college and so they sent you off to Episcopal High School in Alexandria, Virginia to stay for one year before going to college. You did well there, making good grades in algebra and geometry, but you were considered a freshman by the other students and had to suffer certain indignities from the other students who had been there for several years. While there, you developed pneumonia because you ran in the snow when you had a bad cold. You were very sick for a couple of weeks. Fifty-five years later, in Richmond, Virginia, you would meet some of the men who had gone to Episcopal High School with you that you had not seen since the old days. You almost recognized them and did remember their names: Rutherford Moncure, Aldredge Dudley, and Saint Peter Tucker, called "Saint" (his father had been an Episcopal minister). Episcopal High School, called *the* High School, was in truth an excellent school and had some unusually good teachers. Your parents were right—this year did a lot to mature you and prepare you for college. You had graduated from Peacock; you did not have the credits to graduate from Episcopal, but you were accepted at Emory University.

You had always been a Georgia Tech fan and though you wanted to be an engineer and go to Tech, your mother mother wanted you to be a doctor, so you went to Emory University and took premed. At least that is the way that you tell the story. College was a great pleasure—you made good grades, occasionally two A's and a B or one A and two B's, and you managed to have a good time. Your sports were tennis and swimming, and you became the intercollegiate breaststroke champion of the southeast with medals to prove it. You lived at home most of the time; you had an old Franklin (air-cooled) car which you drove to Emory. This was an open touring car that you painted all different colors. You had a coat of arms painted on the door which said, "Cavcat Emptor."

Your major at Emory was chemistry as now you were beginning to take premedical courses. College was a time to mature, and from a thin wisp of a boy, you grew taller and broader. So here you were in college from 1929 to June of 1934, and you really did not know that there was a depression going on except from reading the paper and studying economics at college. You took only the minimal number of chemistry quarters to get your major, but you took many other courses: Latin, Mathematics, Bible, Economics, French, and German. I guess you became a person with an identity while in college. It was a great experience. You were in a fraternity and made some lifelong friends. There was a little group that met in the library and the director of the library would often have to come down to the table to try to make your group quiet down.

Somewhere in college you definitely decided you wanted to be a doctor, but where you got the idea to go to Johns Hopkins Medical School, we are not sure. It may have been from your mother who had known about Johns Hopkins when she lived in Philadelphia and Washington. You applied to Hopkins in 1933, went up to Baltimore for an interview, and heard of your acceptance in the spring of 1934. So you took off by the Southern Railway from Peachtree Station on Sept. 15, 1934, at age twenty-one, just a week shy of your twenty-second birthday. A large group of friends from Emory came to the railroad to see you off, giving you a good send-off to go to a whole new world—the world of medicine.

I think you were pretty much of an idealist as far as medicine was concerned. You saw medicine as a great field of endeavor. You knew that you were at one of the best medical schools in the country, perhaps the best at the time. You felt inspired to study hard. You roomed in a house on Biddle Street off Broadway, and you ate at the fraternity house. You were walking home one snowy night from the library with a big pile of books and papers, warmed by a big coat and hat. Suddenly two girls grabbed you on each arm and said, "Where are you going, big boy?" You protested that you had to go home to do your homework. When they finally decided that you were not a good customer, they shoved you, and down you went, with snow, books, and paper flying everywhere. That was bad enough, but the real ignominy was when they threw snow balls at you. "Hell hath no fury like a lady scorned."

First year: Anatomy with three other students on the same cadaver: Roy Pollock from New York, Menke from California, and Harriet Mitchell from Minnesota. You got to know each other pretty quickly. Chemistry, physiology, pharmacology, and psycho-biology. The latter was under the famous psychiatrist, Adolph Meyer. You had to analyze yourself and write a whole book about yourself. You thought then that you might want to be a psychiatrist. Then, the second year, more of the first plus pathology under the famous Drs. Rich and McCollum. That was the difference between an ordinary medical school and a great one like Hopkins. Almost every one of our professors was a famous man, and we were inspired by these great men. Then those last two years you worked in the hospital, again being taught by great men. A group of you went off to the Massachusetts General Hospital and took your senior medicine there at Harvard Medical School. This was very much like Hopkins. Your attending professor was the famous Dr. Fuller Albright, the parathyroid man. Back to Baltimore for the senior year where you were getting a little more confident working with patients.

You said, "It's fun to be a doctor."

One great thing about Hopkins was that you never got any grades. You had a few little tests which were never graded and then you had your comprehensive exams after two years and again after four years but again, no grades. Man, were you surprised to find that you had made AOA, the Honor Medical Society. Your first year interning was at Hopkins on the Osler Service with nine other great guys; plus assistant residents of fine quality; and Barry Wood, the famous Harvard football star, as your chief resident. What a great year—but you worked like a dog.

It was the fashionable custom to take a year of Pathology then. So off you went to Columbia and did autopsies, research, and teaching—another fun year. Hey, what about the girls—you have been pretty quiet on this subject. There were several in Medical School; one big blonde named Martha lived in Washington; you went by train for a visit. She met you at the train and you hailed a taxi. When you got in the taxi the radio was on.

Martha leaned forward and tapping the taxi driver on the shoulder said firmly, "Turn that radio off; don't you know it is dangerous to drive with the radio going?"

He turned it off. You had a pleasant weekend, but that was the last date you had with her. You never could stand bossy women. Oh, there were several others but none too serious until you got to Columbia and met and dated one of your students—Julia. She later turned out to be too bossy and you dropped her too.

The next year to Bellevue in 1940, and after three months on the chest service, they found a spot of tuberculosis on your lung and shipped you off to Trudeau Sanatorium in Saranac Lake in northern New York to get the cure. As I mentioned earlier in the book, the treatment then was bed rest in the cold air. You had a very small spot; when the war looked imminent, the doctors at the sanatorium went off to the war, so they got you up from bed and made you the doctor for the sanatorium. You stayed almost a whole year, and here the female interest became serious. On your first date you proposed, but she just laughed and laughed. Margaret Gardner, called Margot, came into your life in November 1940 and left you April 11, 1983, with leukemia. Almost forty-three years together, but that doesn't make the loss any easier. You became engaged in the spring and you both went back to New York in the summer of 1941. She got a job with the Inter-allied Information Center, later to become the United Nations, and you went back to Bellevue. On Pearl Harbor Day, you were driving out the Pulaski Skyway to see Dot, Margot's sister, and her husband, Ben. What a shock. You two got married in the spring in New York at the Fifth Avenue Presbyterian Church, and you had the reception at the old Vanderbilt Hotel. She was earning $100 a month, and you were getting $10 a month. You got a little supplement from Momma Emily Neely but still lived in great style. The rent was $62.50 a month. Did you two love New York? Indeed you did and would continue to love it for the rest of your lives. You said it's a great place to live but not much to visit (you can't get enough of it on a visit). Ok, two great years in New York for you Levering—one year on the general medical service at Bellevue and one year on the chest service with the famous Dr. Amberson.

Freddy came into this world on June 20, 1943 at the Presbyterian Hospital in the City of New York, a part of Columbia. Margot stayed in the hospital two weeks as was the custom in those days. Freddy cost exactly $2.00 as Blue Cross paid all the rest.

You, Levering were rejected by the Army because of the recent history of tuberculosis, and you had gone to Atlanta a few months before and found that the internist, Dr. H. Cliff Sauls, desperately needed someone as his associates, Dr. Carter Smith and Dr. Charles Stone, had gone to the Army. So, Levering, you took the job with Dr. Sauls at the beginning salary of $250 a month. This was a low starting salary even in those days, but the opportunities of this position seemed great with an entry into Piedmont Hospital, perhaps the best hospital in Atlanta with a great many of the physicians trained at Johns Hopkins. It was affiliated with Emory Medical School but not controlled by it. Dr. J. E. Paullin was the head of medicine at Piedmont and also the Professor and Chief of Medicine at Emory and Grady Hospitals.

So, Neely, you dove into a very busy practice immediately and would work from early morning until late at night. During the war years you saw very little of your children. Margot came down a few weeks after you arrived with the new baby Freddy. She came by train and somehow daylight saving confused you, and Margot called early Sunday morning, saying she was at the railroad station. Not a great start, but things soon settled down, and you bought the first house on Dellwood Drive for $9,000—two bedrooms, one bath, and a porch.

So here you are, Levering and Margot, beginning a new life—a new career for both of you. You would have six children in a ten year span—three boys and three girls. You, Margot, would become involved in the community early on with the League of Women Voters which preserved your sanity and got you away from those six children at least once or twice a week.

But Levering, we still don't know what kind of a guy you were. How much ambition did you really have in 1943 at age thirty and about to take your first plunge into the real world?

42

Frank H. Neely and Descendents

Okay Frank, we have heard about your birth in 1884. We have some information on you in early school in Rome, Georgia and then a little about your high school days. Then we know that you graduated from Tech in 1904. You went to Pittsburgh with Westinghouse in that same year, came back, and married Rae Schlesinger in 1908. You both went to Pittsburgh again, stayed two years, and then returned to Atlanta in 1910. You did a variety of consulting jobs and finally went to work full time with the Fulton Bag and Cotton Mill in 1915; you were in charge of production. You stayed there for about nine years and apparently did an excellent job of operating the mill.

Walter Rich enticed you to come to Rich's in 1924, and that is really when your career began to take giant leaps. You dove into the job of reorganizing Rich's with a vengeance, and you were the one who initiated the innovations which were responsible for the fact that, within thirty years, Rich's became twelve times as large as it had been in 1924. During these years your personal growth paralleled the store's growth.

One of your greatest admirers was Ralph McGill, editor of the *Atlanta Constitution*. He collected a large mass of information about your accomplishments and efforts in the community. Although there were philanthropists who made huge monetary contributions to certain institutions, there were also many people in your time who held on to their fortunes, and who did not contribute much to the community. But you, Frank, were different. You wanted to dive into the community. Stir it up. Make it move. Give it new direction. And that you did, Frank.

One of your biggest thrusts was in the field of education (your father and grandfather were both teachers). You would not be an actual teacher in a school, but you fostered education throughout your life. You were responsible for the city fathers approving a huge bond issue for the public schools. You were very close to Miss Ira Jarrell, one of the long term superintendents of the Atlanta school system. You helped her when she confronted problems. You helped the schools get a radio station, improving education before television came along. Of course your contributions in time and money to Georgia Tech are well-known. You helped Tech get one of the first nuclear reactors in this part of the country, and when it looked as if there were insufficient funds, you made a massive personal contribution to get things rolling.

You became involved in many aspects of the school, including serving on the search committee to find a new president. You were chairman of the board of the Alumni Foundation and on the Board of the Georgia Tech Research Institute. You gave sixty prime acres of your land at the farm to Georgia Tech before you died. You also worked with other colleges such as Emory, the Tallulah Falls School, and the University System. You had some kind of a compulsion about education. It certainly was in your blood, and if you could not teach yourself, you wanted to help in other ways. Mr. Dick Rich says that you were a great teacher yourself, for any time a little group was gathered around you, you were teaching them.

Education, we see, was one of your great community interests. But there were others. You served on the War Production Board in Washington. You became regional director of that board for the south. You travelled frequently to Washington, D.C. and came to know President and Mrs. Roosevelt. She would later visit you and Rae at the farm. You knew several of the U.S. cabinet members and several of the army generals, including Lucius Clay. When the Lockheed Airplane factory was being planned for Marietta, you expedited it with your firm way to getting things done quickly by cutting through the red tape. Your service to your country brought many official commendations from the government. After the war, you served admirably in converting plants to peacetime activities.

While involved in the national government in 1938, you became Chairman of the Board of the 6th Federal Reserve Bank. As you applied all of your experience in engineering to a department store, so you applied that experience to the bank. Never before had a governmental organization been so completely reorganized: changes in staff, flow of work, product handling, and exchange of information among the member banks modeled on industrial and merchandising operations. Research and statistics as a service to banks and businesses in the district also came out of your planning. You, Frank, made all executives come out of their cloisters and begin to serve the community.

You made it clear (and you carefully followed the precept yourself) that nothing of a civic nature must be planned or supported which had the narrow interest of benefitting only your own business. All of your energy and your executives' energy was directed toward making Atlanta and Georgia a better, a more prosperous city and state. The first question that had to be answered was: "Is it good for Atlanta?"

Ralph McGill writes:

In 1950, as chairman of the Board of the State Commerce Department, he presented to Governor Herman Talmadge the committee report favoring enactment of a "no exemptions" sales tax. Such a tax, anathema to all retailers increasing bookkeeping and employee costs, was necessary to the progress of Georgia, especially the schools.

In 1926, Atlanta was a city in three segments. The railroad yards made gulfs through it. In 1926, as president of the Atlanta Improvement Association, he promoted the building of wide viaducts over the yards as nine main lines came into Atlanta. He was responsible for the organization and establishments of the first Fulton County zoning and planning commission. Later, he had a hand in the merging of county and city planning and zoning activities.

In 1926 he was chairman of the board which spent $10,000,000 improving the city streets, schools, and making growth possible. He was Vice Chairman of the Bond Commission which pushed through the popular expressway system and its necessary limited-access highways which have put Atlanta ahead of her sister cities in this field. Indeed, it is fair to say, that while many men had a vital and valuable part in this, Neely was the first to begin prodding the city officials and the public about the necessity for such a system in a city expecting a population of one million by 1970.

In 1937, during the depression, when city and county were locked in a desperate political debate over welfare programs and costs, he agreed to take the chairmanship of the County Welfare Board. He remained until the problem was worked out and operations proceeded smoothly. He advocated, and caused to be created, a Georgia State Department of Commerce and was its first chairman.

He had a great ability to get things done by indirection. Indeed, in times of civic crisis, his friends were known to fear the ring of their phones or to see Neely approaching for fear they might find themselves on a committee, serving on a board, or doing research on some aspect of a problem. At least two mayors had been needled so much that they have, at times, complained of difficulty sitting down. At times, the city hall and the county commissioners have waved arms and shouted with Frank Neely in red-faced exchanges. But always they cooled off with the realization that they both wanted the same thing; the only difference being that Neely's schedule was always too fast for them. It was his method of the need to get off dead center.

These are but the skeleton outlines of some of the major civic jobs in which he was one of the prime movers for better Atlanta.

—From Ralph McGill, editor of the *Atlanta Constitution* published June 3rd, 1956.

So here is the only one of the Neely tribe who is in *Who's Who of America*. What we would like to know is where his inner drive came from. We wonder again whether it was genes or environment. He certainly had good genes from both his father and mother. Ben, his father, had to have had inner drive and ambition to found and develop the whole public school system in Rome, Georgia, to have learned five foreign languages, and to have been independent enough at thirteen to run away from home and go to sea for two years. Frank's mother is somewhat of an enigma. We know that she was from Augusta where she experienced the Civil War as a young woman and then the long and trying time of reconstruction. She married Ben right after the war and had her first child, Annie, in 1869. There followed eight more children with a total of nine; Frank was the youngest. We can only assume that Henrietta was a great lady and mother to have raised that large and amazing family.

* * *

Let us then look down the generations from Frank and see if we can trace the genes of ambition. His only child was Rachel Eve, born in 1910. She must have shown how bright she was at a very early age. Her mother, Rae, was one of the first women of the south to go north to a famous woman's college, Smith in Massachusetts, around 1900. She undoubtedly was very proud of her own education and wanted the best for her daughter. She investigated the public schools in Atlanta and found the quality poor to her way of thinking. There was one private school, but she did not think it was good enough. While at Smith, she had learned about a method of home-teaching for one's child for grammar school. So when Rachel was six years old, she began this method at home. There was a law that all children should go to grammar school, but Frank fixed that. She continued to teach her through grammar school, but when it was time to go to high school, the public schools had greatly improved. Girls High, one of the few high schools for girls in Atlanta, had an excellent principal who was well-known by the Neelys; Rachel took some tests and was accepted. She did extremely well at high school, and then, like her mother, went to Smith College in North Hampton, Massachusetts.

Rachel matured into a beautiful, dark-haired, and somewhat dark complected young woman. She was still the apple of her father's eye. She graduated from Smith College with honors. She received a good liberal education with a major in art and literature. She had learned French in college and now elected to go to France for a year for further study. She went to the Sorbonne in Paris, learned fluent French, and studied art at the Louvre.

When she came back to Atlanta, the question now was what she should do. She went to work for Rich's in the book department and stayed there for several years. She was very attractive to the opposite sex, but it was difficult for her to find a man bright enough for herself. She finally married a young lawyer from Washington, D.C. whom her father had met while there. He was also a rabbi. Rachel had been brought up in the Jewish faith. The Jewish religion has a rule in mixed marriages; if a mother is Jewish, all of the children must be raised Jewish, but if a Jewish man marries a gentile, this is not necessary. So Rae saw to it that her daughter came up in the Jewish faith. Frank had become a Jew when he married Rae but did not practice Judaism and did not attend synagogue. In the early years, Frank and Rae encountered anti-semitism, and there were still some remnants of it which persisted. Frank could not get into the Piedmont Driving Club or the Capital City Club, the two best private clubs in town. He was later said to have been turned away from the Cloister Hotel on Sea Island. In any event, with his wide contacts and friends among the gentiles, they did not feel any real discrimnation as time went on. So when Rachel married a Jew it simply meant that her religion would now be a little stronger and her children would definitely be brought up as Jews.

The marriage of Rachel and Ben Parker was a big event. It was performed in the synagogue on Peachtree and the reception was at the farm. Frank had purchased his farm in about 1930—two hundred and sixty acres of beautiful land on the Chattahoochee River near Norcross. It was out in the country then, but Atlanta has now grown up around it, and it is an anachronism with a beautiful farm in the midst of developments. Frank had one of Atlanta's best architects, Mr. Henry Toombs, come out and see the farm and plan a beautiful house on a site with large trees and a panoramic view. It was finished in the mid-thirties. There were other houses on the land—one an attractive old farm house near the entrance and several smaller houses. After the marriage, Rachel and Ben restored and enlarged the old farm house, and this would be Rachel's home for most of her life. Ben went to work as the lawyer for Rich's.

Their first child, Daniel, was born in 1941; then came Eve and the youngest child, Nathan. The children all went to public school and did well. Daniel went to Hamilton College and later to law school in Chicago where he met his future wife, Susan. Eve went to Smith College and was thus the third generation to go to this great school. She got a job in California where she met Howard Hoffman, her future husband. Nathan went to college and then married a local girl, Rochelle Galanti. They moved to New York City where he enrolled at Columbia and began working on his Ph.D in education. In the meantime, he got a job in a school in Nyack while he still works on his dissertation for a Ph.D. Each of the three children has three children, so Frank, if alive, would have nine great grandchildren, each with at least one-eighth of his genes.

Rachel remained a lover of the arts, literature, and music, and she and Ben lived on the farm most of the time. As the children grew, they went to the public school like their mother and then went off to college. Somewhere during this time, when they were young adults, Rachel and Ben got a divorce. Frank apparently did not like Ben, but this was not a factor in the divorce. The children remained with their mother. After the divorce and when the children were off at college, Rachel went back to school in Athens at the University of Georgia. She obtained her M.A. in the school of social work; after she graduated she got a job with the Family Counseling Service, one of the best of the charitable services supported by the United Way. She continued to work at this new career for a number of years. The job consisted of counseling individuals and families in trouble, and she was very good at it.

Some time in the 1970's she began to notice pain in her right hip. She was somewhat of a stoic but finally went to an orthopedist. She had degenerative arthritis of the hip. Medicines did not seem to help much, and she lived with the problem for several years; she had to use a cane for some time. She still was very opposed to surgery, but in about 1975, when she was sixty-five years old, she finally agreed to have it done. She decided to go to a small hospital which was close to the farm. The plan was to insert a steel hip in place of the arthritic bone. She was operated upon but had some type of episode on the operating table, probably an allergic reaction to the glue used to keep the new hip in place. She had a severe drop in blood pressure, remained unconscious for several days, and when she did awaken a little, she was found to have a paralysis of the left arm and leg. She now had evidence of cerebral vascular accident or stroke, and she would never regain use of her arm or leg. She hovered between life and death for several days, finally improved, and was transferred to the Georgia Baptist Stroke Unit for rehabilitation. She stayed there several weeks and improved a little but was unable to move the arm or leg. She was finally allowed to go home but had to have three nurses on eight hour shifts around the clock and could never care for herself. She has confined to a wheel chair and would later have a van with a lift for the wheel chair. She would take trips to Florida in the van and would go to New York to visit Nathan by plane along with her nurse. Her mind was clear, but she tired easily, and her span of attention was diminished. She also would go to concerts and parties in her van. She finally moved to the Jewish Home, and when asked how she liked it, she looked around and whispered: "Too many Jews."

Okay Frank, you passed along some fine genes to your daughter, and she did get quite a strong gene of ambition, but a series of circumstances prevented her from following in your footsteps. But she did show drive and determination when she began another career after the divorce. The tragic stroke changed her whole life. The children—Daniel, Eve, Nathan—are all bright and successful at their occupations, but none tried to emulate you, Frank. It will be very interesting to see what happens in the fourth generation; already one of those nine children, Joshua, is showing definite signs of brightness and drive. He, of course, has genes from many other forebears, but it will be interesting to see how he and the other eight develop.

43
Joie de Vivre: Edgar, Jr.

Letter written during summer of 1931. Edgar's Age—21.

Levering, my boy, never pleasure travel alone—to a gregarious beast it's maddening. To a real human who wants to share his thoughts, it's embittering. For the type of person you can pick up is usually a stupid un-understanding bastard who gets in your way.

Levering, my boy, never travel with letters of introduction to people you yourself are not vitally interested in— I shall sink soon into the cesspool of middle aged bourgeois pleasantness—Tunbridge Wells. And then I shall buy myself a coat. And then I shall introduce myself to Mr. Allen, and then I shall see the law-courts, all with strangers, older people whose feeling of obligation will be almost as spoiling as their feeling of condescension. After that, I shall offend everyone by not seeing England (which I dislike) and jumping to Germany because there is a *swell* girl there.

Levering, my boy, never travel at all when you come abroad for the purpose of seeing the show places unless each one that you see has had before some meaning for you. Don't go near the Louvre unless you're really interested in art—Don't go near London Tower unless its history is a part of your emotions from childhood history class. A Paris sky at night is worth seeing—if you have a girl to see it with.

When one is young, Levering, there are only two things important. And those two things are one. The first is to flex the muscles and look at the sky. And the second is to have a woman. And all the talk of education and finer things of life is unmitigated horse-shit promulgated by a sect which is almost as powerful today as the churchmen during the middle ages.

One final word of wisdom—
The only value of a new setting is to give vitality to an old love.

Edgar

Edgar A Neely, Jr.

Hospital Record

Age 73, weight 176 pounds, height 5′ 11″; occupation: trial lawyer, usually for the defense, representing the Southern Railroad, Travelers Insurance Company, and several others. Hobbies: tennis, disco dancing, writing, exercising. Problem causing hospitalization: right inguinal hernia probably brought on by disco dancing, aggravated by coughing, sneezing, and laughing. Admitted to Piedmont Hospital, Suite 242, November 1, 1983. Surgery performed November 2nd. Present illness: noticed pain in right groin about October 15, 1983. He went to Piedmont Hospital Emergency Room October 18, at night; examined somewhat superficially by a female emergency room physician, who informed him that he did not have a hernia. Continued to have pain in the groin, and noticed a definite bulge in the right groin. Examined by his brother, an internist who diagnosed a right inguinal hernia. He continued experiencing almost constant pain when on his feet, which seemed only to be relieved by lying down. Saw Dr. John Akin, surgeon, who recommended surgery and arranged for admission to hospital for November 1, 1983.

Past history: General health very good. No recent serious illness. Appendectomy many years ago. Slight weakness of right arm due to mild birth injury. Left-handed. Plebitis of left arm in 1960, following slight injury. Recurrent back pain, hospitalized in 1980 because doctors suspected intervertebral disc disease. Myelogram advised but declined. Following steroid injection into back by orthopedist, symptoms subsided; left hospital, with no serious back or leg pain since. No history of heart of lung problem. Once a heavy smoker, but has not smoked for some thirty years.

System Review: negative.

Family history: Father died of cirrhosis at age 73; mother died of natural causes at age 83. One brother living and well; one brother deceased of a ruptured aneurysm of abdominal aorta.

Social history: Married for second time. Wife living and well. First wife divorced, deceased. Five children living and well, ages forty-four to twenty-nine. Three boys and two girls.

Physical Examination

Vital signs: Temp. normal. Pulse 76 and regular. Blood pressure: 170/90. Well-developed and nourished seventy-three year old white male with white hair. Very active. Looks younger than stated age in spite of white hair. Nose slightly stuffy. Throat clear. Lung examination, normal. Heart regular; sounds of good quality. No murmurs. Abdomen, soft. Right inguinal hernia easily reducible. Extremities—no edema. Slight weakness and decreased coordination of right hand. Laboratory data: Complete blood count, normal. Chemical profile—including blood sugar, cholesterol, liver function, all normal. X-ray of chest; no heart enlargement. Lungs clear. Cardiogram, normal.

Impression: 1. Right inguinal hernia.
2. Mild hypertension.
3. Mild incoordination of right hand due to birth injury.
4. History of phlebitis, left arm.
5. History of appendectomy.

Operation: November 2, 1983. Right herniorrhaphy under general anesthesia.

Course in the hospital: No complications. Up walking in hall on second day. Discharged on third hospital day to return to Dr. Akin's office on November 9, 1983 for removal of stitches.

The hospital summary fails to mention that Edgar Neely Jr. constantly bantered with the residents, nurses, laboratory technicians, and x-ray technicians, and they all agreed that here was a man who plainly enjoyed life, one who relished living life to its fullest. But these people who came in casual contact with him did not realize the active life he was leading. His primary occupation was work, with most other activities of secondary concern. When healthy, his daily schedule consisted of getting up early, at least by 7 a.m., taking a half-hour of exercises, then a bath, breakfast, followed by a drive to work. Arriving at the office, he began the day's work with dictation, reading briefs, and preparing for the trial cases that he'd soon be facing. The work typically related to the next trial. He meticulously prepared for each trial: this surely was the secret to his success as one of the best trial lawyers in the city. He learned every detail, every fact for each case. He thus experienced very few surprises while arguing a case. He was now close to the zenith within the courtroom. With fifty years of expereince, he was no longer nervous but radiated an enthusiasm and excitement over new trial cases. He mastered the art of questioning witnesses without going too far and being called out of order by the judge. An actor by nature, he knew how to use theatrics effectively in front of the jury. He commanded complete facility with the English language, both written and spoken, and his diction seemed flawless, his argument persuasive.

He toiled diligently throughout the day, stopping only briefly for a lunch he often had delivered to the office, and worked on into the night—either at the office or at home. His brother, Levering, a physician, said that he was glad he was not a lawyer as lawyers worked days, nights, Saturdays and Sundays.

Dr. Neely, making the rounds in the hospital one day, visited one of his patients ill with diverticulitis. The patient's daughter stood by. She was a lawyer in Augusta, Georgia. She asked Dr. Neely if he was any relation to Edgar Neely, Jr.

"Yes, just my brother."

She said, "Well, he is the talk of Augusta."

"Why?" asked Levering.

She went on: "Mr. Neely tried a case in Augusta, where he represented a ten-year-old girl who suffered injuries in an automobile accident. She had a broken leg, and one badly injured eye had to be removed. During the summation, Mr. Neely became very emotional and began to cry. Moved, the jury began crying. The opposing lawyer called for a recess and settled the case with Mr. Neely right away. He knew whom the jury would favor."

Edgar has a large office on the top floor of the Summit Building in the northern section of downtown Atlanta. The view from his office is striking. Both Stone Mountain and Kennesaw Mountain may be seen in the distance. The expressways can be followed from the base of the building for miles to the north and the south. Part of downtown Atlanta is visible. But the most amazing sight is the blanket of greenery from the large number of trees extending from the outskirts right on up to the city. His office is decorated in a modern style with a comfortable sofa, a huge desk, and a great deal of dictating equipment. He enjoys having at least two dictating machines within easy reach and sometimes has three. Dictating with him is a way of life. He may have a mass of papers before him, but once he dictates, the words flow. He stops and listens from time to time. If in the middle of dictation he suddenly thinks of something else, he stops, picks up another dictating machine and says, "Memo, remind me to write Mrs. Machery about her will." Then he will proceed with the original dictation. Although there are often phone calls in the middle of his dictating, he is always able to switch his mind back, after the call, to his oral letter. He expresses himself with grace and clarity, despite talking rapidly. Rarely does he have to change a word or sentence much less a paragraph. He had, in addition, three secretaries, each working forty hours a week.

His law firm consists of twenty-five lawyers, including his son Ned, who is forty-three years old. Edgar, the senior partner, is "Big Daddy" to the other members of the firm. He is being encouraged to slow down and draw some retirement pay, but one doubts that he will do this soon. He may change directions slightly and do a little writing. He's already penned one short book, *How to Gain and Maintain Vibrant Drive From Twenty-five to Ninety-five*. He is in the process of having it edited by a former editor for the *Saturday Evening Post* and will have it published soon.

He is plainly obsessed with staying active and young, and this obsession forms the thematic nerve center of his book. A great believer in working out, he does his own exercises every morning—including pulling in his abdominal muscles about 500 times as well as exercising his arms and legs with push-ups and the like. A tennis player throughout his life, he still wields formidable left-handed chops and cuts. He also enjoys dancing and does it every chance he gets. He really did win the disco award at the Atlanta Bar Association meeting. His younger partner had a hard time keeping up with the old guy.

He always diets, and when he makes up his mind, he reduces with a passion. He often takes exercises at the Piedmont Driving Club, often betting with some unsuspecting member that he will lose a certain amount of weight within a certain period of time. His trick is to make a small bet the first time and purposely lose, and then make a much larger bet; this time, he sticks to the diet and wins.

There are five children, all bright and active in this world. Edgar's IQ was once measured at 140, just below genius level. As a pure guess, the children's IQs would measure from about 125 to 140. All were intellectually gifted, but none was brighter than Edgar. Ned went to law school at the University of Virginia and is a successful lawyer in his father's firm. He became a nationally ranked tennis player at age fifteen. He held the number one spot in the country for his age group at fifteen years. He played several times at Forest Hills in the U.S. open tennis tournament and played at Wimbledon too.

Michael, the second son, is the most charming. He has a sly wit and a charismatic smile. After graduating from Yale, he studied poltergeists at Oxford. He returned to the U.S., became a helicopter pilot for a large farm in Florida, later made movies, and sold Christmas trees. He finally settled down, went to law school at Virginia, and is now married. Living in Seattle, he fathered three children, practices law, but his main occupation centers around entrepreneural interests: buying and selling buildings and docks. He is part-owner of a firm that makes door handles for airplanes. He has a beautiful home overlooking Lake Washington with a striking view of Mount Rainier. His wife is bright as are the three attractive and most unusual children. Michael also was a nationally-ranked tennis player, although never of the stature of Ned. With a tennis court on the house grounds, all of the children play tennis.

Little Claire, often called Sister, went to Bryn Mawr in Philadelphia and now teaches in a special school in Boston. She is now thirty-eight.

Alan, thirty-seven, went to Choate Preparatory School, then to the University of North Carolina where his father went. He presently is in an executive placement firm and is quite successful.

The youngest child, Constance Levering Neely, went to the University of Tennessee, and later attended agricultural school, and married a classmate. They now farm, although she has a job with the Department of Agriculture of Alabama.

Genes of equal quality were passed into these children by their mother, Claire Graham Neely, now deceased of a brain tumor. Bright, oversensitive, hardworking, and a great mother, this was Claire. She and Edgar divorced a few years before her death.

Edgar, we see, had strong genes of ambition. But that drive or ambition drove him to be a lawyer. He was fulfilled and was regarded as one of the best trial lawyers in the city, just as his father was. He was so busy with law that he had no time for other things and had no desire to get into politics or community affairs like his Uncle Frank. However, he clearly passes some strong genes of ambition to his offspring.

Okay, Edgar, lets look back in time to see if we can trace your ambition back to any specific occurrences in your life. Born in a hack on the way to the hospital, you were delivered by the black nurse from your mother Emily. After about twelve months, they noticed that the coordination of your right hand wasn't quite normal. You were taken to several specialists; the final diagnosis was that you had a slight birth injury.

You have attributed many of your drives and directions in life as being influenced by this injury. You couldn't enter into all the athletic games as a young man, so you compensated for the injury in several ways. Since you could not play football, basketball, or baseball very well, you took up tennis and became quite good. Also, since you couldn't play the major sports, you had more time to devote to studies and you made good grades. You were widely read by high school.

You went to 10th Street Grammar School, then on to O'Keefe Junior High, and then to Peacock School for Boys. You moved on to Episcopal High School in Virginia (your mother's idea) where you did well. Then it was on to the University of North Carolina. College was, as Hemingway would put it, a moveable feast for you. And the story about earning three A's one quarter and three F's the next was true. However, the story now comes out that during the quarter you were secretly in New York spending your college money and having a great time. In any event, you were, as we came to expect, a little different.

We knew you were a little unusual as a small child, with that thick, brown, long curly hair. You were great at tantrums. I personally remember the one you had when I got my first bicycle at age five. You had not received yours until age six. You screamed and cried for over an hour. Then you suddenly wiped away the tears and went off to play with some of that Ansley Park gang. You made very good grades, when you wanted to, in grammar school. But the mood swings were there and if you didn't want to do something mother suggested, you would do anything to get your way. Only the threat of a whipping by your father would allay your anger.

In high school you began living life to its fullest. Your Bar Mitzvah was on your sixteenth birthday, when you obtained your driver's license. You didn't get your own car, so you wanted to drive the family car at every opportunity. At the Peacock School of Boys, which was for flunkies from the public schools and unusually bright students (you fell into the latter category, of course), you were encouraged to develop your unusual traits. You still had this drive to excel at school. You contend it was due to the fact that you could not really excel in sports because of the right arm, but is that the case? Everyone seemingly worked together to make you a little different from most of your peers. Your genes, your family, and now your school were fostering your individuality.

You had the sixteen-year-old syndrome of adolescent rebellion. You rebelled against the family, and especially your mother, who was trying to keep some control over this wild buck straining at the bit. You sneaked out with the family car several times at night, and when caught, you had the keys taken away for a month or more. We don't know just when it was that you discovered girls, but when you did, it was like a whole new world. We are not sure if you lost your virginity in high school or college, but we do know that in college, the pursuit of the opposite sex was your number one avocation.

And it was in college that your real rebellion came along. The University of North Carolina had some very radical professors in those days and you espoused socialism and even communism for a short time. But you swung away from it before you left college. You used to love to bring your ideas home and shock your parents. It was in college that your flare for dramatics was encouraged as you became very active in the Carolina Play Makers which was the best college theatrical group in the South. You would be a great actor for the rest of your life; your stage was the courtroom, and the jury was your audience. You took public speaking and debating in college and excelled. You always had great respect for your father and his courtroom ability and the role of the lawyer seemed to suit you to a tee.

So, after graduation from Chapel Hill (University of North Carolina), you enrolled in Emory Law School in Atlanta where you graduated in 1934—in the depths of the depression. You went to work for your dad's firm for $100 a month. Starting at the bottom of the ladder, you did routine tasks for the other lawyers. It was a disappointment at first after that freedom in college, but, after all, you were just getting started in the adventure of life. Your progress as a lawyer, so slow at first, soon developed momentum. And your rise over the last fifty years has been nothing short of spectacular.

So what star did you ride to ascend to your zenith? It really was many stars that carried you. You were a mental astronaut, leaping from star to star, sometimes falling and nearly crashing. But no: somehow you held tight, finding new rocketry to propel you onward and upward. So here's to you Fuzzy Wuzzy. You're a diamond, you're a daisy, you're a ducky. You're an India rubber *idiot* on a spree.

You are Edgar.

44
Ben's Second Voyage

Fall 1848. Ben had been gone from home almost one year. Now he embarked on another long journey, one taking him to several other foreign lands. After Charleston and Savannah, the good ship *Nerrisa* would sail south along the Florida coast, the peninsula having been recently purchased from Spain by Mr. John Forsyth, a U.S. agent, for $5 million. Some fancy houses and estates in Palm Beach now sell for that sum; but in 1848, Miami was not even established. The ship proceeded to Puerto Rico, the Virgin Islands, then south to Barbados, on to the mouth of the Amazon in Belem. After that, it crossed the south Atlantic, arriving in Dakar, the Senegal, in Africa, and then, to the north, Portugal and Spain. In Brest, France, the ship and crew took a much-needed breather. Here they refurbished the boat and then sailed on to Liverpool via Dublin, across the Irish Sea.

Would the first journey have any lasting effect on Ben's character and future life? At age thirteen (really almost fourteen), he was thrown into the hard life of the sea. Fortunately, he was on a clipper ship that had a very good captain and first mate. But in addition, he was among a group of tough, though friendly sailors. Ben had seen and experienced a wide variety of life's problems at a very early age. He had seen drunkenness, violence, and death. He had visited a number of countries with entirely different cultures and languages. Had all these things matured him that first year away from home? It had to have had some effect on him, but it would be difficult to measure it at that time. And here he was embarking on another journey to foreign lands.

He was now much more relaxed with the crew. He was somewhat of a novelty for them on the first trip. Now he was accepted as part of the group. He had grown fast that first year, and he had become more muscular. He was still the cabin boy, but the captain and first mate were giving him more small jobs on the bridge. He could write well and the captain kept him busy making copies of invoices and doing some secretarial work. He still had full use of the captain's library and was an avid reader. The captain had an Old World Encyclopedia, and Ben was determined to read the whole thing. So he was growing mentally as well as physically. He came to realize the value of education and how important it might be for his future life. He soon found that the sailors working on the boat did not have much hope of doing anything more than being sailors for the rest of their lives. He was thus getting an opportunity to mature early. Although he did not recognize it at the time, the experiences were giving him a drive or ambition to do something in this world. Also, he would be able to cope with the problems of a teenager much better and push on to fulfill his aspirations with ease and self-confidence.

* * *

So in early April, 1848 on a clear but windy day, the *Nerrisa* weighed anchor, dropped its ropes, and took off once again from Baltimore, this time with a plan for another long journey. So here was the good ship *Nerrisa* sailing off again almost a year after the start of the previous trip. The crew had changed some, but the camaraderie was still there. Ben and the cook still got along well together. The cook was conscientious and Ben made every effort to help. The daily preparation of food and the dishwashing for eighteen people was no easy job, and it kept them both quite busy. The cook had by now full confidence in Ben. The crew had come to accept Ben by now, and Ben felt quite comfortable with most of them. The crew was still a very international group. The cook was Norwegian, and there were two more Norwegians on board. There were two French-Canadians and one Frenchman from Brest. One Brazilian was aboard who spoke Portuguese and Spanish and three Italians from Baltimore. Many immigrants had settled in Baltimore, and when they could not find work easily, they went to sea. There were also three Pennsylvania Dutch from Philadelphia who spoke German.

Thus at the mess table, there were often three or four different languages being spoken at the same time. Often at sea when there was not much to be done, the sailors sat around the table, drank coffee, smoked, and talked most of the afternoon. Ben had always been fascinated with languages. He had studied French in school for one year, and when the Frenchmen talked, he could understand a little, but it was spoken so fast that it was a jumble of words at first. However, his ears had a fine tuning for language, and he sometimes listened to the French sailors talk for a long time. The captain had a French-English dictionary, and Ben began carrying it around in his pocket. He soon was able to talk some to the Frenchmen in their own language. For anyone to make an effort to speak French charmed them, and they would often give him informal lessons. It wasn't long before he could talk with the Frenchmen quite well.

He next began listening to the two Italian sailors talking in Italian with each other, and soon he was beginning to understand a little of what they said. He also would ask the cook to talk Norwegian to him and name things around the galley in Norwegian. It wasn't long before he had a small vocabulary in Norwegian. He also listened to the Brazilians speaking Portuguese and Spanish as the sailors could speak both. Soon everyone was competing with the other to get Ben to speak their language. This would continue for the entire trip which would last for months. Soon Ben was able to say simple sentences in French, Norwegian, Italian, Portuguese, and Spanish. In addition, Ben enjoyed mimicking the way the broken English was spoken by the different persons on board. As time went on, he could communicate more and more with the various nationalities in their own language. At first he had no idea of how the words that were spoken were written, but in Charleston he went into a bookstore and bought small dictionaries in Spanish, Portuguese, Italian, French, and German.

It is a little difficult to believe that a fourteen-year-old boy should take on such a huge project. Would one call this ambition? I don't think so. It was more of a combination of a spark of interest that was lit up to a fire by a gradual accelerating desire to learn languages. Then there was the very unusual opportunity to practice these languages with people of the various nationalities on the ship. So there was some unusual drive within him which was different. It was not done as an urge to do more in this world; rather it amused him and delighted him to hear a foreign language and almost understand it. He had had a good start in languages in the first year of his voyage, but now he began working in earnest in trying to master several foreign languages.

Their first port of call was Charleston, South Carolina. This was already a thriving cosmopolitan city with considerable history, and it would make more history later. It had been established by the British in 1670 and became the main seaport of the South very soon. During the American Revolution, it was defended and fell to the British who held it for two years; it was then retaken by the revolutionary forces and was incorporated as Charleston in 1783. Later Charleston would be the site of the first shots fired in the Civil War when the Confederates attacked Fort Sumter in Charleston's bay. It was an educational center with several colleges and a medical school by the time Ben arrived there about April 15, 1848. Little did Ben know that he would be coming back here to college in about six years.

The *Nerrisa* came into the harbor and docked along the Battery as the waterfront is called. They would be here only a short time as they had a small shipment of textile machinery to deliver. Ben did get a chance to go into town with the cook to buy a few more supplies in the marketplace, and he had a chance to wander around in this beautiful old city, now almost two hundred years old. It was also the largest entry point for the slaves who were brought over from Africa, especially from the largest exit port for slaves—Dakar in the Senegal. Ben had a chance to see a slave auction with men, women, and sometimes children being put up on the auction block and bid on by the plantation owners who came here from all over the South. It made a deep impression on him to see a human being sold for a price.

Next day the ship sailed for Savannah, Georgia, another historic site; another city that Ben would later visit. It had been founded by James Oglethorpe in 1733; some have argued that the settlers were a large group from debtor's prison in England, but recent historians say that although there were a few debtors from prison, the great majority simply wanted to try out the new world. This city, although not quite as big as Charleston, was quite similar being a big port. Savannah was held by the British for almost four years during the Revolutionary War. In the Civil War it would be taken and partially destroyed by Sherman in his march to the sea. But in 1848, when the *Nerrisa* arrived, it was a flourishing seaport which brought in slaves for auction. The first steamship to cross the Atlantic was named the *Savannah* and sailed from there to Liverpool in 1819. Ben had a chance to make a quick trip into the city before their departure to San Juan, Puerto Rico.

Here was a large island under Spanish rule with a city named by the Spanish explorer, Juan Ponce de Leon, in 1508. Sir Frances Drake tried to take the city from the Spanish but failed. It remained under Spanish rule until 1898 when U.S. General Nelson Miles landed and took over. It was ceded outright to the U.S. shortly thereafter. In 1848 it was another busy port. The *Nerrisa* delivered wheat here and picked up sugar and rum as these were the main exports even then. The ship stayed there two days; again Ben had a chance to go into the city. He was immediately struck by the crowded conditions, the apparent poverty, and the huge Catholic churches. It was fascinating to wander around in the city and Ben had a chance to try his Spanish a little.

From Puerto Rico they made a quick trip to Charlotte Amalie on Saint Thomas, the one Virgin Island that belonged to Denmark at the time. Here was a beautiful harbor with transparent aquamarine water. Here they picked up another cargo of rum and bay rum for England. Then on down through the Caribbean into the Lesser Antilles and the Windward Islands, through St. Vincent's Passage to Barbados Island with the Port of Bridgetown. The trip was great with warm weather, almost tropical, and a gentle breeze, although in September hurricanes begin in this area. Barbados was under British rule and had been for many years. It was much more orderly than San Juan, and poverty was not evident. Another load of sugar was added to the cargo.

Then for a long journey of 1200 miles from Barbados to Belem, Brazil, above the eastern tip of South America and just above the equator. It took about twenty days to make the trip. Even with all the cleaning and repairing that Mr. Rumans thought up, there was still a lot of free time for the crew. Ben stayed busy feeding the crew and cleaning up after them, but he also had a good bit of free time. He continued in his avocation of languages and had ample time to study and practice talking. Belem, meaning Bethlehem in Portuguese, was the port of the whole Amazon River basin. Though not on the Amazon River itself, it is on the Amazon Peninsula area, and any products from up the Amazon would come here. Coffee was being grown in Brazil. Around the middle of the nineteenth-century coffee was becoming very popular in France, Spain, the Netherlands, and Germany. Tea would remain the the main drink in England, although coffee for breakfast was becoming more popular there also. So the *Nerrisa* planned on taking on a big shipment of coffee beans to take to Europe.

Brazil was under Portuguese rule in 1848 and, although their rule has been contested from time to time, it remained a Portuguese colony until it got its independence in the twentieth century. Belem was a huge port in those days as it was a starting point for explorers and adventurers going up the Amazon—that mammoth river whose headwaters begin in that rim of high mountains along the west coast of South America. So here was another world and another language to which Ben was exposed and again he was amazed.

Time raced by. It had taken about a month to get to Puerto Rico, and the Virgin Islands and another month to arrive in Belem, Brazil. So it was now June 15th, and the good ship was being prepared to cross the Atlantic to Dakar in the Senegal. The eastern point or tip of South America is considerably south of the rounded bulge on the west of Africa; the *Nerrisa* would have to sail northeast and the journey would be about twenty-four hundred miles and could take from fifteen days to a month depending on the weather. They had been lucky on the weather in the Caribbean, but the Atlantic was considerably more unpredictable and changeable. In truth, they did go through a fairly bad storm, but the Baltimore clippers were sturdy ships as well as fast; they weathered the storm very well except for taking in a little water which had to be pumped out by hand. The storm did delay them, however, and it took almost a month to travel those miles.

The captain and Mr. Rumans were both experienced navigators and with the sextant and their experience they rarely were off course, and if they were buffeted around a bit, they could shoot the stars or the sun and get back on course. It was only when the sky was cloudy for several days that they had any problem, but usually the compass would keep them on the approximate course. The captain asked Mr. Rumans to show Ben how to shoot the sun and the stars with the sextant and Ben learned rapidly. Mr. Rumans would often test him to see if he had done it correctly.

It was July and the weather was hot and humid with thunder storms and squalls. They sailed along with the coast of Africa visible in the distance and then arrived at Dakar, the capital of the Senegal. This was under French rule. French was the official language. The population was black, but there had been schools established and the Africans could speak French as well as their native tongues. The main crop was peanuts, and that is what the *Nerrisa* came to pick up to take to Europe. Ben tried his French again. In 1848 Dakar was one of the largest slave exporting cities in Africa; they had huge compounds in which to keep the slaves. They were brought in from neighboring states by slave hunters. Ben witnessed them bringing slaves shackled together through the streets and putting them into the slave compounds to await shipment to the southern United States, which was the largest user of slaves at the time. The ship remained in Dakar several days before departing for Cadiz, Spain just east of Gibraltar. This was another two thousand mile trip and took twenty-five days. On arrival at Cadiz, it was August 15th.

Cadiz was one of the largest ports of Spain. It was especially important as a starting point for the new world. Here the ship took on a cargo of wine to go to England. They had several days here so Ben explored the town with the cook; they had to get food after the recent long journeys. They now had enough food to take them to England. However, there would be a stop at Lisbon, then on to Brest, France where there were shipyards for refurbishing and repairs.

The trip from Cadiz, Spain to Lisbon, Portugal took only six days. They hugged the shore line, had a good trade-wind breeze, and got into the huge beautiful harbor of Lisbon on the 25th of August, 1848. Here was a huge city and here was a fourteen-year-old boy, almost fifteen, arriving at the waterfront for a short stay. Lisbon's history goes back to Phoenician and Carthaginian times and was under Roman rule for a long time; there was still ruins of the old Roman rule for a long time; there are still ruins of the old Roman fortifications. The important date for Lisbon was 1147 A.D. when King Alfonso I, with the help of the crusaders, took the city back from the Moors. The *Nerrisa* left Lisbon for Brest, France and a refurbishing and overhaul at the shipyards which would take two weeks.

Ben's second journey from Baltimore was in many ways different from the first. In one year Ben had grown about four inches. Also, he had become more muscular and broadened his shoulders. He was now as large as many of the sailors. He was much more self-confident than on the first trip. He was now able to joke with the sailors on a more equal basis whereas on the first trip, he was the one who was being teased and kidded. He was well-liked by the other crew members as well as the captain and grudgingly by the usually cold first mate, Mr. Rumans.

But more importantly, he had now really seen the world, and all of these experiences would be clear to him for the rest of his life. Not only that, he would be able to profit by his experiences; he would be able to get along with people better, and he would develop that sense of humor which is so important for life in this world. So Ben, you were very fortunate in many ways to have pushed your stepmother down the stairs.

45

Family—Early Years

Can we cover forty years from 1943 to 1983 on twenty pages of a legal pad? Obviously not, but we can hit the high spots, and if it takes another twenty pages, so be it.

1943 in Atlanta, Georgia—summer time. U.S. participation in World War II had been going on for 1-1/2 years and the tide was beginning to change in favor of the allied forces both in Europe and in the Pacific. The Germans had surrendered in Tunisia; the allies had just captured Sicily and were about to invade Italy. German submarine attacks in the Atlantic were being controlled by the British and U.S. Navies. In the Pacific our navy had two great naval victories over major parts of the Japanese fleet, one in the battle of the Coral Sea and one in the battle of the Bismark Sea. Guadacanal was a costly but important victory for the U.S. After the shock and paralysis of Pearl Harbor, the forces of good were beginning to prevail.

War was not very obvious in Atlanta that summer. To be sure, coverage constantly appeared in the papers and the radio (remember, there was no television then). There was gas rationing, but the streets were crowded with cars. The rationing really only prevented one from taking a trip to Florida, but there was enough gas for ordinary activities such as going to work and to the grocery store. True, most of the young men had volunteered and gone off to the armed forces. The Bell Bomber Plant had been built in record time and was now beginning to turn out some of those huge planes that helped win the war. No unemployment. Wartime factories had taken up all available workers. In England Fleming had just discovered penicillin. The armed forces used it first; soon after, the public was able to recieve small quantities. Sulfa drugs were available for infections.

So Margot and Levering and little Freddy, age six weeks, were beginning life anew in this city of 250,000 (which would be two million in forty years). They had just left New York City where they had been for two years while Levering was in training at Bellevue Hospital and Margot had been working at the Interallied Information Center, a forerunner of the United Nations. Levering had been rejected by the army because of his minimal case of tuberculosis and had preceded Margot and Freddy to Atlanta by one month. He had begun the practice of internal medicine with Dr. H. Cliff Sauls in the Medical Arts Building.

The plan was to live with the Neely family until an apartment or house could be found as housing was scarce. The family home at 4 Barksdale Drive in Ansley Park was adequate and comfortable with three bedrooms, three baths, and a big enclosed sleeping porch. There were two black servants and Mancy Brown would come when free from his job at General Motors. Margot fitted into the household remarkably well and got along fine with mother-in-law, Emily.

Levering went off to work every morning but usually had breakfast with the family, served by the servants. He would not get home until about seven o'clock at night and again there would be a family supper. Margot spent most of her daytime hours looking after the infant, Freddy, or being driven around looking at houses or apartments. Levering was making only $250 a month, so there were some limitation on what they could find suitable to their finances.

Levering alternated working on weekends with Dr. Sauls. When on, it meant making rounds on about twenty patients at Piedmont Hospital on Capital Avenue, Sunday morning admitting new patients, and making house calls. In those days house calls were taken for granted, and there might be three or four house calls a day as well as an occasional one at night. Dr. Carter Smith, Sr. and Dr. Charles Stone had gone off to war as had many other doctors, so Levering was busy from the start. It was mostly caring for Dr. Sauls's patients at first, but he was slowly building up a practice on his own. He began working and teaching at Grady Hospital in the Emory Medical School very soon, as this was the exciting side of medicine. He worked in the clinic and made attending rounds with the interns, residents, and students. Dr. Eugene Stead was the Professor of Medicine at Emory at the time. A brilliant teacher and researcher, he had already made quite a name for himself, attracting a large number of young researchers and teachers to work with him. He believed in work, work, work, work, and he had his grand rounds on Sunday morning. This early association with Emory Medical School would continue for the rest of Levering's medical career; he enjoyed teaching medicine the rest of his life. Perhaps the teaching genes of his grandfather, Ben Neely, and Ben's father, had jumped a generation in Edgar and had been passed along to Levering.

Piedmont Hospital, although small, was probably the best private (but non-profit) hospital in Atlanta at the time. Many of the doctors had been trained at Johns Hopkins. The administration was very good and the nursing was excellent. There was a very good nursing school run by Miss Genivieve Garren—a martinet of a woman—who produced superb nurses. So Levering fitted into the family of Piedmont Hospital very well. He would remain a loyal member of the staff for the rest of his medical career. Piedmont would continue to be one of the best hospitals in Atlanta even after its move to Peachtree and Collier. It had a phenomenal growth from one hundred sixty beds on Capital Avenue to five hundred on Peachtree.

Finally, after five months of living with the family, they found a small house with two bedrooms, one bath, a porch, small front yard, and a fair sized backyard on Dellwood Drive. The price: $9,000. Miss Emily helped with the down payment of $3,000 and there was a mortgage of $6,000 at four percent interest. The monthly payments were $72. Levering's salary had been increased to $400 a month and so it was affordable. They moved in right after Christmas. Their furniture from New York had been stored and was now brought out. Fixing up the house was quite a job for Levering and Margot. She had to do the day-to-day care of the house, look after Freddy who was now six months old, and cook for Levering when he got home at night, which seemed to be later and later. The first real friends were neighbors. The house next door soon had the Dickeys and later the Olmsteads, both of whom would be lifelong friends, although the Dickeys would move away in about ten years. In another year the Isons would move in across the street, and although their interests were different, they would be very close friends for life.

Margot, having adjusted extremely well to a sometimes difficult arrangement of living with her in-laws, now began to blossom as a housewife. But her mind was too alert and too active to be satisfied with only the role of housewife. She began to look for a little more intellectual stimulation. She picked the League of Women Voters, an organization which was already strong in Atlanta. Some of the brightest young women, and older ones as well, were in this group in Atlanta. Margot had part time help so it was not difficult to get someone to care for Freddy while she would go to meetings at least once a week. She would often go to the League office and work during the day. This gave her a volunteer job which was interesting, and she could do as much as she wished. She would visit the Georgia legislature as well as the county commissioners and city council and then report to the League. This gave her a real insight into Atlanta and the state and how they were run. Civil rights for blacks had not yet begun in those days, but later the League would take an active part. The League was considered too liberal by some, but it had the support and respect of the business community, which gave it financial help. The League's report, distributed before all elections, was a great help to voters.

Margot had majored in American literature in college, and she continued to be a prolific reader on almost all subjects. Although she did not play the piano anymore (having studied it as a child), she knew classical music well and would play the classical records at home and would go to the concerts when they came to Atlanta. She also had taken art in college and kept up her interest in art at every opportunity. She later would paint. Atlanta had a good musical community; her mother-in-law, Emily, introduced her to some of the musical and art lovers of the city.

One family of friends that were seen fairly often were the Distins. Bob and Jean were from Saranac Lake, had grown up with Margot, and he had been brought down to help plan the Bell Bomber Plant. They both were very tall—he six feet four inches, and she a full six feet. They had small children when they came to Atlanta, and the two families got along well together. Bob was an unusual person—very bright technically but a little different. The story goes that when he was sent to Atlanta to talk to the city fathers about planning the bomber plant, they were amazed that Bell would send a man so young (only twenty-three). They called Bell and asked for a more mature person. The Bell people answered that this young man could do it. Later it was said that the only mistake he made in the overall plan was insufficient parking.

Bob neglected to buy new Georgia license tags for his car for several years and when caught was jailed; Margot and Jean had to go down and get him out. The two families became very close and would remain so the rest of their lives even though the Distins later moved to Columbia, South Carolina and much later to Washington, Connecticut. The Distins definitely were a part of the Neelys' lives. Not only were they great friends, there was a deep abiding love between the two families.

When Freddy was nine months old Margot became pregnant with Virginia who was born in December 1944. After that Margot was on an eighteen month schedule and had Gardner in June of 1946 and John in December of 1947. There was a hiatus of five years and the "little girls" would appear. Sara in 1952 and Betsy in 1954. But we are getting ahead of the story. Virginia McKenzie Neely was born at Piedmont Hospital on a cold and windy night, December 11, 1944. She was the child who never caused any problems. She cried very little, was not a whiner, accepted life as it came, and later would help with her younger siblings. When the third child, Leroy Gardner, came, Fred was only three and Virginia eighteen months. Leroy Gardner Neely born in June, 1946 was named for his grandfather; he was a summer baby and seemed to cry all summer. Some said it was summer colic and much advice was given from the neighbors, but the doctor pronounced him healthy and he gained weight and thrived. By fall he had stopped his crying.

But in the fall of 1946 tragedy struck Margot's father. Although there was no history of heart trouble, he developed severe chest pains one evening and apparently died immediately, at home in Saranac Lake. Autopsy confirmed an acute myocardial infarction as the cause of death. He was only fifty-seven years old and had been a tuberculosis and silicosis expert since 1919. He was now world famous and in *Who's Who*. Another life snuffed out at its zenith by a tiny one millimeter blood clot blocking the main coronary artery.

Margot had always been very close to her father, and his death was very traumatic. Since moving south in 1943 she had seen him only occasionally. In the summer of 1945 the two families had met in Jacksonville Beach for several days. Levering, Margot, Freddy, and Virginia had driven down; Leroy and Carabel had come by train. It was a pleasant reunion and vacation. Viola, the black maid, had come along to help with the children. It was their first introduction to the beach, and the Neely family would be coming back to Florida beaches for the rest of their lives.

People and physicians from all over the country attended the funeral of Dr. Gardner. Long obituaries in the *New York Herald Tribune* and the *New York Times* appeared. Slowly Margot recovered emotionally but it had been a big blow. Leroy Gardner Neely was only four months old at the time. Levering and Margot went to the funeral by train, which took almost twenty-four hours as there was little commercial air travel then and Margot did not like air travel in any event. Leroy was buried in the graveyard of a small but historical church close to the village of Saranac Lake. Carabel elected to stay in the house at least for the present time but would spend the winter in the South, coming to stay in Atlanta with the Neelys for Christmas and then going down to Pass-a-Grille, Florida, a pleasant little town on the Gulf. The Neely family would often come and visit her here in late winter or early spring.

Shortly after John was born, Levering developed a bad sore throat with a high fever for a few days. He apparently recovered but two weeks later awoke with severe abdominal pain. He realized that this was not a simple stomach ache, and Margot drove him across town to the hospital. He was seen by several physicians, including Drs. Carter Smith and John Akin; they felt that he had an acute abdomen and should be operated on right away. He was taken to the operating room, laparotomy was done by Dr. Akin, and the only finding was a little inflammation of the peritoneal surfaces. It looked as if he had had intussusception or telescoped bowel which had corrected itself just prior to surgery. His appendix was removed but was not inflamed. He apparently did well that night, but the next day was found to have a very rapid pulse. A cardiogram was taken and found to be very abnormal; he was thought to have either a myocardial infarction or myocarditis, called post-streptococcal myocarditis, as he had had the severe sore throat two weeks previously. Margot was told that he was in a very serious condition, might not recover, and to be prepared for anything. Here she was with four small children—the oldest four and one-half and the youngest two months—and this was only four months after her father had passed away. But she kept a great spirit throughout.

Levering remained in the hospital for one month and gradually improved. He then went to his parents' house as it was thought that Margot had enough on her hands with the four small children in that small house. He stayed at the Edgar Neelys' for several weeks and then came home, but he still was not able to return to work. After a couple of weeks there, Margot and Levering took off without the children for Florida, leaving a brave baby-sitter at home. They drove first to the Gulf and stayed at a small hotel in Indian Rocks, but they did not like it. So they went further south through St. Petersburg to Fort Myers and stayed on Captiva Island. They happened to find a place called the Tween Water Inn. In the front was a beautiful beach lined with Australian pines and in the back was the tide river with boats for fishing. It was an ideal spot at which to recover; they discovered the pleasure and serenity of the beach as well as the fun of fishing.

Now, after three months, Levering seemed to have completely recovered, and he dove back into the practice of medicine. After the war was over in 1945, Carter Smith and Charles Stone had returned from the army and had resumed practice with Dr. Sauls. The bookkeeping office was moved down the hall and the office was a little crowded but ample. In 1947 Dr. Sauls had had a major stroke and became completely disabled and had to stay at home. The practice was busy for everyone, and when Dr. Sauls was forced to retire, another young doctor was added to the group—Dr. Edward Russell, a conscientious young Johns Hopkins graduate. So the late forties and early fifties were busy times for this group of young internists with Carter Smith being the senior member now, and he was only about fifty. Charles F. Stone was about one and one-half years older than Levering who was forty in 1952. Both Drs. Smith and Stone trained at Harvard's Peter Bent Brigham Hospital. Group practice for busy internists was a very pleasant way to work because one could take off on a vacation and leave the practice with those remaining. One doctor of the four worked only one weekend a month.

Piedmont Hospital moved from Capital Avenue to Peachtree and Collier in 1956. The new hospital was modern in every respect; it still had excellent facilities and superior staff. It was still associated with Emory Medical School. Almost ninety percent of the physicians were on the volunteer teaching staff at Grady, and there were students, interns, and residents from Emory working at Piedmont.

The Neelys now were living at 1885 Anjaco Road, which was half a block from the hospital. The office in the Medical Arts Building had been moved to the modern Sheffield Building next to the new hospital. More young doctors were taken into the office: Carter Smith, Jr., trained at Cornell, joined the firm; James Brawner, a Hopkins graduate, also came into the group; and several others who stayed a short time. Then Brantley Burns joined. Carter Smith, Sr. and Jr., Charley Stone and Jim Brawner all did Cardiology as a subspecialty to Internal Medicine. Levering was a lung specialist and Brantley was a gastroenterologist. The largest part of the practice was straight internal medicine.

The house on Anjaco was large but modest. There were four bedrooms upstairs with two baths and a small sun porch downstairs in the front. A large playroom for the children was added in the back of the house as there were now six children by the mid fifties. Betsy, the youngest, was born in 1953. The children now were over a ten year span. Margot remained extremely busy with the children but continued to be active in community and cultural affairs. The League of Women Voters was still her number one project. Sometime in the fifties she took up tennis, and this became a very good outlet from caring for the children. She had a group of women friends with whom she played, and she would also play mixed doubles with Levering on the weekends. She continued to play tennis the rest of her life, and the children all played tennis, except Gardner.

A new lake had been made north of Atlanta and in the early fifties the Neelys bought a Lyman fishing boat with an outboard motor. Many a summer Sunday was spent at the lake with swimming, picnicking, and exploring. The only problem was the ride home with six irritable children; ghost stories told by Levering were a salvation.

In 1951 Virginia went to bed with a cold. She awoke the next morning with a severe headache and had trouble raising her right arm. Dr. Richard Blumberg, by now an old friend of the family, came over to the house and made the diagnosis of poliomyelitis. She was taken to Grady Hospital's infectious disease unit, had a spinal puncture, and the diagnosis was confirmed. She remained at Grady for about two weeks. Her fever came down, but she still couldn't raise her right arm. Warm Springs Rehabilitation Center, founded by Mr. Roosevelt, was advised for her. She went and spent six months there with weekly visits by the family. She improved, but the deltoid muscle of the right arm never came back. She never complained or cried throughout the illness and took things in her stride as she would for the rest of her life. Take life as it comes and make the most of it was her motto. She never let this handicap of a weak right arm brother her.

When Fred was ten, he was playing with a group of children in the neighborhood; they were throwing long stick spears at one another. One struck him in the eye. The family was out of town, but the baby sitter called the doctor. They rushed him to Emory, but the eye could not be saved and had to be removed. Freddy also took it in his stride. When Gardner was about eight, he had a high fever, sore joints, and was hospitalized and found to have rheumatic fever which required rest and a wheelchair for several months. He also took this infirmity well.

So the early years for Margot, Levering, and the six children had their ups and downs, but Margot and Levering felt that they were most fortunate as all of these illnesses could have been so much worse. Also, all of these problems had the effect of making the family ties stronger. You, Margot, lived through these vicissitudes on a day to day basis. Caring for six children could have shaken a weaker person, but it made you stronger.

46
Family—Middle Years

Elizabeth Eve Neely was born August 17, 1953, the last of six children. Her sister Sara Margaret Neely was born August 15, 1951. John Longstreet Neely was born in December, 1947, over three years before Sara came along. The first four had been on an 18 month schedule, so why the pause? Margot had a miscarriage or rather a hydatiform mole which is similar to a miscarriage, but is actually a tumor which can become malignant. So good old Dr. Upshaw, the obstetrician watched her carefully, and as there were no signs of malignancy, he gave her the go ahead signal, which she took and became pregnant with Sara.

It was about this time that the Levering Neelys began going to Fernandina Beach, Florida on a regular basis. They had kept in touch with the Frank Neelys, had visited them in Fernandina, and had been invited to spend one or two weeks a year there. This became a regular occurrence. Margot was pregnant with Sara on one of the early trips. Margot and Levering had been to the west coast of Florida several times, but the quiet peaceful waves of the Gulf were no match for the big towering rollers of the Atlantic. So it became a yearly "ausflug" and many happy times were had there as the children began to grow. Fernandina was a small fishing and factory town then with no big developers interested, mainly because of the smell from the paper mills. Later the pollution would be corrected, and the developers moved in on the lower half of the island with the huge condominium complex—the Amelia Plantation. So here was the yearly recreation spot that became such a big part of the lives of all of the Levering Neely children.

What about school for the children? They all started at public schools. It was E. Rivers for the four older ones and Morris Brandon for the two youngest—"the little girls." The public schools were all very good in those days. There wasn't any integration in the schools in the fifties. However, the public high schools had a rather poor reputation in the late 50's and 60's and when integration did come in, there was a white flight to private schools.

Westminster School was already organized and running in the 50's. It was a private school which began in the first grade and went on through high school. It was an outgrowth of the well-known private school for girls called North Avenue. Dr. William Pressley had been brought down from McCalley's private school in Chattanooga. He was an excellent organizer and administrator. The school was given a huge tract of beautiful land on the northwest side of Atlanta in the poshest part of town—the prestigious West Paces Ferry area. Sufficient money was raised to build three beautiful schools—one for the grammar school, one for the boys' high school, one for the girls' high school. Dr. Pressley insisted on the highest standards of teaching for his teachers and learning for his students. His purpose seemed to be to get as many seniors as possible into northern colleges, although he did not come out and say this. He hired only the best qualified teachers and he did not hesitate to expel any students who did not do their work. The school became the finest school in the city, the state, and probably the southeast. It did, in truth, send many students to Harvard, Yale, Princeton, Smith College and Bryn Mawr. As integration progressed, it became very difficult to get into Westminster as the demand was so great from people taking their children out of public schools.

All of the Neely children went to Westminster, with most beginning in the 8th grade and staying through high school. However, they were there to get a good education, and it had nothing to do with integration as Margot and Levering were both on the liberal side, and were opposed to segregation. So as one child followed another to Westminster, it became more and more ingrained into their daily lives. Freddy was the only one that did not graduate from Westminster. He was the one who had a little teenage rebellion and did not think study was important. His parents were once called in the middle of the night by the police and were told that they had Freddy at the police station.

Levering told them: "That is impossible; he is downstairs in his bed. Hold the phone and I will see."

Sure enough, he was gone and had slipped out in the car with the neighbor's children. He was only fifteen at the time. He was put on probation and taken to a psychiatrist, but everyone finally weathered the storm. His grades fell, and he was sent to a boarding school called Darlington in Rome, Ga. He did very well there, graduated, and went to Emory University. He became a mortgage broker for the Trust Company. He married and has two fine children.

Virginia did very well at Westminster. She earned good, though not outstanding, grades, ones that got her into her mother's college, Middlebury in Vermont. She would later go into the Peace Corps in Africa, come back to the states, move to Arkansas, marry and have two children. Gardner got along well at Westminster; he made satisfactory grades, graduated and went to Sewanee to college. He was the football player of the family, loved body-contact sports, and was a great wrestler. However, once at a gathering in his fraternity house, a girl from Texas who knew Karate challenged anyone in the room to a short wrestling match. Gardie volunteered. After circling around each other for a short time, she threw him over her back into the corner. His back hurt immediately, but he got up and got her in the bear-hug until she surrendered. However, Gardie has had trouble with his back ever since; he stills wears a corset and now 15 years later had to have a myelogram to be sure there was not a ruptured disc. Gardner is now a research librarian in Cincinnati, Ohio and is divorced.

John went through Westminster with no trouble. He was the tall one, eventually hitting about 6 feet, 2 inches. He became a good basketball player for the school team, and he also was on the tennis team. These two sports gave him good (or bad) self-confidence and were good for his ego. He later went to the University of North Carolina. At one time, he thought he might become a doctor, but changed his mind, majored in history, went in the Peace Corps in Brazil, came back, took a city planning degree at the University of Georgia, and now is in Savannah, Georgia doing economic development for the city. He is married and has two children.

Sara did extremely well at Westminster, graduated with honors and awards, went to Smith, then Dartmouth Medical School, transferred to Columbia Medical School, graduated, did her internship and residency at the Billings Hospital at the University of Chicago Medical School; is married, lives in Los Angeles and is on the faculty of the University of Southern California. Her specialty is internal medicine and hematology. No kids yet.

Betsy, the baby, age 30, also did well at Westminster, but did not do quite as well as her next older sister. She went to Bradford Junior college in Massachusetts, then transferred to Colorado College in Colorado Springs. She got her degree in Botany, went in the Peace Corps in the Philippines for two years, came back, took a master's degree in range ecology at Utah State University in Logan, Utah. She had a job with the U.S. Forestry Service in Logan doing research. She is not married. She is the outdoor girl who loves the West.

So Westminster was the diving board for the lives of most of the children. It provided a good environment for learning. The children all had the same genes, but the genes were manifest in many different ways in each child. Whether there were still some genes of ambition from their great grandfather Ben, playing a role, we don't know. We do know that all of the children were influenced by Westminster School in those formative years. They were all exposed to excellent teachers. They all learned to study, which they all did to various degrees. Did Westminster affect their lives more than home life at the time? That is very difficult to say. They had the concern and interest of their parents and they had a pleasant home life, but how to decide what factor influenced them is difficult to determine. One thing is certain—they are all different; different interests, different drives and varying amounts of ambition. So in spite of the same genes and almost the same environment, they are each individuals with very few similar characteristics. Could one have predicted at Westminster what the future would hold? Clearly not.

When the family moved into the new house in 1960, that was a watershed—or at least a turning point—for the family. Fred was seventeen and Betsy was seven years old. The land had been purchased a year prior to 1960 from the de Golian family. They had divided up their large land estate into sixteen generous sized lots, and made a dead-end street. The area was beautifully wooded with tall pine and deciduous trees. The Neely lot was sloping up and the house was built high close to the rear of the lot. They found an architect specializing in building houses for families who had six children. Levering and a neighbor's child drew a picture of a house they thought would be pleasant, and the architect came up with a plan close to what they had drawn. The house was modern, and yet traditional with three bed rooms up stairs, a study, a large living room—dining room combined, a big picture window in the front and big sliding glass doors in the back with a nice-sized terrace. There were three bedrooms and a bath down stairs and this served as the boys' dormitory. There was a large playroom and a messy playroom for storage.

But with all the pleasant amenities for parents and children, it was really the neighbors that made the difference and a whole new, small world. The Felix de Golian family, whose father owned the original land, built a house and had fourteen children.

When Mrs. de Golian became pregnant with the last child, she called up her obstetrician, and said, "I am pregnant again."

He said, "I am sorry, Helen, I am retiring this month."

"Oh, no you are not, you are going to come out of retirement and deliver my last child."

And he did just that. Several of the Neely children were friendly with their children. Then there was the psychiatrist, David Kahn and his wife, Carol, whose backyard was adjacent to the Neely's backyard. There was a fence between the two yards, but the Neelys consulted Dr. Kahn so much that they tore the fence down. They raised two bright children, younger than the Neely children. The older, a girl named Cynthia, was very beautiful and bright. The younger, a boy named Beecher, would come over and discuss life in general with Margot when he was about six years old. He would go home only when a great cry would come from his mother. Then there were the Amisanos across the street—he, a fine second generation Italian architect and she a former school teacher from Mississippi. They had three younger children, all adopted. Delightful people those Amisanos. Then the urologist and spouse—Sam and Betty Ambrose. Finally, the Curries built a beautiful home next door with a swimming pool—he a lawyer and she a great mother of five. Oh there were others but friendships formed and ties were very close with the Kahns and Amisanos.

In those middle years as the children were growing into adults, Margot became increasingly active in community affairs. Her first interest had been the League of Women Voters, and she was active in this for about ten years. But now new interests came along—she began to work at Grady Hospital as a volunteer in the Emergency Room. It was one of her jobs to get the information for the record on the patients as they came in. She had a chance to work with the hospital staff in the emergency room. She had some interesting experiences there, and would continue this weekly job for almost twenty-five years. She became more active at the church, and was elected the first woman elder. She took on the help and support of several people working in the poor black community of Vine City. There were several black ministers helping the community to survive and the John Kennedy Center—both a school and community center had just been built. One of the black workers in the senior citizens program was Mrs. Lundy, whom Margot helped with old clothes and a stipend from the church. She would drive down into this black community several times a month with old clothes she collected. She served on several juries and also the Grand Jury, and later joined the Grand Jury's association.

She continued her interest in cultural affairs, attending all classical music concerts. She joined a music group taught by a great retired musician, Anabelle Wood Mansfield. She and Levering joined a Great Books group which met every two weeks. They continued this for about ten years. She was elected to the History Club of 1886 in which each member presented a paper on history. She took up painting, took lessons, and from primitive, she progressed to classical art and has given many of her friends and children some of her paintings. She took up tennis at about thirty-eight and would continue to play the rest of her life.

It was somewhere in the early sixties that Margot and Levering took their first trip to Europe. They flew to Italy, saw Rome, Florence, and Venice, then traveled by train to Munich and picked up their first Volkswagen. They drove through Switzerland, into southern France, visiting the caves of Lescaux with their famous cave paintings. They drove on to Paris, then across the channel by car ferry to England. They drove by Canterbury and saw the famous church near Dover. Then they took off to

London, arriving at night, and knowing nothing about the city. And, of course, they were driving on the left side of the road with a steering wheel on the wrong side. They made it to the hotel that night. After a few days in London, they headed southeast through King Arthur country, ending up in Bournemouth where an international tennis tournament was going on. They watched this for a day, then drove the car back to London, where it would be shipped to the states. Then the trip home was by boat, the Queen Mary, sister ship of the old Queen Elizabeth. What a great and jolly time aboard for five days. They met some very pleasant people on board. This was the first of a series of European travels.

All of the countries were interesting, but Italy seemed to have the greatest charm. They had gone from New York to Rome on Al Italia, and had a chance to observe many high spirited Italians: they had pushed and shoved to get on the plane (although they all had seat reservations). So not only were the people delightful and interesting to watch, but the Italian towns and cities were charming, and there was always some amazing new sight to see or place to visit. Rome and Florence remained favorites for many years to come. On the first trip to Rome, their friends, the Isons, recommended a guide, Mr. Thomaso Paollicelli. He was a typical Italian who loved Rome, and knew it like a book. He not only showed the Neelys the city, but he invited them to have a meal at his house with a wife who could speak no English and a daughter who was a lawyer, and could talk and write English very well. So for many years to come, the Neelys would drive through the countryside of Italy. They covered most of it, including the tip of the boot and Sicily. Art, Greek and Roman ruins, old churches, little picturesque towns, these were never ending wonderment to Margot and Levering. Their very last trip to Italy was in March of 1982. This time they visited places that they had not seen before—Naples and Reggio Calabria, near the toe of the boot. They never were bored in Italy. It was always a joy to visit.

The middle years of medical practice for Levering were satisfying. The practice of medicine is fun if one gives it a chance, and one can relax when not working. Practicing internal medicine requires a great deal of reassurance to the patients. Any practicing physician needs a nurse to act as a buffer. Levering had such a one. She is bright, intelligent, sympathetic and patient with the patients. Beverly Baugham, you deserve a special seat in Heaven. You are and always have been a jewel. His goal was to combine practice, teaching, and research. Practice is never in itself dull; however, it can still become tiresome and routine if one lets it get that way by too much work. Teaching residents and medical students is a very gratifying occupation. Research in medicine is a real challenge. Trying to prove some simple fact in a laboratory or what is called clinical research, that is working with people rather than laboratory animals, can really keep one's interest high. Usually the practice of medicine becomes so demanding that teaching and research can be interrupted and sometimes interfered with badly. Levering was fortunate—he was working with a group doing internal medicine and thus was able to get certain days and times when his practice was covered by some other member of the group, and thus could pursue these other activities.

His first research was in the second year at Johns Hopkins Medical School when he and his pathology partner, Roy Pollock, were quizzed by the famous professor of pathology, Dr. Arnold Rich. Levering thought that he would ask Dr. Rich a question before Dr. Rich asked him one.

So Levering asked, "Why do people have a tendency to bleed with jaundice?"

Dr. Rich said, "I don't know, but would you like to find out?"

Roy Pollock and Levering both said yes.

"Well," said Dr. Rich, "come to my office Friday."

They went and very soon began a series of experiments on dogs. They would operate on the dog and sew up his bile duct. The dog would get jaundice and then blood would be taken for several general tests. No definite results were obtained but it was the beginning of a series of experiments for many years to come.

At Columbia Medical School while interning in pathology, he had given guinea pigs asthma and tried different drugs on them. In the early years at Grady, he tried to find how to measure oxygen from the blood by a finger stick sample. He made no real discoveries, but it was interesting work. Then in his middle years of practice, he began a long series of experiments, stretching over several years on the best liquid replacement for athletes. Gatorade had just come out. He got a grant of $20,000 from Coca Cola. He would have athletes run on a treadmill in a controlled hot and humid room, and collect blood and sweat at intervals. His conclusions after several years were that for strenuous exercise for a period of about one hour, water, Coke, Gatorade, and another drink called ARF (Athletes Replacement Fluid) were all about equally good. Coke did not get into the field of a drink for athletes possibly on this basis.

Teaching at Grady on the wards and in the clinic as well as at the Veterans Hospital added much to keep medicine interesting. This continued into the late years, but then it was more at Piedmont Hospital as there were residents and students from Emory Medical School there.

Office practice in the middle years was extremely busy, but there were many advantages in practicing with a group of great doctors like Carter Smith, Charles Stone, and many others. Hospital practice at one of the best hospitals in Atlanta became so much a part of life that Levering would often call the hospital phone number when he thought he was calling home.

So life was great for the Neelys in those middle years; no financial worries, no children worries (they all did well in college and were not into drugs), good travel, good friends, good tennis, good music, good art and good literature. Life for Margot and Levering in the middle years was a rich experience and no one had any idea that it would ever end.

47

Family—Later Years

Dogs do become part of the family. Charley Brown, Boots, and Buddy; the three successive Levering Neely dogs were definitely in that category and should all have the surname Neely. Each had his own personality. Each was loved by the whole family and each returned that affection. Each was a loyal faithful friend to all eight Levering Neelys.

Oh, there were other dogs before these, but Charley Brown was the first to be really adopted by the family. Somewhere in the early sixties he arrived on the scene, a puppy from the Humane Society, and he would later become a large brown dog with medium long hair. His ancestry was clearly a mixture. He was a good firm dog, friendly to the family but suspicious of intruders, especially delivery men and the like. His only great fear was thunder, and Sara, as a young girl, would take him into a closet during the clap of thunder and play on the guitar for him until the storm was over. Charley Brown remained with the family for over eleven years. He became a little more crotchety as he got older but still loved the family. Gardner had his wedding at the house and Charley Brown thought he should be part of the wedding. He had always had free access to the house, came in just before the ceremony was to start and lay down close to where the bride and groom would be questioned by the minister. The Neely family thought nothing of his presence and interest in the wedding. The bride's mother, however, a rather bossy woman, insisted that he be removed. So Freddy and Johnny got him out of the house as best they could. Charley Brown obviously did not like this removal and was hesitantly led and pushed outdoors. He did not come home that night and was never seen again. The neighborhood was searched, notices were put in the papers but he was never found. The Neelys always blamed the new mother-in-law.

Charley Brown was followed by Boots, another product of the Humane Society and also of uncertain parentage, although it obviously was a combination. He came full grown, was moderate-sized and fitted into the family well. He too had a good, firm personality and did not like any other dogs invading his territory. He was a good traveler, loved to go to Fernandina Beach with the family, and he greatly enjoyed the beach. He was part of the family for about five years. He met his Maker on the road in front of the house at Fernandina. He was following his mistress across the street and did not see a fast moving car. We buried him, appropriately, at Fernandina Beach behind a sand dune. Hail to thee blithe spirit, bird thou never went, but you were a hell of a fine dog.

That brings us to Buddy. Buddy is nine or ten years old. He is still in the family. He is a true and beautiful Golden Retriever although his papers have been lost. After Boots left this world, an advertisement was seen in the *Fernandina Beach New Leader* that a Golden Retriever was for sale. We located the address and this beautiful, very active, but full grown animal was playing in the yard and chasing a frisbee for a child. The family was asked why they wanted to get rid of him and they replied that when he wagged his tail, he would knock over the smallest child and was just too much for them. We got him for $25 because he had no papers. They said he had a tendency to wander but would usually come back. He especially liked tennis balls and would continue to retrieve them as long as anyone would throw them.

He was taken back to Atlanta and did in truth show that wanderlust. One night during the first week back in Atlanta, the phone rang and a voice asked "Do you have a dog named Buddy?"

"Yes."

"Well, he is down here on Peachtree at the Singles' Bar, and I saw your phone number on his collar."

"Thanks very much, we will come to get him right away."

A few nights later another person called: "Do you have a dog named Buddy? He is here at Peaches Record Shop listening to the music."

"Hold him and we will get him immediately."

Several days later Levering came home in the evening and said to Margot, "Where is Buddy, did he wander off again?"

"No," said Margot, "I took him to the vet."

"Oh," said Levering, "You didn't have that done to him?—that's against my religion."

"Yes," said Margot, "the vet said that is the only way to keep him at home."

And, sure enough, he never was a wanderer after his operation. Never was there a more loving dog than Buddy. Anyone that would pay any attention to him was his friend, even burglars, as we shall see later. He would goose people from the front as a gesture of his good nature. Often at a party the ladies with the new skirts and dresses did not seem to appreciate this show of friendship, especially when it came as a surprise.

There never has been a sweeter personality in a dog and he also was loyal, faithful, a great companion in town and at the beach. He loved to retrieve tennis balls. His assistant was Tullulah, another beautiful Golden Retriever from next door, who retrieved the ball, gave it to Buddy who returned the ball to the thrower. His other great pleasure was riding in the car, and he would go anywhere, anytime. He lay quietly on the back seat all the way to Fernandina, a seven hour drive.

These animals were all known well to every member of the Levering Neely Clan, and though living away, the children all want to know how he is and say that they miss him. So, these great spirited dogs added to the richness of the family life. Although it seemed a small part, it was an important part.

The children were by now (in the eighties) either out of college or almost out of college although several went to graduate school. Virginia took an M.B.A. at Arkansas; Gardner an M.A. in Library Science; Johnny, a degree in City Planning at the University of Georgia; Sara, an M.D. at Dartmouth and Columbia, and Betsy, a late Master in Botany in Utah State University. Three of the children went in the Peace Corps—Virginia, to the Senegal in Africa, Johnnie to Brazil on the Amazon, and Betsy, to a small rice paddy village high in the mountains of north Luzon in the Philippines. Each had a two year stint and their Peace Corp service was a tremendously broadening experience for all of them. It also began them on some vocation which was either related or connected with their work in the Peace Corps. Virginia taught in a small community in the countryside of Senegal, Johnnie worked with the farmers in the Amazon basin, and Betsy taught nutrition in the Philippines. Gardner worked in Vista for a year. All of the children did interesting things and none were ever blighted by the drug culture. They would all become good citizens and contribute to their world.

Freddy was the first to be married in 1968 to Shelby Mills, whom he knew at Westminster. He had a job with Adair Realty which was taken over by the Trust Company Bank. He says he is a mortgage broker and when asked what that is, he replies, "What I do is try to convince my clients to borrow money at high interest rates." After a long lean period he is now becoming very busy and rarely gets home before 7 PM.

Shelby fitted into the family extremely well and in addition to helping raise two fine children, she has a business of her own making quality little girls dresses.

We have talked about Emily in some detail but Lev is an equally fine boy. He is quiet, studious, and is intrigued by computers. He now has an Apple of his own and can do many things on it in addition to the usual games. His mother, father, and grandfather all are taking a course in computers in order to keep up with Little Lev.

Virginia married in Texarkana, Arkansas to James Jackson. He had also been in the Peace Corps, in India. He is a quiet, studious, friendly man with a good sense of humor. He is a lawyer in Fayetteville, Arkansas. They have two children, Keefe and little Margot. Their wedding, in contrast to the large wedding of Fred and Shelby's, was in a home of one of their friends, a simple ceremony followed by eating and drinking into the night. Virginia works as a fund raiser for a mental institution. She had a bumper sticker on her car that says "Support Mental Health or I will kill you!" Keefe is a most unusual child as we have seen. Margaret Elizabeth Eve Neely (but called Margot for her grandmother) is the infant's name. It combines the names of Virginia's two sisters as well as Virginia's mother, Margot.

Gardner married at the Neely homestead, as we have described earlier. He was married to a fellow librarian and they both went to Cincinnati to work in different libraries. They bought a home but the marraige was not destined to last, ending in divorce. Fortunately, no children. Gardner now has a fine girlfriend, Clare, who is the mother of four children all grown. Gardner still has trouble with that back which began when he was thrown by the lady wrestler in college.

Johnny married at the Druid Hills Presbyterian church to Carolyn Turner, whom he had met at the University of North Carolina. They have two bright children, Jordan and Thomas Longstreet. Carolyn is a social worker and a good, solid, well adjusted woman. Johnny is still with the government as an Economic Developer for the city of Savannah, and Carolyn still works part time as a social worker. Johnnie is beginning to get restless in his job and is looking at alternatives in the private sector.

Sara married in July, 1983, several months after her mother's death. She had been a resident at the University of Chicago Medical School for several years and then a Fellow in hematology there. She met Steven Newman there and after their marriage in Chicago, they moved to Los Angeles where Steve is in the practice of Oncology (cancer), and Sara teaches full time at the University of Southern California. No children yet.

Betsy has a job with the Forestry Service in Logan, Utah. She and her friend, Allen Carpenter (a Ph.D. student at Utah State) are thinking about getting married but are not in a hurry to do so. She loves the outdoors, likes to camp, hike, climb mountains, ski, especially the cross country skiing, and obviously enjoys the wilderness of the west.

In the late sixties, Margot and Levering went to Afghanistan on a teaching trip at the hospital in Kabul. This was a program of CARE—Medico. It was an amazing experience for both Levering and Margot to work in this undeveloped country. Levering would round with the residents in the morning and then usually work in the clinic in the afternoon.

All kinds of diseases were seen, such as polio, lockjaw, rheumatic fever, as well as many cases of tuberculosis and malaria. Margot worked in the hospital every day as a volunteer. All of the interns, residents and students could speak a certain amount of English. All were eager to learn, although the facilities were quite primitive. Several short side trips would be interesting—one to Pakistan and several within Afghanistan. There was a small group of physicians in various specialties from all over the U.S., and there was one lifelong friendship formed with Dr. Bella Strauss. The trip to Afghanistan was for a one month period and there was a visit to Lebanon before getting to Afghanistan, and a trip to Moscow afterwards. A car was at their disposal in Afghanistan. They were told that they should not stop if they should strike a child. Recently, they learned, a German woman injured a child, and when she stopped in the small town, the natives had stoned her to death. Usually the people were pleasant and friendly. The entire Afghan trip was a rewarding experience for both of the Neelys.

They would travel to Europe almost every spring or early summer in the seventies. They would usually rent a car and drive through the countryside and stay anywhere they could find a place at night—England, Ireland, France and Switzerland. All were charming places and the people were so friendly and helpful. Greece and Egypt had been seen previously. Vienna, Austria and Budapest, Hungary were especially great and interesting cities.

Probably the highlight of all the trips was the one to China. Prior to this there had been a trip to the Philippines to visit Betsy. Hong Kong, Thailand, Singapore and Malasia were seen. From Hong Kong a short trip was taken by bus to look over into Red China but one could not easily visit them. However, that view over into the mysterious country of Red China gave the impetus to go later. So, in 1981, we arranged a trip with the Linblad Travel Agency. The group met in Hong Kong, took off by train to Canton, thence to Kwelin with the usual mountains, and to Xian to see the ancient warrior statues. Then Shanghai and finally Peking. A totally different world of millions of people but no poverty was seen in contrast to India.

Travel is said to broaden one's outlook and I think this is true. One sees that people live and seem to enjoy an entirely different lifestyle from that in America. But also for the individual who travels there is a deep feeling of a treasury or bank of memories that add to the fulfillment of one's personal life. All travel inspires more travel. So it was with Margot and Levering, and most of their children, who had an opportunity to see the world; Virginia to Africa and Italy; Gardner to Florence to clean up the mud; Johnny to Europe and Brazil; Sara to Europe several times; Betsy to Europe, later to the Philippines, and then around the world. Freddy has a trip to Europe coming to him but hasn't take it yet. Is wanderlust a gene? Who knows?

Since 1968, Fernandina Beach, Florida loomed as a major factor for the entire Levering Neely Family. We purchased the house from an unusual retired sea captain from World War II at a ridiculously low price when one considers today's market. Uncle Frank had already given our family the lot next to the Captain's house but there was an access walkway between. The house was comfortable, spacious enough, and right on the beach. There was a sand dune in front of the house, and the captain had built a large deck just forward of the sand dune. This was called the roost and many an hour was spent relaxing on the roost, watching the sea, the birds, the shrimp boats, the porpoises and, unfortunately, the cars driving up and down the beach.

So as the children grew older and moved away, the house at Fernandina provided a wonderful place for a reunion as well as spot for individual or groups of family to come. When there was a reunion, the family would take on a project such as painting part of the house. Great was the camaraderie of such an endeavor. The house was called Frogmore because a small, black town in South Carolina really is called Frogmore, and Levering and Margot had been amused by the name. Later they would learn that Frogmore was burial place for the English Royalty in London. The house at Fernandina did become truly Frogmore as all of the guests would send a present with frogs on it. The ultimate was the toilet seat with a big frog on it. The house at Fernandina Beach would always remain the "Ultima Thule" of each and every Levering Neely child, no matter how far away they lived.

Margot continued all of her multiple activities. Each was a small community unto itself, and Margot's charm was affecting all of these groups. She worked at Grady Emergency Room every week for about twenty-five years and the clerks to the supervisors all felt her magic. She was greatly beloved at the church, having been on the session, involved in Christian Action and deeply interested in the music of the church. She had a history class mainly composed of the older intellectuals of Atlanta, and she fitted right in, presenting great essays which she researched and wrote. She had a tennis group who greatly appreciated her company. Probably her favorite organization was The Child Service and Family Counseling. She remained on the Board of this for many years and served them well but in addition to her service, she was respected and loved by one and all. She had that indefinable quality that a rare person has of being able to talk with anyone from the yardman to the mayor, from a child to an elderly person, from a pauper to a millionaire, and make them laugh.

What is it—who knows? Is it passed down as an inherited characteristic? All of her children have their own charm but none has her special, engaging personality.

Levering in the late years made a gradual medical evolution. His specialty has always been chest disease and although he did more straight internal medicine he had continued with his special interest in chest diseases. He worked as a consultant and teacher of chest diseases at the

Veterans Hospital with the Emory Medical Residents and students. He worked in the Chest Clinic at Grady for many years. He attended Pulmonary Meetings throughout the country. He organized the Pulmonary Intensive Care Unit at Piedmont Hospital and helped select its first medical director. Sometime in the seventies, the chest specialists in internal medicine began doing bronchoscopies to help in diagnosis. He had not been trained in this and did not think he could take the six months out to learn this procedure. So, he was not used as a chest consultant as he had formerly been. He was already developing two subspecialties, one, Geriatrics, as most of his patients were getting older and two, Vascular Disease. He had been sued for advising an arteriogram and, although the suit was dropped, he and a vascular surgeon created and developed the Non-Invasive Lab, which grew like a weed over the seven years since it started.

The heart of the Vascular Laboratory was Mary Hartz, later Mary Brakebill. She was the first and only technician of the lab when it started. She is now the chief technician with four workers under her direction. A graduate nurse, a conscientious and creative worker, a sparkling personality, a bright inquiring mind, a deep loyalty, a beauty and a joy to behold are a few of her attributes. Hail Mary! You have added to my life.

Levering and Margot had reached that state in their lives where they both were slowing down a little but still remained very active in projects. They had no financial worries, they had a host of friends, a beautiful home and a great house at the beach. The children were all settled with four married, one divorced and one considering marriage. True, five of the children then lived out of town but this gave Margot and Levering a good excuse to go visiting. There were five grandchildren and one on the way. Life seemed as if it would go on forever.

Then tragedy struck with vengeance.

Margot was feeling a little short of breath on the tennis court. She went to Dr. Stone, her regular physician, who found a fairly marked anemia which required blood transfusions. Then off to Duke to see a hematologist, who diagnosed pre-leukemia. Two weeks later she developed a severe pain in her back, was hospitalized, found to have acute myeloid leukemia. She began strong chemotherapy which brought down the high white count but was followed by a severe bacteremia, and then an overwhelming fungus infection. She was gone within three weeks from admission to the hospital.

All of the children had come to see her in the hospital and now they all came for the funeral. Trinity Church is a large church but it was completely filled with her friends and admirers. Her ashes were spread at her most loved spot, beside the sea on the beach of Fernandina.

"I must go down to the seas again, to the lonely sea and sky . . . I must go down to the seas again, for the call of the running tide is a wild call and a clear call that may not be denied."

48
Three Great Mamas

Mary was thirteen years old when her father Ben Neely died in 1892. Annie was twenty-four, a mature grown woman and a teacher, having studied mathematics and the classics with her father and been taught to teach by her father. Hannah was between the two. She was sixteen and in the Tubman High School in Augusta. So there was great turmoil when Ben, Sr., died.

The boys were all working. Allen was working in Atlanta. John was eighteen years old and had finished the Neely High School, and was now in college at the University of Georgia. Ben was working as an accountant which would stand him in good stead later. Edgar was fourteen and still in high school but he too was working, delivering papers. Frank, the baby, was only eight and he felt his father's death less than the other children who had known, loved, and respected their father. Frank was beginning to blossom. He had known enough about his father to begin to worship him, but he was still young enough to adjust to almost any situation, even his father's death. Although the children knew that their father was terribly sick, the final event left them devastated. Fortunately, the older children were all working and all contributed to the financing of the household. Henrietta Eve Carmichael Neely, though severely wounded by Ben's death, nevertheless continued to be a great mother. Bessie would die in childbirth and never get an opportunity to be a mother. But the other three girls would all have large families. Annie—seven children; Hannah—four; and Mary—four. Every one of these children will attest to the fact that their own mother was positively the ideal mother.

O.K., let us look at Mary, the youngest girl, born in 1879 and thirteen years old when her father died. She finished high school and went to Shorter College, and learned how to teach. She finished Shorter College in 1898 at age nineteen. There were no jobs in Rome at the time but Brother John had written from Bainbridge and said that there was a great shortage of teachers there and to come on down. Mary, a resourceful type, took the train to Atlanta and thence to Bainbridge.

She arrived in Bainbridge, stayed at Brother John's house, and began teaching the fifth grade in the public schools. There was very little unemployment at the time and a good, well-trained teacher was difficult to find. Bainbridge was a mill town with large lumber companies cutting and sending much lumber to all parts of the state. Many blacks were employed and there was a large black population. Mary had to walk or ride a horse through the forest for several miles to the school in town. The blacks in 1898 were aware of the freedom that had been awarded them but were still somewhat resentful of the way they were treated. There were several instances of attacks and rapes of white women. John insisted that this beautiful red-haired sister of his carry a pistol and know how to shoot it. And, that she did, and became quite a good marksman with her pistol. This was fortunate as once a black man jumped out from behind a tree and grabbed her horse's reins. She did not hesitate to draw her gun, point it at him, and he took off back into the woods. She continued at this job for two years but now decided to go back to Rome and teach. Annie, her older sister, had taught in Rome for several years at the Neely High School, but was now married to Augustin Young, and had moved to Cedartown.

It was back in Rome that Mary met Reginald Grinell, who had come down from Ohio with his family as his father was working for a rug mill in Rome. Reginald's family was related to the famous Grinell family that invented and marketed the Grinell Sprinkler System to put out fires. Reginald's own family, however, had not been involved in the sprinkler industry and were not wealthy by any means. Reginald finished the Neely High School, and went to work for the copper mines at Copperhill, Tennessee fifty miles from Cedartown. He would come back to Rome on weekends and stay with his parents. He met Mary at Sunday School and was smitten by her beauty and personality. She in turn was intrigued by this handsome Yankee man who was a little different than the run-of-the-mill Southerner.

Reginald had done a little gold panning in North Georgia and when he heard of the gold rush in Alaska, he decided to take off. He and Mary were now going together and became engaged before he left. So off he went by train to Seattle and then by boat to Alaska. He found a job in a large commissary for the miners in a small town in Alaska and did his panning on the side. He had collected quite a little gold within a year and so had enough money to return to Rome to marry Mary. Sister Hannah, already married and living in Augusta, invited the whole family to come there for the wedding. Edgar had passes on the railroad for the whole family and off they went and were married in 1902. Off to Alaska they went as Reginald had found a small house for them near the mines. He continued his job at the commissary which was now feeding hundreds of miners and he continued to pan for gold.

Life in Alaska was fairly primitive in those days. There was no running water, but there was a wood stove for cooking and heat. Mary adapted well to the life and very soon became pregnant. She went into labor in mid-winter. Reginald had to go far for the doctor with his dog sled over the snow, and brought him back to the house. The doctor thought she needed Ergot to hurry the delivery. It was rather dark in the room with only a kerosene lantern. He poured out a brown liquid which he thought was Ergot but was really Tincture of Iodine. She swigged it down, immediately felt a terrible burning in her mouth and throat and the doctor realized she had got the wrong medicine. However, it did bring on the baby and Adarine was born, a beautiful healthy child with red hair, even at birth. Mary suffered with her mouth and throat for two weeks but there was no permanent damage.

They remained in Alaska about two years more and then came back to the states. Reginald had several thousand dollars from his gold panning. They went to Florence, South Carolina to try farming and finally bought a large tract of land in Ebenezer, South Carolina. Reginald had bad luck with his crops and barely eked out a living in the early years between 1905 and 1925. They had three more children in a twelve year span, all girls—Mary, Allen, and Hannah. They had a large mortgage on the land and were unable to pay it off. Brother Edgar came to visit, found their plight and suggested that they take over his farm in North Georgia. So that they did and this became their home. Big Mary began teaching at the Baldwin High School and added to their income, which had its ups and downs. Adarine moved to Atlanta and became a secretary. Mary and Hannah went to Nursing School in Augusta and Allen worked for a business firm in Atlanta.

Throughout her married life, Mary remained cheerful and a great mother to those four girls. In spite of the hardships in Alaska, the periods of poverty in South Carolina and the ups and downs at the farm in Alto, she kept her spirit. Reginald, at times, bitter at his fate, was difficult to live with but Mary could cope. Reginald's health began to fail in the 1950s and he finally died in 1955. Mary, plagued with diabetes, followed in 1956. Adarine, Mary, and Hannah have all lost their husbands but all have children and grandchildren. Allen moved away and died some years ago. All of the girls still talk about what a great mother they had. She never lost her sense of humor, could mimic anyone in a hilarious way, was highly religious, and a musician with a beautiful voice. She taught for many years in the Baldwin High School and there was no question that she carried on her heritage of fine teaching.

Annie Wyatt Neely was the oldest of Ben's girls. She was born in 1869 in Augusta, right in the early part of Reconstruction. Federal troops were still stationed in Augusta and there were many instances of clashes between the Federal troops and the citizens of Augusta. Ben was twenty-seven when the Civil War began. He had finished college at the University of South Carolina and was teaching at Richmond Academy when the war broke out. Following the war, he came home, began teaching again and soon thereafter married Henrietta Eve Carmichael, and began the family. Allen was the eldest child born in 1867 and Annie came in 1869.

Annie, being the oldest girl, soon began taking the responsibility of helping care for the younger children and was a great help to Henrietta Eve, who had her hands full in a very short time. Annie remembered Frank as a baby (she was 16 when he was born) and when she would hear of what great things he was doing, she would say, "How well do I remember Frank at about two when his diapers would fall to the floor immobilizing him and how he would bawl until he got some help."

So she partially raised all of the younger children and it made her a mature person at a very early age. She taught in Rome at the high school and then married Augustin Young in 1897. The story goes that Augustin was courting the younger sister, Bessie, at first. Once Annie came into the store where Augustin was busy in the back of the store and someone said to Annie, "Your sister's beau is in the back!"

"Who is that?" said Annie.

"Augustin Young."

"Oh, he is no good," she said.

Gus heard this and came out to see who was running him down. They had a lively conversation and Gus then began courting Annie instead of Bess. After they were married, they moved to Young, Georgia and later to Cedartown. Augustin came from a fine old family in the area. He began farming and later had a country store in Young, Georgia. He later moved the store to Cedartown, where it was quite successful. He became a member of the Board of Trustees of the bank. He was one of the outstanding citizens in the area, had large farms, employing as many as forty hands. He developed the first good hardware store in Cedartown and had his son, James, run the store.

The family continued to grow and there was finally five girls and two boys. All of the girls went to Shorter College in Rome and Augustin, Jr., went to Sewanee, and James went to the University of Georgia. They had a large house on Walnut Street. Augustin had a big family of brothers and sisters who lived in the area. There were five bedrooms and a big sleeping porch in the house. In those days, families would come to visit and often stay a few months to a year. There was a steady procession of family and Annie had to feed from nine to twenty people every day for dinner, which was in the middle of the day. She had several servants including one named Mamie Lou. She would come out on the veranda at about 12 noon and ask, "How many for dinner, Miss Annie?" Annie would count them off and somehow there was always enough food for everybody.

The house in those days was surrounded by woods and fields. They had a large barn with several horses and cows. There were carriages around the early part of the century but these changed to automobiles right after World War I. There were several servants like a cook, two maids and a man to look after the place. Augustin had obtained the house by a swap with one of the tax assessors. He did it on the spur of the moment and then came home and told Annie what he had done. She wasn't happy with this at first but later came to love the big old mansion. There often were family members living or visiting. One of Augustin's sisters stayed there for years and often a child of Hannah's or Mary's would come up for the summer or sometimes for a year, and go to school in Cedartown. Annie would have thirty nieces and nephews on her side and Augustine had forty-three on his side, and so the Young House was very much like Grand Central Station as practically all of the nieces and nephews came to visit at one time or another.

Annie was very active in the Episcopal Church in Cedartown. It began as a very small church but gradually grew. All of the children were baptized in the church. Augustin, however, would not go to church. He had the excuse of business and had to visit his farm on Sunday. Someone asked Annie about this and she replied, "Don't worry, when I die, I am going to heaven and I will take Augustin right along with me. The Lord will be delighted to have such a good man." The Episcopal Bishops, Neilson and Michael, would come up to visit and always stayed at Gus and Annie's house. It was always a jolly household. Annie had a wonderful way with the children, either her own or other people's children and they all loved her. She could quote from the Bible and also from Shakespeare. The children all grew up and many of them taught, as was the family custom, at least until they got married. Most of the girls married and began families. Annie was the disciplinarian for the family as Augustin was very easy going and would rarely make the children do anything that they did not want to do.

Augustin became ill in the late thirties with what was diagnosed as cancer of the lung. He had been a smoker for many years. He finally passed away in the late thirties. Here was the end of an era. No one could take his place. He left sufficient funds to take care of Annie's needs. She had been ill with a chest condition for several years and would continue to suffer with it until she passed away suddenly in 1947 at age seventy-nine.

The children scattered over the state but many remained in that area. Eve married Roy Hand and moved to Philadelphia but they would come back frequently and Roy bought a farm there. Mary married a geologist and they moved all over the country. Ida never married but she had a most interesting job, working for the NASA, and doing editorial work for their publications. She had a very responsible and interesting job which she kept for about thirty-four years. She is now retired and does picture painting. James ran the hardware store for years and Buddy was in the dairy business. All of the children think their mother (Annie) was the greatest mother in the world. They say she died suddenly while eating ice cream.

Hannah Longstreet Neely, the sixth child of Benjamin and Henrietta Eve Neely, was born November 19, 1876 at "The Cottage" six miles south of Augusta, Georgia. She was a bright, happy child and, like the rest, did well in school. Her father, Ben, taught in Richmond Academy, the high school for boys, and he was asked to be the first principal of Tubman High School for Girls. However, in about 1882, Ben took the job of organizing the first high school in Rome and the family moved there. Ben's brother, John, succeeded him as principal of Tubman High. Hannah had begun Tubman when the family moved, and it was decided that she would stay in Augusta with Uncle John and Aunt Mamie and finish high school, which she did and graduated in 1896. She taught school in a small town close to Augusta. The Plumbs lived next to Uncle John and she met William Alpheus Plumb about the turn of the century. She would travel back and forth from Augusta to Rome, but her roots were really in Augusta, and after she met William Plumb she stayed in Augusta most of the time.

They were married May 20, 1902 and lived in Augusta. William Plumb was a widower. His wife had died and left him with two young children, Adeline and Katherine. Adeline had a twin, William A. Plumb, Junior, but he died at age six. Thus, Hannah started with two stepchildren but very soon presented the family with her first born, Bessie, in 1903. Mary Frances Plumb came soon after that in 1906, followed by William Warren Plumb in 1908, and finally Benjamin Neely Plumb (called Neely), born in 1912. William A. Plumb developed a severe headache in September, 1916 and died in December 1916 of a brain tumor.

Hannah was left with two stepchildren as well as her own four. The youngest, Neely, was only four years old. William's death came unexpectedly and although Hannah had a home there was a very little income. She tried selling real estate at first but this was not very successful. She then bought and operated the Tea Shop which was a restaurant on Eighth Street off Broad Street. It became quite popular and was a success, providing enough income to support the family. It took a great deal of hard work by Hannah; the children would help out in the kitchen and waiting on tables. Hannah insisted that the children do well in school. William and Neely had numerous jobs such as paper routes, and the like. Their father had been interested in music and played the flute. William and Neely both were good at music and played the saxophone in the school band. This early musical training was to have an effect on their lives. They both went to Georgia Tech Co-op School, where one would work four months, go to school four months with no summer vacation, and one would get an engineeering degree in five years. Both William and Neely did this but they found that for the work period they could make much more money playing the saxophone in a jazz band than they could at the various technical jobs that were available.

Hannah worked from early morning until 10 or sometimes 11 at night as the Tea Shop served early breakfast and a late supper. There was always work after the last customer had left. Hannah kept going without a complaint and served as a good example to the children, who all pitched in and helped in any way they could.

Bessie, the eldest, graduated from Tubman High just like her mother and went to work as a secretary for the Otis Elevator Company. She married Frank Conlon, who died relatively young. Mary also graduated from Tubman, then went to Nursing School at The University of Georgia. She later went to New York and became a psychiatric nurse at the Psychiatric Hospital at Columbia. The two boys finished Georgia Tech and were now so good on the saxophone that they got jobs with big name bands that traveled over the country. They were with the same orchestra for a few years and Hannah, who had retired from the Tea Shop, would travel all over the country with the boys. Home was in Augusta, however, and she lived with Bessie until her death in 1958. William returned to engineering but Neely stayed with music and gave up engineering. He ended up in Hollywood heading a company that made music for the movies.

Three great mammas—Annie, Hannah and Mary. They fulfilled the role of a mother whose purpose was to take care of their children and that they did. All three were good disciplinarians and the children would obey. All were great comforters and psychiatrists for their children. All could give good solid advice. All had a great sense of humor. If you should ask any of their children what kind of a mother was she, the answer would come from all loud and clear, "She was the greatest mother in the world."

Their ambitions were to be friends to their children and yet bring them up right. They all attained their ambition.

49
Ben's Homecoming and What Followed

The good ship *Nerrisa* stopped briefly at Cadiz before proceeding on to Brest, France. Brest was one of France's largest shipping bases for overseas trade. It had many piers and also many repair docks very much like the busy port of Liverpool. The *Nerrisa* needed some repairs after its long, circuitous world travels.

Ben had a good chance to practice his French while in Brest. He learned to converse with the two French speaking sailors and read several books in French while on duty. So, when he and the cook went into town to buy supplies, Ben did the talking in French with the men in the market. After about four days, they took off for Dublin, one of Ben's favorite cities as he knew his father had gone to Trinity College. After spending three days in Dublin, he went one morning to see the college. It dated back over one hundred years and was already one of the older colleges in Ireland. Ben wandered around the campus and watched the students walking about with their green coats. He wished that he might come back here to college but that was not in the stars.

Finally, the boat took off to Liverpool, its final destination before returning to Baltimore. They always stayed about two weeks in Liverpool because the ship's company was based in Liverpool and Baltimore. The captain would consult with the company and plans for the following trip would be made.

Ben and several of the shipmates went into town to try the food at one of the seafood restaurants. There was a group of about five from the ship who were sitting around a table drinking beer, waiting for their supper. There was an older man sitting at the bar having his supper. He looked at this group of sailors idly at first but then his eyes lit on Ben. Finally, Ben looked up at the man. Ben thought the man looked very familiar but he did not place him at first.

The gentleman then came over to the table and said, "Aren't you Ben Neely from Easton, Maryland?" Ben now recognized him and said, "Why you are Mr. McGregor! What are you doing here?"

Mr. McGregor lived on the same street in Easton, Maryland, as Ben's family and knew Ben's father, John, quite well.

"I am here buying sewing machines to take back to Easton for my store. But I am more interested to know what you are doing here. I knew you had run away from home almost two years ago. Your family has almost given you up for dead. They thought you might have gone out west. They were all heartbroken when you left. Your stepmother and your father have forgiven you a long time ago and I know they want you to come home. Also, Mrs. Neely is ill with consumption. So, you are needed at home. Why don't you come with me? I am going on a fast Baltimore clipper that should get there in about two weeks."

Ben's mind raced. He knew he wanted to get home. Here was the chance brought about by this coincidental meeting.

"I'll talk to the Captain and First Mate and will let you know tomorrow. Where are you staying?"

Mr. McGregor replied, "At the Mariner's Inn, as it is close to our boat, *The Viking*. It will be taking off the day after tomorrow morning early. Why don't you come and stay with me tomorrow night and we can leave early the next day?"

Ben ate his supper, went back to the ship, told the First Mate and the Captain of his plans and began packing his belongings.

The Captain and First Mate both said they were sorry to lose Ben, but they understood and they had two weeks before sailing and could replace Ben. The next day, Ben moved his belongings to Mr. McGregor's room at the hotel, but the crew was going to give Ben a farewell party on the boat. The whole crew had become very fond of Ben and Ben really thought of most of the crew as an extension of his family. So he came to the party that night and sat at the head of the table, and they all waited on *him*. They gave him presents and then all sang "For He's A Jolly Good Fellow." That night when he left he had to hold back tears. Ben had grown almost five inches in two years. He was still thin but had much more muscle. His voice had changed and he had a stubble of a beard. He had begun the transition from boyhood into manhood.

Ben saved his money and his passage back on *The Viking* cost only $25, as he shared the stateroom with Mr. McGregor. The weather was great—spring was in the air. It was about May 8th when they sailed and they got to Baltimore May 21, in almost record time of thirteen days.

What a surprise it was for the family when Ben came into the house at suppertime that spring day of 1850.

The whole family was overjoyed, though at first they did not recognize him. When they did, there were whoops and hollers and girlish screams. Everyone asked questions at once and Ben could scarcely answer one, when there would be several more. Father John was delighted and kept shaking his hand. Euphemia was now a young lady of eighteen. John, Jr., Ben's younger brother, was twelve and he had grown a good bit. The others were young girls. Mary, born in 1836, had died in 1845 at age eleven. Elizabeth, born in 1840, had died of scarlet fever in 1845. Maria was born in 1841 and was nine when Ben came home. Annie, born in 1843, was seven. Julia, born in 1844, had died a few days after her birth. So, there were only four children to greet Ben that night. His stepmother was very ill with consumption and was in the hospital.

School was almost over and Ben had a long conference with his father as to what he should do. Ben told his father about his studies on the boat and his interest in languages. They decided Ben would study hard for the summer and then take exams in the fall with the hope of getting into the tenth grade, as he had been out of school two full years. Euphemia was to be his coach but his father did a lot of teaching himself. When fall came he took the exams and was able to skip those two years. So he had caught up with his class, had three more years of high school, and planned on going to college to learn to teach like his father.

He did well in school and would work at various jobs to earn a little money. Isabella McAlister Neely, his stepmother, had died in 1851 and John married (for the third time) Rebecca Somerville Dorsey in September, 1852. About this time, Father John got an offer to come to Augusta, Georgia to organize the high schools there. Richmond Academy was a school for boys which had been in existence for about twenty-five years, but they needed a girl's high school which was being built. So, John went to Augusta by train, liked the people and the area, and moved the whole family to Augusta, in 1852. Ben was now eighteen and became a senior at the Richmond Academy, and did well. The next year he went to college at the College of Charleston, where he went for four years and took a liberal arts course with a special training in teaching. He graduated in 1857 and returned to Augusta to teach at the Richmond Academy teaching French, Latin and History.

In 1857 there had been considerable animosity developing between the North and the South. No one knew that war would be only four years off. Georgia was an agricultural state and there were numerous large plantations around Augusta with many slaves. Father John was a minister as well as teacher. He believed that slavery was wrong but he also felt that the North should not be telling the South what to do. As a teacher he was really not personally involved, but he thought of himself as a Southerner. Ben also thought that slavery was unjust but the four years at the College of Charleston had made him more of a Southerner and he, like his father, thought the Yankees were going too far in trying to change the South.

The four years from 1857 to the outbreak of the Civil War was a period of rapidly growing ill-will between the North and the South. The Abolitionists in the North made forays into the South, preaching about the evils of slavery. They were often run out of town by the Southerners. In Augusta, there were some who sympathized with the Abolitionists but these were very few and the greatest number of people were strong supporters of the Southern point of view. Meetings were held, there were many speeches, and the idea of pulling out of the Union was born. This idea, mild at first, soon became the strongest opinion and there sprung up a new patriotism whose battle cry was "Secession!" The South was becoming more independent. Plans were made for getting guns and ammunition for the South as the industrial North had most of the factories to manufacture goods and arms. Meetings were being held all over the state. Ben continued to teach but he felt now that a civil war would come sooner or later.

When the shots were fired at Fort Sumter, everyone realized that this was the start of the war. There was great elation and enthusiasm at first. The war was slow to get going. The North and the South were not prepared for war and they both began frantically getting prepared with arms, ammunition and supplies. The Confederate Army began to spring up all over the South. A new uniform was adopted and made in several factories. Meetings were held, everyone was enthusiastic, and the strong feeling of adventure had begun.

Ben was more realistic. He knew that war was going to be terrible, no matter who won. He had seen the effects of war in his travels. However, he felt obligated to join the southern army in 1861. He conferred with the army people from Richmond who had come down to enlist for the Confederate Army. When asked what skills he had, he answered "languages." The colonel asked him if he could talk like a Yankee and he said as a matter of fact he could as he had been imitating people from all over the country. The colonel then asked if he could shoot a pistol and rifle and he answered yes because he had been a hunter most of his life.

"Well," said the colonel, "you are going to be one of our spies and I want you to come to Richmond to take our course in spying."

The Civil War caused 500,000 deaths—more Northerners than Southerners. There was hardly a family on either side that had not lost someone. The South was devastated; and in addition to loss of life, there was a great loss of material goods, food, and also spirit. After four years, Ben came marching home, also broken in spirit. The homestead was still standing near Augusta and Ben vowed that he was going back to teaching. Augusta was not on Sherman's path from Atlanta to the sea and was not completely laid waste. However, large parts of it had been destroyed and now there was a fairly large Federal Troop installation to enforce the freeing of the slaves and keep law and order.

Soon Reconstruction began. It was very slow with many readjustments for everyone. The schools did not start for about one year. Ben pushed for the schools and got permission from the Federal Troop Commander to begin the grammar school and the high school. There was some food at the family house and when spring came, there was much more.

Ben met Henrietta Eve Carmichael right after the war. She was a teacher also and now was teaching the fifth grade at grammar school. As Ben became more confident that he could get the high schools going and as he was getting a little income, he proposed to this attractive young teacher, several years his junior. They were married in 1867; Ben was thirty-two years old and Henrietta twenty-five. They both continued to teach in the schools. Their first child, Allen, was born in 1868, followed by Annie in 1869, John in 1871, Elizabeth Eve in 1872, Benjamin, Jr., in 1874, Hannah in 1876, Edgar in 1878, Mary in 1879 and finally Frank, born in 1884. All of the children were born in Augusta except for Frank, who was born in Rome. Ben had gone there to organize the first high school.

Those early years in Augusta were happy years. Henrietta made a great and loving mother to her children and Ben was a good father. They both insisted that their children do well in school and most of them did. As other children came along, the older ones would help the younger ones. Annie, the oldest girl, helped raise the younger ones and she also acted as a second mother when Henrietta became too busy.

They had their own house in Augusta and had a large garden for growing vegetables and fruits as well as a smoke house for the hams, bacon, and chickens. It was a very happy home. Reconstruction had been very painful and slow, but in the late 1870's the town of Augusta began flourishing. There was plenty of food and the Federal Troops had finally left. Ben and the family were pretty well settled in Augusta, but in 1882, Ben suddenly decided to move to Rome. We think it had something to do with Ben trying to upgrade the black schools which were segregated at that time. So the next ten years for Ben were in Rome, where he organized and became the principal of the first high school in Rome, the Neely High School.

Ben died young. He was only fifty-eight years in 1892, leaving his wife with nine children; the oldest, Allen, was twenty-four and the youngest, Frank, was eight years old.

So here's to you Ben—you lived a full life. Your career probably was greatly influenced by that one incident when you were thirteen and pushed your stepmother down the stairs, prompting your escape to the sea. So here's to you Ben—adventurer, cabin boy, world traveler, linguist, teacher, spy, family man, organizer of schools, and a great father. You certainly had the genes of ambition and you spread these genes across our great land of America.

50
Leroy Gardner Neely

Hey, Gardner. On an overlook perusal of the book, we can't find much on you. This certainly was not intentional. We did the same to my father, Edgar, and had to go back and write about him.

You are, and always have been, a most unusual guy. Definitely not the run-of-the-mill. All of the Levering and Margot Neely children are different from their peers. They still fit well into society. Some are economically successful and others not as well off financially, but fortunately none are motivated primarily by money. And so it is with you, Gardner.

We know your parents well and we know and have pictured your grandparents, siblings, and even several of your nephews and nieces. We begin to get fuzzy when we talk about your great-grandparents, but we still know a number of facts about several of them, especially Benjamin Neely Sr., our chief protagonist. Thus, we are in a position to make some conjectures and possibly some simple conclusions in tracing your characteristic genes back to your forebears. Also, we have a wonderful opportunity to see how your environment may have affected you.

So here goes! You were number three on the hit parade. Two siblings older than you, a boy and a girl; Freddy and then Virginia. Three younger were John, then Sara and Betsy. Thus, your position in the family was not unusual. You did not have the possible pressure of the eldest, nor the extra attention usually given to the youngest. You were, perhaps, closest in early interests and activity with your younger brother, John. He was a year and a half younger than you. You two generally had a close and pleasant relationship and still have, even though living in different cities for the last fifteen years.

Let us see if we can characterize you when very young, then follow you in the procession of life and watch for traits of character that might have been inherited and might somehow have rubbed off on you from events or other people in your environment. We have already mentioned that you were born in June and seemed to cry most of the summer. It was called summer colic and the doctor could not find any trouble; physically you thrived. The neighbors next door were concerned and offered suggestions so they could get some sleep, as the houses were close together. By Fall, you settled down and were a very husky, muscular youngster.

In the preschool years, we realized that you had a very strong will. If you set your mind on something you wanted, it was fairly difficult to distract you. You were very definite in your ideas about food and your diet was certainly restricted. Hamburgers and ice cream were fine but all vegetables were a no-no for many years. One Memorable Incident—the whole family would go to the grandparents for Sunday dinner in the middle of the day. Once, or perhaps several times, at age four or five, you would look at the food brought in by Mancy Brown, the butler, and a disdainful look would come over your face. You would then get under the dining room table and stay there throughout the meal. In spite of exhortations, you would remain there. Soon it was taken for granted that you would stay there but perhaps come out when ice cream was served.

You became enamored of Indian lore early on. It seemed to fit your personality. This interest continued for many years and was manifest as a great love of the outdoors. You collected arrowheads, and you went to Camp Sequoyah where Indian customs and history were taught. You made an Indian headdress of eagle feathers and later made a large one with a long, flowing decoration down the back. You had hatchets and tomahawks. You personally built a genuine Indian teepee in the front yard of our house on a busy street. You would often spend the night in it and I think would occasionally cook in it. It attracted considerable attention from the neighbors. You developed a great interest in the woods and talked seriously of becoming a forester when you grew up.

Your health had many ups and downs. Even when you were young, at about age two, you developed what appeared to be a bad chest cold but it would not clear up and you would wheeze badly, especially at night. Finally, a chest x-ray was taken and they found an enlarged thymus gland. You were admitted to Emory Hospital under the care of a well-known chest surgeon. He bronchoscoped you for possible foreign body in the lung but nothing was found except the enlarged thymus. You received radiation therapy and your symptoms disappeared.

In about the fourth grade, in grammar school, you developed a severe sore throat and two weeks later developed a high fever. You were seen by the pediatrician, a cardiogram was done, and the diagnosis of acute rheumatic fever with myocarditis was made. You were admitted to the hospital again and stayed until the fever subsided, then went home and stayed in bed for several weeks. You were coached at home so you would not get behind in your school work. After they let you up, the doctor recommended a wheelchair for a time. It was meant to keep you from exercising too much because of the heart involvement with rheumatic fever. You learned to move around in the wheelchair very well and if you came to some stairs, you would jump out of the wheelchair, carry it up the stairs, and get back in it and off you would go again. You had to be watched very closely for a recurrence of the rheumatic fever but none ever developed. You did have considerable contact with doctors due to your health problems.

From grammar school you went to Westminster School, the best private school in Atlanta at the time. You did all right on your studies but were not outstanding. However, you went on and graduated in the middle of your class. While at Westminster, you developed two new interests—wrestling and football. You were definitely a body-contact man. Your skill in wrestling was increased by your drive and determination to overcome. Football really was your favorite sport. Although you were never very tall or heavy, you had developed your muscles and were fearless when it came to tackling some opponent, no matter how large he was. You had a few injuries, including a broken ankle. You were too small for college football but you continued the wrestling and this interest may have been a factor in the change in your young and early middle life.

Sewanee was your college of choice. It was a small but good college in the Tennessee mountains. It had foresting courses which you thought you might take in preparation to becoming a forester as this still was a strong interest for you. You joined a fraternity and one evening at a party, one of the dates of a fraternity brother said that she would challenge any of the boys there to wrestle.

Gardie volunteered and the room was cleared. The girl was slight but obviously muscular. Gardie thought he could easily subdue her. They walked around each other several times, sizing up the other. Then suddenly, before Gardie knew it, she surprised him with a quick movement and threw him over her head and onto the concrete. Gardie's back hurt immediately but he got up and ended the fight. She was a judo expert and teacher and knew how to use the principles to throw a much heavier person onto the floor. From that time on, Gardie had trouble with his back. First there would be short periods of pain, then larger periods of discomfort. He had x-rays taken showing that he had an extra lumbar vertebra (six instead of five). This made him much more susceptible to injury and disability.

This problem caused him to lose interest in forestry. He found that he could no longer go back-packing in the woods. The back troubles would plague him for years, coming and going. He finally (20 years later) became almost totally disabled and had to have a major operation with a back fusion. He was not able to get into the army because of his back.

Gardner's family were somewhat on the liberal side and were staunch Democrats. Gardner's older sister had gone to Middlebury College in Vermont and had joined the Peace Corp and spent two years in Africa. During college, she became involved in the civil rights movement and came down with a group of fellow students and marched at Selma, Alabama when Martin Luther King Jr. was pursuing his non-violent activities. Gardner, in college, was widely read, majoring in history and was active in the blacks' movement. He also participated in some of the civil rights marches in Atlanta. Following graduation from Sewanee, he joined the Vista Volunteers, a group similar to the Peace Corps that involved working in the black slums of our own cities. He worked at an economic development project in Atlanta and lived some of the time in the black community as well as helping with recreational activities of black adolescents.

About this time there was the terrible flood in Florence, Italy, and a distress call went out for volunteers to help with restoring the art work and homes which were damaged by water and mud. Gardner was fired up by this idea, volunteered, and carried a message from Ivan Allan, the Atlanta mayor, to Bargellini, the Florence mayor. Gardner traveled by freighter, this being the cheapest way to travel. He finally arrived at La Specia, Italy; then on to Florence. The volunteers lived in youth hostels and came from all over the world. He met some very interesting people including several from Australia. One in particular caught his fancy and she was his companion, along with several others, on trips around Italy. Gardner found a job as a student houseman for a physician, Dr. C. Ciarugi, and thus was able to stay longer. He remained and worked in Florence for four months, cleaning not only artwork but poor people's quarters. He was called "Angelo de Fango" (angel of mud) as were all of the volunteers who worked in mud. It was a great experience for him and, following finishing his work there, he travelled around Europe, staying in youth hostels. Finally, he arrived home, coming to Atlanta from New York by Greyhound bus. A large delegation of his friends went to meet him with large signs and placards welcoming him back.

After looking around for some suitable work, he finally decided to go to library school. His back still bothered him and he knew he could not become a forester. He had developed a great love for books, had a fairly good library of his own and after investigating the Emory University Library School, applied and was accepted.

While in library school he met one Sally Moffett, who went through the year's course with him. Following graduation, he found a job at the public library in Cincinnati, Ohio and she at the University of Cincinnati. So right after graduation, they got married at the home of his family. That was the time when the new mother-in-law-to-be ordered the family dog out of the house during the wedding. Charlie Brown, as he was called, left and never came back in spite of much searching. Later, this was thought to be a bad omen for the marriage.

The job was going along O.K.—he was a branch manager in the biracial neighborhood of Bond Hill. He got a chance to help in the education of black and white school children and to do community work.

After years of sporadic back trouble, his symptoms began to get really bad. He would lose days and sometimes weeks from his work and was walking with a cane. He consulted orthopedists, neurologists, internists, and psychologists but was unable to get any permanent improvement. Surgery had been recommended after a myelogram. He came back to Atlanta, consulted another internist, neurosurgeon, and orthopedist. A myelogram was done which showed disc disease and spondylosis—an unstable back. The decision was made to operate in Atlanta at Piedmont Hospital. Surgery was performed, the disc was removed, and a fusion done to stabilize his back. Long, slow recovery with very gradual improvement. While awaiting complete recovery, he went back to library school and took computer courses. He learned how to retrieve information via computer. Here was the wave of the future. In about six months, he applied for a library position at Georgia Institue of Technology; he was accepted and began his job.

So, Gardner, what manner of person are you? What makes you tick? What do you want from this world? Several characteristics stand out. You are a charmer, just like your mother was. You can talk to almost anyone; you make them feel good and they like you. You have had a good education and generally keep up your reading and know what is going on in this world. You love music and have sung in your Cincinnati Church Choir and now in the Trinity Presbyterian Church Choir in Atlanta. This characteristic has given you entre to a large number of interesting people. But you have had your ups and downs—you tend to get depressed at times. You were very emotional as a child and had fits of anger. This seems to have subsided but you have shown some evidence of stress. You still have somewhat of a rigid personality and you like to make suggestions, particularly to your father, who doesn't like suggestions. In fact, your father is somewhat paranoid about this.

But Gardner, you carry within you some great genes, both from the Neely side and from the Gardner side. You are getting along well with your fellow man, you have a good sense of humor, and a good intelligence. If you haven't quite found your niche in life, you will soon. In looking over your life, you have done many great things and we think that you will continue to do so.

51
To Levering from son Gardner

What can I say to my father about life? There is a man who loves life shamelessly. A man who would not dwell on the set-backs, but pay the devil his due and then blast ahead for some vast new empire to explore.

Neely, I like your spirit!

I saw your stuff when you lost Margot, a loss that all six kids felt, but that only you had to deal with from morning to night. A forty year marriage; a damn good marriage from what I can tell.

You and Margot had the gumption (or the foolhardiness!) to produce a half dozen kids without reservation—today's parents feel we must fill out a 1040 form five years in advance and get bonded for daycare and tutoring payments.

You and Margot led us through the great procession of life—brought us into the world, showed us where the green pastures were, and with your love and care inspired us to begin our own unique spiritual journeys.

You gave us that marvelous mix of traits us kids have to draw our character from (warrants a novel, you say?): southern medical student stricken by tuberculosis goes to New England mountains for treatment, recovers enough to court Doctor Gardner's daughter and bring her home to Atlanta after the war, to raise a tribute of half yankee/half southern children.

Mother would sometimes hum the tune of "John Brown's Body Lies A-mouldering in the Grave" (claimed her father had re-assembled the bones of John Brown on the dining room table after recovery from Harper's Ferry for proper burial).

While picnicking at Kennesaw Mountain, Johnny and Freddie would shoot make-believe cannon shot and minie balls of watermelon rinds at Gardie and Sammy and Wright—the neighbor's kids—during a not-so-historic re-enactment of the Battle of Atlanta.

Mother thoughtlessly fussed with the black maid about wanting to come to work through the front door, but then changed her tune a few years later and hosted thirty Freedom Riders overnight on the way to a Selma, Alabama civil rights protest with Martin Luther King. She inspired Dad to start a clinic in Vine City in the racially troubled 1960s, helped lobby for open housing in Atlanta, and together they prodded the social conscience of the church, the city and their well-fed children.

What I marvel at about my folks is not so much the worth or earnestness of their accomplishments, but rather their ability to have fun while marching into the thick of battle. Dad seems to be a man who not only can sing in the rain, but a man who can tap dance in a hail storm as well!

Stormy weather you have had—first with Margot's death two years ago, and then the loss of your grandson, Little Lev, in May 1985. But I see now your bounce, your strength, Big Dad. You grieve, but you do not despair. You see life's end not as a curse, but as a natural sequence. Perhaps these losses have intensified your quest for achievement, and for sharing the love of life you have always known.

Since these losses you have energetically gone on to pursue your vision of the world as a grandfather, a doctor, a citizen, and as an artist, musician, romantic, and poet.

You founded a non-profit Atlanta Coalition against Tobacco to fight the scourge of preventable heart attack, emphysema, lung cancer which you see all to frequently in your patients; inspiring, irritating, and recruiting friends, neighbors and colleagues, including pre-med student D. Beecher Kahn for this notable cause. You saw a threat to international peace, so you unflinchingly joined a church-sponsored peace mission to Russia.

You gave countless hours to senior projects, including the geriatric department at Georgia State University, Northside Senior Citizens at Trinity Presbyterian Church, and now planning the development of a multigenerational retirement village with architect Joe Amisano.

Your humor and love of dance transformed your Wallaby shoes into Fred Astaire's tapping torpedoes, as you gathered friends and stray children into a singing and dancing routine called the Half-Baked Hams that was cheered on by an eighty-three-year-old piano player, Miss Irene, who took us to nursing homes to entertain.

Your talents seemed to expand exponentially, as you would come home from a sculpture class with a bust of Eve, or a likeness of Adam you made in clay, only to set down at the piano and compose a friendly tune such as "Ice Cream, Ice Cream, More and then More" and "Its Fun to be Married," or a lament such as "Fernandina Blues."

While shy at first, you became less bashful and made the effort to start dating ladies. To find a lady with less difficulty, you would say to your friends' wives . . . "Find me a gal as nice as you are . . ."

This *Renaissance Man* was overflowing with energy and creativity, so much so that I became exasperated and jealous of his dynamism . . . and would open the kitchen door and wonder aloud to the heavens . . . "Deliver me, St. Peter, I cannot keep up with this man."

And when Dr. Cooper found a lump in your stomach, and family physician Sara insisted on operating, did it slow you down or sting your spirit? Would you let your sons pamper or fuss over you? Hell no! You insisted not only on cheating St. Peter, but also on driving to work the Monday following surgery, and were playing tennis a few weeks later!

As your surgeon, Johnny Akin, says, "Lev, looks like you dodged the bullet on this one, you healthy rascal, you!"

Yes, Neely . . . I like your spirit! . . .

St. Patrick's Day
1986

52
Finale and Prayer

Sitting in church, one's eyes tend to wander. There in the next pew in front and to the right is a family of four, a typical American family: husband, wife, one son, and one daughter. Noses seem to be the first thing one notices: the profile of the nose with the forehead and the indentation between the forehead and the nose. The son is about fifteen years old; the father perhaps forty-seven to fifty years old. The profile on those noses is almost identical and also is characteristic or a little different from anyone else's nose. The mother has a slightly different nose. The boy obviously got it directly from a dominant gene of the father. What about the girl, about age twelve, sitting beyond her mother? Her nose is different also, perhaps more like her mother's. But the shape of the head and the curl of the hair are definitely quite similar in mother and daughter. Then look to the left. A brother and sister in their thirties sitting together. They look different but look at those ears: a definite similarity; the girl's are a little smaller, true, but the overall shape, the little sinking-in on the upper-outer side are very much the same and yet different from other people.

Now look five pews up and to the right. That family has four children sitting between father and mother. They all have a certain similarity but also a certain difference. The shape of the head, the forehead, the hair, the hands: they all share many of the same physical characteristics. But what about the mannerisms? There is a family down the street of a man, his wife and fourteen children, all of the same parents. The father has a famous laugh, extremely loud. No one else has that exact laugh, but several of his children have a very similar laugh, so similar that everyone remarks on it.

So what about the same mental characteristics, are they passed along like the physical ones? We have already talked about intelligence, an inherited characteristic at least to a certain extent. However, environment can certainly affect intelligence. Emotional instability is certainly greatly determined by inheritance. Manic-depressives definitely run in families, but what about a little emotional instability, as in the Italians or Irish? The Germans, English, and Scotch, in contrast, seem more stable emotionally. Again, we are faced with the question—is this characteristic in the genes or determined by the environment? There is no easy answer. How about musical ability? The geniuses seem to develop very early without much parental pushing or help. Some families have many excellent musicians but the family lines of good musical ability can stop abruptly. Finally, what is this thing called *ambition*? We touched on it in chapter one and we have tried to trace it through the generations. We know that Frank Henry Neely had ambition and fulfilled his ambition. We know that his father, Ben, had the ambition to be a good teacher and we have seen that he fulfilled his ambition. Ben's father, John, was also a great teacher. He also became a minister when he was forty-five years old and continued to have the two professions of teacher and minister.

We now have some new information on John's father in Ireland which has only recently come to our attention and which goes back to 1778, when Benjamin Neely was born in County Lisburn and operated a school known as Neely's Academy from about 1805 until his death in 1838 at age sixty. His son, John, born in 1804, was in America teaching when his father died. Three generations of teachers—all very good. They all did well at teaching—but then the next generation, Ben's children, four girls and five boys, failed to produce a single male teacher, but then most of the girls taught at one time or another. However, the five boys certainly got their share of ambition with the youngest, Frank, having the most. The next older, Edgar, the lawyer, mixed in the love of life with his ambition. There was Ben, who supported the others with his hard work, and John who became a doctor after his leg had been cut off. He stuck to his profession and did it well in spite of this handicap. Allen, the oldest, is somewhat of an unknown quantity as he went to South America and died at age thirty-two in Colombia. Whether he went there because of ambition or lack of it we don't know.

So in four generations there was in truth an abundance of ambition. The girls, Ben's children, may have had it but had little opportunity to show it except by being great mamas. However, in four generations, we must remember that there were the genes of equal quantity from the female side and Ben's wife, Henrietta Eve Carmichael Neely, had forebears from the Longstreets, two of whom were a General and a College President.

O.K., we now have four generations. What about going down the generations and see what happened? Frank's daughter, Rachel, certainly got a lot of Frank's ambition but could not use it as well as a man might have. One of her grandchildren, Joshua, seems to have a spark that definitely showed at his Bar Mitzvah when he conducted himself like a mature man, while only thirteen years old. So after seven generations, there is no indication that ambition has burned out.

There are many other families of Ben's progeny, but we will just mention a few. Edgar, Sr., produced three boys and their children and grandchildren. All are achievers of various degrees. Take Keefe Jackson, Virginia Neely's child. Here is an unusual little guy bursting with ambition at age six. Then Edgar, Jrs.' children—three boys and two girls—all well-adjusted with one a teacher—possibly from great, great, great grandfather's genes? Who knows? And Michael Neely's children in Seattle. Those kids bear watching. There is Heather, age thirteen, who wants to be a brain surgeon, which is quite possible with her ability and easy but not ostentatious self-assurance. John, the doctor, had two children. One had ambition, the other, very little. Ben Jrs.' family, mostly girls, are all dead and the grandchildren and great grandchildren are all spread out. Then the girls—Annie, Hannah, and Mary—all had big families and ambition popped up here and there in all of them.

So, unfortunately, there is no clear answer. It is more of a feeling, a judgment upon a large group of human beings with very diverse environment but yet somewhat similar genes. Ambition flows and ebbs, high here, low there, but it seems to continue and probably will keep on through the next generation.

We have discovered what kind of people our forebears were. We have some old pictures and we have some brief descriptions of their activities. We have taken the liberty of trying to make Ben a living person. We have taken a number of facts and fictionalized them somewhat to make him more real to the next generations. Great gaps in our knowledge still exist, of course, for we really often know very little about our first cousins—people who have two of the same grandparents as we do. I have looked up a number of my first cousins—many are dead, but it is interesting to talk to some that I have not seen for thirty and forty years.

There should be some way of preserving more about our forebears. Perhaps with the advent of the videotape there will be more information left to our offspring and their offspring. If videotapes could be made once in awhile and stored away for the future it would help our understanding of ourselves and each other.

Will computers be a big factor in our children's lives? All signs point to this. One of the IBM officials says that in ten to fifteen years, it will not be necessary to go to the office more than about once a week. Fifty percent of all jobs have to do with information and that is communication. This could be done by computers at home and people could come to the office once a week to check-up and have personal conferences, which are so necessary to personal satisfaction. But will computers add to our lives? Will they foster creativity or will they stultify it? Creativity in man is one of his most important characteristics, and we must be extremely careful not to do anything to change it or hold it down.

Johns Hopkins Medical School became great because they fostered creativity, something which had never been done in a medical school before. One inventive mind brought others and encouraged creativity in anyone connected with the medical school. So in looking back over the giant strides in medicine over the last fifty years, as recently summarized in the *Social Transformation of American Medicine* by Starr, Johns Hopkins Hospital and the Harvard Medical School played a major role in the great advances of medicine. It was at these schools that bedside teaching instead of didactical medicine was born. No medical school is much greater than the others now, but great men with ideas inspire other men and women to let their imagination wander rather than being shackled by mere rote learning. So creativity may be the most important single characteristic to let man fly rather than crawl.

So back to those genes and chromosomes. New concepts in genetics have to do with genetic engineering. This is how to change heredity. Inherited characteristics have always been a mystery, but now they are beginning to explain how they are passed down, and it seems that DNA or deoxyribonucleic acid is in the genes, but that RNA—ribonucleic acid—forms the template which replicates and produces more genes. And now they are changing the genes in bacteria and animals and producing some substances that can't be produced any other way.

The whole process of conception is one of the wonders of the world. One cell from each parent joining and mixing up the genes and then producing a real live person with characteristics from each seems simple and yet is so difficult to really take in. It's a kind of metaphysical experience. This process makes one wonder about evolution and how true it is. The survival of the fittest is a very important theory in evolution but it fails to explain that there seems to be a purpose and a goal. It is difficult to see how the evolutionary process is all happenstance. The whole reamarkable intricacy of the body with all of its checks and balances to keep the body stable seems to have some purpose. The eye, with its myriad of cells working for a common purpose to see and to record in the brain, is much more fantastic than a computer.

Thus it would appear that there is some hand behind these things, what the contemporary playwright Sam Shepard would call an "unseen hand." The atheist says that this is false, the agnostic says, "I don't know," and the believer says, "There is a Lord, a Supreme Being that is governing the universe." When early death comes to someone the believer tries to make an excuse for it happening. A young person dying with a heart attack, a child dying of leukemia, a young mother dying in childbirth—these things are difficult to explain. Some say it is the Lord's will. Some say it just happened but we really can't explain these tragedies any better than the amazing conception of life. However, it gives one more comfort to have trust in the Lord than to take these problems to heart and to continue to brood over them.

This book is almost finished and we have not said one thing about the sex urge. Without it, there would be no progeny, no generations, no animals. This basic urge on the part of both man and woman is a very important part of history. In fact, there would not be any history if this basic urge were not a strong part of life. So the sex urge had to be present in all those generations to produce so many progenies. We usually take this for granted. Freud was one of the first to bring sex out into the open and also to make us realize how important it is in everyday life. Freud thinks it is a factor in all of our behavior and it plays a larger portion in mental illness. Some are questioning Freud now and discounting his theories, while others claim that his theories are disarmingly sound.

One will have to draw his own conclusions about heredity and environment and the question as to which is dominant. We have tried to give facts about six generations. There are no clear answers, especially about ambition. We do see it ebb and flow through the generations and that is as far as we can go.

A Prayer

Dear Lord,

What a wondrous organism thou hath made, Oh, Lord. Man is a strange but wonderful animal. Strange that he can function so well in this world. Strange that his body is so beautifully constructed and able to function in such a remarkable way. I don't mean, Lord, that your other creations are not equally amazing. A rose is almost as remarkable as man. The beauty of a rose—did you think about the rose long before you made one? A rose is a rose is a rose, as Gertude Stein said. I am not sure what that means but I do know that the delicate color and beauty of a rose is something that could not have just happened. Someone, Lord, had to conceive the rose, and I think it was You, Lord. Man cannot make a rose but You did, Lord.

And when people begin tampering with genes, Lord, they should remember that they did not make the genes. You did that, Lord. So let them tamper, let them change what they can, but let us see them make genes, Lord. Let us see them make a rose.

A teacher once asked the kindergarten class, "What does *create* mean?" They all said "To make something."

But one child in the back of the classroom said, "No, that is wrong; it means to make something no one has ever made before."

And that is what You do, Lord. And all of these atheists and agnostics say, "I don't believe in anything, I don't think there is anyone watching over me. Why do children die? Why does a young father or mother die? That shows you that no one is running this world." But I say, Lord, that You can't look after every little detail in the world. You created this organism called man, and that Lord is a great accomplishment, but to continue to look after each and everyone of them is totally impossible.

Perhaps you, Lord, worked along with evolution over the millions of years. A barren world millions of years ago had no reason to produce life. It should be still barren, but something started it all and even with evolution look what we have: the plant and the animal kingdom. It's said to have all begun because the right chemicals came together at the right time and then evolution took it from there. But this is inconceivable; this is preposterous; this does not make sense to man. Surely there was some direction, some balance—something.

And that is where religion begins and why it continues—because there are just so many things that we do not understand and will never understand. Almost every civilization has had some form of religion—mainly because they don't understand life—its origin, its development, its function.

So come on agnostics, come on atheists, let us see you create something. Let us see you make a rose, let us see you make a lowly weed. Let us see you make a functioning human being. And while we are picking on you atheists and agnostics who criticize the church as being hypocritical, let us see how much you can do by yourselves to help this world. "What can I do about this nuclear war problem, what can one person do?" you ask. You atheists and agnostics are mostly individualists. You are loners, you stay by yourselves and with one or two you discuss your common feelings. But do you ever do anything as a group? No, you just criticize.

O.K. the church makes an effort to help mankind. It may not succeed but it works as a group to try to change the world. You Ag's and Ath's, you do nothing but deny—you are negative on everything. And so when it comes to doing something about the nuclear bomb threat, the A's and A's will discuss but that is all. The Christians, Jews, even Moslems can band together and make some common effort. So this getting together as a group or groups of groups is the only hope against the extinction of mankind. They say that if the Russians strike first and annihilate our entire country with atomic bombs there will be a great nuclear cloud which will go up into the sky and gradually cover the earth and this nuclear cloud will slowly settle over the world and will either slowly kill or sterilize the rest of the people in the world. The same thing would happen if the United States struck first and laid waste Russia. So, I am saying let us Christians, Jews, Moslems, Buddhists, or whatever, all join together and make a real effort to stop this insanity of a nuclear war.

Lord, hear us, as we have not listened to You. We have done those things which we ought not to have done, and we have not done those things that we ought to have done and there is no help (or health) in us. But thou, Oh, Lord, have mercy upon us miserable offenders. Man against man. Wars continue on. Is there no end? War in South America, war in the Middle East, war in Africa. This is the story of mankind, but now war is no longer war if it involves the big powers. Now war is extinction. So let us be informed and let us inform others. That is what one person can do.

Lord, help us.

53
Postlude—Edgar Adams Neely, Sr.

Hey, Dad!

This book is finished and we haven't said nearly enough about you. You were one of the most interesting members of the clan of Neelys: Ben Neely's son, father of Edgar Adams Neely, Jr., Frederick Levering Neely and Randolph Carmichael Neely, as well as grandfather of sixteen—nine girls and seven boys. Also, you are the great grandfather of sixteen—eight girls and eight boys.

You had one burning ambition and you attained it—to live life to its fullest, to use all of your talents, and, like Abou Ben Adam, you loved your fellow man. You had no desire to be a great man but you attained it. Some say you were a greater man than your brother, Frank, who was nationally recognized for his achievements. But you were recognized as great within your wide circle of friends and acquaintances. You were the human one. You had the magic touch of making friends almost immediately on meeting someone. You had the respect of your fellow lawyers as being one of the finest trial lawyers in Atlanta and the state. You had the ability to communicate with the jury.

But it was a wide variety of activities you pursued that rounded out your life and made it interesting for you, your family, and other people. Yes, trial lawyer but also poet, singer, composer, dancer, golfer, lover, repartee expert, big drinker but not drunkard, card player, world traveler, big spender, Shakespearean scholar, good father, pillar of the church, and actor in plays.

Yes, Pops, you would get up in the middle of the night, go downstairs, poor yourself a half tumbler of straight whiskey—usually Bourbon and write poetry for an hour or two. True, the poetry was very flowery and usually had to do with love and sorrow; but you were an emotional guy. You could cry very easily if something struck you as sad. For instance, when you gave your second son, age eight, a good whipping with your razor strap because he stole money from church mite box, you cried the whole time. Somehow, most of that poetry can't be found. But we do have some songs that you composed and they were set to music by your friend, George Lindner, who ran a school of music. And you could sing quite well yourself.

You had other unexpected talents such as tap dancing. We never knew where you learned that but your second son got the gene and did a lot of tap dancing too. You played golf for the camaraderie but you were not very good at that. You played cards in the late afternoon at the Piedmont Driving Club with a group of old buddies.

True, you used alcohol to excess but you were never seen drunk. You consumed it before supper and in the middle of the night but it never interfered in your work. It was more a part of your *joie de vivre;* it was something you enjoyed. You died at seventy-five of cirrhosis of the liver, which is usually caused by excess alcohol, and your daily consumption was undoubtedly very great.

When your second son was fourteen and you found he was beginning to experiment with beer and wine, you called him into the pantry and said "Son, if you are going to drink, I want you to drink with me."

Putting down two glasses, you filled them half full of straight Bourbon. You drank yours down. Your son drank. He choked for about five minutes and finally got his breath. He did not drink any whiskey for three years.

Now, Big Edgar, you always had an eye for the ladies. While in high school in Rome (The Neely High School) you were said to have chased Miss Lucille Moses around a haystack three times. We don't know whether you caught her or not. We don't think you began to stray until you were about 45 or 50. But stray you did and although you were still a good family man, the reports were that you had a series of girl friends. You just had that exuberance that bubbled over when you saw a pretty woman. You always were discreet. Mama probably knew but would never admit it. You were as charming to the ladies as they were to you. You still were a good father to your children; played and joked with them at night and traveled all over the country with the entire family either in the touring car or on the Southern Railroad.

And, speaking of travel, you and Emily took the boat to Europe many times, more to England and Ireland, and you looked up family connections in Ireland. So in spite of being born a little country boy, you were not provincial by any means. Once you and Emily took the three boys out west and to Canada on the train. It was a great experience for the three boys, age twelve, ten, and eight. You saw Pikes Peak, Los Angeles with Catalina Island, San Francisco, Seattle, Victoria, Banff, and Lake Louise. Then on the Canadian Pacific across the Canadian Rockies, then finally to Winnipeg, Chicago, and home. The children still remember that trip well with pleasant memories. There were, of course, some sibling quarrels such as the time the three boys were out in a small boat on Lake Louise and began fighting. Mom and Dad on the shore were pretty upset until the boat finally came to shore and they all got a whipping.

Your first trip to Europe was when W.W. I started. You were going over on business by yourself. The boat got word by wireless that the war had started and the boat turned back to New York. During W.W. I, you had your three children and were not eligible for the draft. You became a "Minute Man" who made speeches to raise money for some good cause connected with the war effort. So you never did serve in the army.

You were a great supporter of the church. You and Emily joined St. Luke's Episcopal Church early, and you were very active in the community service of the church. Old Dr. Wilmer was one of the first ministers, and later there was a Reverend Hi Moore who was very dynamic and brought many new ideas and activities to the church. He was very active with the young people and is remembered by some as the minister who took them to the wrestling matches. You, Edgar, worked right with him and he was often over at the house at Four Barksdale Drive.

Incidentally, the house on Barksdale played a big part in the lives of Edgar, Emily and the children. It was built in about 1911 and was one of the first houses in Ansley Park. Not a pretentious house, it had a big living room, dining room, three bedrooms, and a large sleeping porch upstairs. It had gas fixtures at first for lighting but these were changed to electricity when that came along. There was a small lawn in the front, a fair-sized garden in the back, a servant's house in the back, a small barn for a pony, and a chicken house. There was a garage. The house looked out on a large park in front called "The Hollow" in those days, a great spot for the children to play in. That house was home for almost fifty years for Edgar and for another ten for Emily after Edgar died. Ansley Park slowly built up and there were many friendships made there that continued for life.

Edgar and Emily entertained frequently in the house. They had a large Steinway grand piano and there were many musicals there in the old days. They also entertained friends, relatives, international travelers, civic groups, church groups, and a great variety of cultured people. Edgar and Emily knew and were friends with many of the Jewish community. They never had any Jewish prejudice.

They always had one or two servants and when the children were small they had live-in nurses from Highlands, North Carolina. They also had a live-in French governess at one time. She taught at Washington Seminary and taught the children French and would be there when Edgar and Emily would travel. But the servants there the longest were Addie Fitshugh and Mancy Brown. Addie stayed for many years but did not live on the premises. She would come by streetcar or be brought by her husband, Aaron. She would be there from seven o'clock in the morning until about eight o'clock at night, six and one-half days a week. She received $5 a week for a long time but it finally got up to about $15. The most loved servant was Mancy Brown (real name—Emancipation Brown), who was a faithful servant off and on for thirty years until his death. We have already talked about him but it was his relationship with you, Edgar, which kept him so faithful. He thought that you, Edgar, were just about the greatest thing in the world. You two could talk to each other on anything. He kept that great respect for you, Edgar, as long as you lived.

You would usually drive the children to school in the morning in the Buick touring car and then you would go on to work. Frequently, there were ladies standing on the corner waiting to go to town on the streetcar and if they were pretty, you would invariably give them a ride to town. In those days, no one was afraid to get a "lift to town," especially when there were several children in the car.

There were other relatives living in the house. The two nephews from Augusta lived there while going to Georgia Tech. Many other nephews and nieces would stay there and at one time, an Englishman, Major Ferguson, stayed a long time while teaching at Oglethorpe University. No. 4 Barksdale Drive became almost an institution with so many people coming and going.

Edgar, you were generous to a fault. When your sister, Hannah, became a widow and had some financial trouble, you assisted her in establishing her tea room until it began to make a good income for her. You helped sister Mary when she and her husband lost their farm in South Carolina during the depression. You sent them to your farm in North Georgia and I think finally just gave them the farm. Any relation that needed something could rely on you, not to mention many people who were not relatives.

You made a good salary but you spent it all as fast as it came in. In late life, you would go into debt and have to be occasionally helped through a crisis by a child. Emily had a fairly good inheritance, something around $100,000, big money in those days. You spent some of it without telling Emily and when she found this out, she kept the rest of the money in her name and managed it through an advisor. Edgar felt guilty but she held on to her money, and left a considerable amount to each child. Edgar had some extra money in the twenties before the depression. He invested heavily in real estate with a Mr. Weyman in East Point, a suburb to the south of Atlanta. Atlanta, unfortunately for Edgar, grew to the north and not the south. When the crash came in 1929, he lost his entire investment. But, fortunately, his salary from the law practice continued. There still is a very long street in East Point called Neely Street, named for Edgar.

You loved your children and they loved you. You would get up early every morning in the summer, and take the whole family to the Driving Club to swim. It was just about two blocks away and was a chance for the family to play together. You enjoyed the Driving Club, not in a snobbish way but as a place to entertain and to be with people and family.

* * *

But, your greatest accomplishment in life was becoming a great lawyer, perhaps one of the greatest in the city in your time. This did not happen because of the kind of burning ambition your brother Frank had. Rather, it came to your naturally by adding little bits and pieces until there was a great statue which simply appeared—Edgar Adams Neely. Edgar has become an old-fashioned name and even a somewhat comical name. Very few people are named Edgar any more. We are not sure where the Adams came from but it gives a nice ring to the name.

You started as a court reporter, were inspired by the theatrics of the courts, and then went to Georgetown in Washington to study the law. We think that Mr. Hughes Spalding had something to do with getting you to go to Georgetown, a Catholic university. You graduated, married Emily Levering Eckfeldt, and came back to Atlanta. You joined the firm of Black, McDaniel and Alston, three great lawyers. You must have impressed them for them to take you in, but you had ability and personality and I am sure both helped in getting you the job. You must have worked hard. Your particular talent was in trying cases and slowly you built up your reputation. You developed that great court presence which is so important. You learned how to make the jury laugh and you found out how to appeal to them. You made them feel that you were a down-to-earth lawyer. Your command of the English language was great but you talked simply so they could understand.

You began winning cases regularly. You were usually on the defense for the Southern Railroad and for the big insurance companies like Travelers. Later, there would be big companies like the International Paper Company and many others. You were asked frequently "How did the case come out?" Your reply would be "In the usual fashion" which would mean that you won it.

The firm grew. But Mr. Eugene Black left to go to the Federal Reserve Bank. Mr. Alston left because the firm could not take in his brother. Mr.

Harry Green and Mr. Rembert Marshall joined. The firm ultimately became Neely, Marshall and Green, after Mr. McDaniel died. Then young men came in—Neal Baird and Edgar Neely Jr. The excellent work they did continued and the young added to the prestige of the firm.

You, Edgar, were not only a good speaker in court, you prepared your work very carefully, knew the facts and how to present them. You once had a case with the opposing lawyer, Mr. Smythe Gambrell. You would call him by his first name before the jury. He objected to the court and then you began calling him Mr. Gambrell with emphasis on the Mister. The jury would laugh every time you did it and this may have influenced them in awarding you the winning verdict.

One of the young lawyers who worked with you on a case asked, "How do you stay so calm during the trial and while the jury is out?"

Your response was quick as a wink, "Well, I'll tell you, Hugh, I may look calm but I pee in my britches a little every time the jury comes back in."

You were always quick on repartee in a very humorous manner. You had many expressions which you were known for and your children still remember them. "I heard you the first time" was one when someone made a good suggestion but it had not been suggested before. Another was, "Oh, doctor, struck a nerve" when something was done or said which surprised you or hurt you emotionally. Then there was "It knocked the living pee water out of me." Also, "Mean as Catshit." And "Oh, Dr. Munion, I have got an awful bunion."

Once when you were acting in a play with the Shriners, the curtain would not come down at the end and there was a bad moment until you said in a loud voice, "Friends, let us repair to the anteroom." The entire troupe walked off the stage and you saved the day.

So, Edgar, you need immortalizing. This brief description does not begin to do you justice. Your oldest grandchildren barely knew you and you died long before your youngest grandchildren were born. You were a great character. You lived life to its fullest. You were beloved to hundreds, even thousands of people. You had that wonderful gift of the human touch. May it pass down in your great genes. You proved that the genes of ambition are not so important. Your stature just came naturally.

ABOU,
BEN EDGAR

May your tribe continue to increase.
Please pass along some of your great genes.

54

Margot—Summation and Death

Hail to Thee Blithe Spirit or

—To Margot with Love—

Postlude written on her 65th birthday in 1982

1940—42 years ago—
How many can remember it—
That's two score and two years,
  Think on it.
Saranac Lake, New York—a little town of 7,000 souls
Nestled in the snowy Adirondacks,
40 below in winter (maybe 75 above in summer)
Margaret Gardner back home from New York for an indefinite stay,
 Possibly the winter.
Tired of New York City for a while—Now here,
 the skiing great, work for Dad,
  Just regroup a little.
This guy Neely—where did he fit in the picture?
A doctor being doctored at Trudeau for tuberculosis acquired at Bellevue Hospital.
His mother came to visit from Atlanta—was standing on the corner waiting for a bus to her hotel.
Along comes Mrs. Thompson in her car—"Can I give you a lift?" she asks.

The two ladies get to talking as ladies will.
Yes, I'm visiting my son at Trudeau—he's got T.B. but not very sick.
Mrs. Thompson, "Could he come for dessert some night to meet a very nice girl who lives next door to me?"

And soon there they were at Mrs. Thompson's for dessert and bridge—
Margot just happened to have on a pink angora sweater,
And girls if you want a man, go out and get a pink angora.
Later that night they drove over to Lake Placid in the Model A. Ford.
WELL—he says he proposed the first night. She just laughed.
Two years later they were married and she just laughed again.
She worked for Interallied Information in New York—the beginning of the United Nations.
She made $100 a month and he $10 and a towel—Believe it.
They lived great—well maybe Mamma Neely helped a little.
Freddy was soon on his way and in June, '43 arrived at
The Presbyterian Hospital in the City of New York.
She stayed 2 weeks in the hospital and the bill was $2, thanks to
Blue Cross and no doctor's fee.

Lev started practice in July, 1943. She came down with Fred on the old
Crescent Limited in August. He got confused on daylight saving and was
one hour late to meet her—
Not an auspicious start but after that things settled down or rather
they settled down to having children.
One right after another, Fred, Virginia, Gardner, Johnny, then a pause and then the little girls, Sarah, Betsy.
How did she do it? We don't really know, but she did it and did the League of Women Voters and other things. She kept her cool—most of the time.
But what she really did keep was her sense of humor.
That pulled her through.
And now they have all gone and left us.
But we have Buddy, a very fine dog that gooses people from the front.

So here's to you Margot.
You are the one with that indefinable quality—charisma?
Charm—neither word tells it all.

You are the one who makes people feel good just talking with
you.
You are the one who can get along with a prince or a pauper.
You are the one with the bright quick wit.
You are the one with the active mind—
You have intelligence, and you use it.
You know music, art, and books.
You work at Grady, the Family Service, and the Church—
You raised six children, but do you rest on those laurels?
Not by any means—
Your energy just keeps on bubbling—
Hey, I almost forgot—
You are and always were
Cute and attractive.

So, hail to thee blithe Margot
Bird thou never were—
But you were and always will be
A hell of a fine wife.

* * *

Yes, it's harsh, tough, mean, shocking, soul-grinding, an earthquake to the equilibrium, devastating, incomprehensible, and unbelievable. It can't be taken in—it doesn't quite register—it seems unreal. One says no, I don't believe it. And yet one knows it is true. She's gone, she's not here anymore. But where? Where has that spirit gone to—to Heaven? Certainly, if there be a heaven.

All civilizations considered life after death—the Indians with their "Happy Hunting Grounds," the Greeks with their Elysian Fields, the Romans—well maybe, but nothing clearly defined. The Chinese, the Arabs, the Jews all have their special interpretations of life after death. It is just very difficult for man to understand death, although the human spirit makes a great effort to come up with plausible, credible ideas of immortality.

Margot was born February 7, 1917. She died April 7, 1983. Sixty-six years is a fairly good life span. We have looked at her early life. We followed her life through college at Middlebury, Vermont. We have taken her to New York to get a job at a time of relative depression. We have seen her get married and begin begetting a total of six children. The oldest is now forty-one and the youngest is thirty-one—three boys and three girls. Four married, one divorced, and one messin' around. Margot lives in the hearts of all those children and all those grandchildren who really knew her. But she must stay alive to the children of the grandchildren and on down the generations and so here goes.

Margot was indeed a blithe spirit. She was different from the usual child in grammar school with whom she came up. She would take on a job as a young girl and would usually carry it out. She could run as fast as the boys, and as she went into high school and college she became a great skier and could keep up with the boys in their ski escapades, whatever they were—climbing mountains, coming down steep slopes, or cross country skiing for miles. And yet though a good athlete in skiing, she was still very feminine and attractive. But there was beginning to develop an unusual personality or spirit even when in high school, which flowered while in college. It was a personality which could reach out and make other people feel close to her on even a slight acquaintance. Her personality, with its easy wit, bubbled over. It charmed strangers and old friends alike. It was fun to be with her. This made her unusually popular in college. Although she went the full four years at Middlebury, she spent a lot of time visiting at Williams College where she went for dances and other social functions. They called her the first coed at Williams. This ebullient spirit persisted throughout her life and was even felt by the nurses who took care of her in her last illness.

What is this "personality?" How does it evolve? And is one born with it? Margot had a sister—two years younger than she was. Both had the same parents, thus they had the same chromosomes but the genes were mixed. Margot was the more outgoing, the more emotional with wild swings from joyousness to periods of depression and moodiness. But she never stayed depressed for long. Dorothy was the more even-tempered. They both had a great sense of humor, Margot's the fountain type, Dot's the dry wit. Alike and yet different. They had the same environment—at least in the early years, but their personalities were entirely different for the rest of their lives.

Let us look back and see how Margot's spirit had a part in shaping her life—how it was felt by different groups—and then let us try to see if she had genes of ambition beneath this delightful spirit. Ambition usually means there is a purpose or plan in one's activities in life. We have said that great men seem to attain their stature by some natural process and not from a burning drive to accomplish something in life. A personality or spirit may be a help in getting along with people and thus arriving at some satisfying station in this world.

So it was with Margot. She had no clear cut goal, but her spirit and friendliness made it easier for her to achieve more than onc with a dull nature might. So what were some of the things that she did in this world? Many of these activities were not recognized except by the small groups with whom the activities were carried out. Thus she had a special circle of friends both black and white at Grady Hospital who knew her and loved her, while there were many other circles who did not know each other. There was the church group, the Vine City group, the Family Counseling group, the tennis group, the music and artistic lovers, the League of Women Voters, the history classes, to name a few.

She had a great deal of friends. She was very close to several of her contemporaries who were very different from Margot. They still would get along extremely well. Margot's interests were broad—music, literature, art, the theater. Although many of her friends had very narrow interests, she conversed easily with them and had a great affection for them. They clearly felt the same way. She kept many of her friends made in college throughout life. These she would see only every few years and yet the old ties persisted and they would pick up their attachment immediately upon seeing each other. There were friends like the Distins who came to Atlanta during the war. They had come up in Saranac Lake with Margot. He was six feet four and she was six feet. Their four children were all quite tall. The two families had great times together, often at the lake where the Neelys had a boat. The Distins moved away after a few years but there would be frequent visits over the years even up into the 1980's. The two families always kept in touch at Christmas. The children of both families moved to different places all over the country but they too would keep up the contacts when possible. The Distins were thus a part of the Neelys' lives. In addition to being tall, the Distins were most unusual and perhaps this was what kept the relationship alive over the years.

Then there was the more intellectual group of Margot's friends whom she met at the League of Women Voters and through the musical groups. These were the really interesting groups. Margot joined the League in her early years in Atlanta. She was very active at first and often went to the state legislature as well as to the Atlanta City Council to observe and report. Her children had come very rapidly. She was on an eighteen month schedule for the first four children and thus at one time had four children under five years of age. So although she had some help at home, the care of these small ones took up a great deal of her time. There was a brief pause in the birthing after the first four, and then came the "little girls"—Sara was born five years after John and then came Betsy two years later.

The relationship between mother and children was interesting. She loved them all dearly and the love was returned in kind. Probably that abiding shared love was one of the factors that kept all the children out of any major trouble. Oh, there were problems, but a secure child is one that is loved and knows it. When the pressure to stray comes along there is a firm bond that prevents really bad behavior.

Then there were all those illnesses the children had. Gardner had an enlarged thymus gland and had to be hospitalized when very young. He later developed rheumatic fever and myocarditis requiring a rather long stay in the hospital, plus time in a wheel chair. Virginia developed poliomyelitis at age six and was sent to Warm Springs for six months. Freddy lost an eye in an accident with a neighbor boy while they were playing with spears. Margot took these problems in her stride.

Margot kept up her own education. She was an avid and rapid reader; she read books of all types, but rarely did she read trashy books. She and Levering joined the Great Books which met every two weeks throughout the school year. This group was actively attended by Margot and Lev for a period of about ten years; they stopped when they arrived back at the first book again. Then Margot had her history class of 1886. This was a group of ladies who would each present a paper on some author or subject. It was a very cultured and erudite group of southern ladies who were all well educated. Margot would work for months on preparation of her paper, doing extensive research for her presentation.

She attended the symphony regularly and had special music classes by Anabelle Wood, a music teacher who had retired from piano teaching in New York. She knew art well and on the Italian trip was a great student of all the art that she saw. She loved art and later became a painter in oils. At first her paintings were primitive, but they soon became more sophisticated; some of her last paintings were quite good, especially those of the beach and Italy.

Italy was her country. She could speak French fluently but never was able to talk Italian well. But she could communicate with the Italians in one way or another. She loved the Italian people with their jolly spirit, talkativeness, and love of life. Thus on every foreign trip (and there must have been fifteen of these trips) she would make it a point to get to Italy for either a short or longer period. The Neelys usually rented a car and would travel from the far north throughout the land. They often had their friends, the Londons, with them, and they would drive to remote towns in the country. By the time they had finished they had been up and down the east coast, the west coast, the Italian Alps and lake country, the tip of the boot, all around Sicily, and to the hill towns in the central area.

There was a guide in Rome that they met on the first trip to Rome—Thomas Paolicelli. He was about sixty-five years old on the first trip, and he charmed the Neelys. He really loved Rome and was always very enthusiastic about showing it. He would be contacted on almost all trips to Rome and always had something interesting to show. He invited the Neelys out to his house for meals, and although his wife could not speak English, his daughter, a lawyer, could speak it well. He soon became a friend and Margot would always send him a check at Christmas as he was of modest means.

Then those other trips—Egypt and up the Nile to the Aswan Dam. The Greek trip with the boat to the Greek Islands and to Istanbul was a great one. Margot always read up avidly before a trip and knew what to look for. Of course England and France were greatly enjoyed but not as much as Italy. Probably the highlight of all the trips was the one to Afghanistan in 1967 which I described earlier. All of the trips were great experiences, and wherever they went, Margot was able to communicate easily with the natives as well as the fellow travelers.

Of course the house in Atlanta was an extension of Margot as well. It was built in 1960 with room for all six children. Margot loved the patio and worked in the yard with spectacular results. She had life-size statues of Neptune as well as Aphrodite on the embankment; they were lit up at night. Of course there was a fountain in the center of the patio and Margot insisted that it be turned on whenever anyone was sitting there. The neighbors were all friendly and the Kahns, who lived behind the Neelys, had a nice swimming pool. "Don't be rich. Have rich friends," Margot would say. The Neely house was set back from the street in the middle of a forest of towering one hundred and fifty foot pines, and oaks and hickories almost as tall. The house was wood painted dark brown with large overhangs from the roof. It sprawled across the one hundred and fifty foot wide lot. The living room had a high ceiling, a huge picture window in front, and sliding glass doors into the back. There was a large fireplace that was used frequently. There was a Steinway grand piano on one side and book shelves on the other. The room was decorated with paintings from all over the world such as cave paintings from Lascaux to prints from Italy and a Khyber rifle from Afghanistan. Small statues from all over formed silhouettes high above the picture window and glass doors. There was a large kitchen and breakfast room area. The hallway to the bedrooms was covered with photographs taken through the years. Downstairs was the boys' dormitory with three rooms and a bath. It was a comfortable house, lived in, and there was easy access to the outdoors. The house was a great memorial to Margot's taste and discernment of how one's character is shown in their own home.

There were bird feeders in the back yard, and Margot fed the birds daily. She knew all of the kinds of birds and many of the birds were like old friends. The patio or garden had banks of azaleas, many gardenia bushes, roses, and holly. Lilacs were planted but never bloomed. Margot's favorite flower was lilac, which are so abundant in the Adirondacks. Gardenias were also a great favorite. There was a small greenhouse in the back yard which Margot tended in the winter. She wanted something flowering all winter long. It was a great idea, but it didn't always work out, especially when Levering tried to do hydrangeas. Margot definitely had a green thumb and sultanas (impatiens), poinsettias, begonias, dusty millers, marigolds, and bulbs were some of her favorites. Roses she loved, but that was Levering's department.

Illness struck first in 1978. Margot could not walk the width of the beach without her legs cramping, causing her to stop and rest. She went to her physician, Dr. Charles Stone, who found that she had almost no pulse in her feet. She was referred to the vascular surgeon at Emory, was admitted to the hospital, and arteriograms were performed by an Afghan physician, ironically. She had terrible pain with the arteriograms and vowed to never have another one. But the findings showed that her main artery to the legs, the aorta, was almost blocked. Surgery was performed and the arteries cleaned out like a roto rooter. She had a tough postoperative period, but the surgery was a complete success, and back to tennis she went with no trouble. One thing happened after this operation: she had smoked for about forty years and stopped cold.

The final event, the leukemia, was fast. She was sick only about three weeks. It was a terrible illness, but it was relatively short.

So, Margot, we are not interested in your genes of ambition. What we want to know is did you pass down your charm?

Epilogue

Margaret Gardner Neely died April 12, 1983 of acute leukemia after an illness of about three weeks. Her spirit lives in the hearts of her many friends. She has that indefinable warmth of personality that was felt by anyone who came close to her sphere. She was animated.

On Sept 2, 1984 her ashes were scattered into the shallow waters in front of her beloved summer home, Frogmore, at Fernandina Beach, Florida. Her husband, all six of her children, and five of her grandchildren participated in the brief ceremony in her memory. It was fitting that the "Hymn to Joy" by Beethoven was sung and also "Sea Fever" by John Masefield was read as well as "To a Skylark" by Percy Bysshe Shelley.

"Teach me half the gladness
That thy brain must know,
Such harmonious madness
From my lips would flow,
The world should listen then as
I am listening now."

TO LEV WITH LOVE

A rising star shot down
just as it was beginning its ascent
It had all the indications
that there was no limit
to its rise, infinity perhaps.

There were no limits to its grace—amazing grace.
No limits to what it would do in this world.
And no limits as to how it would function in time and space.

Like a rocket perhaps—on its way up—everything working perfectly until, until, until—

But no need for tears
Oh yes—you will be missed—
missed with a soul-aching grief
You leave an empty spot that can never be filled.
But we do have great memories
and we will cherish them.

Gentleness was one of your many attributes—
The ability to love and be loved.
We don't think of this as one of those great characteristics but it is.

Your parents loved you
and you returned this love,
Sounds simple doesn't it?
But that can make all the difference in the world as to how a child looks at life and gets along in life.

Love in the family is a security—you knew you could depend on it—
solid as a rock.
That one fact motivated you,
kept you striving for good things.
Children can be influenced by their peers
for good things, but also for bad.

But we never really thought that you
would step far out of line.
Oh simple mischief perhaps
boy's fun—

But if there was anything bad that was really tempting, you had a power to resist.
You did not really know that it existed—

You simply knew that if your parents wouldn't approve,
you wouldn't approve and you would not do it.
You carried that bag of love
wherever you went—

Why did your rocket malfunction?
Why did it happen to you?
Of all the millions, why,
Why, Oh Lord?

Lord, you try our faith.
Just when we think we know what is going on—
Bang, crash, thunder and lightning, a devastating tornado comes through, killing and maiming.

You were taken by a personal tornado
and we still don't understand.
Oh, one can rationalize
but still there is no real answer to why, why, why?

Perhaps we must just accept.
Help us Lord with this acceptance.
Anger does not help—
Crying may help a little.
Harder on your sister than us older persons.
But to sister just turned sixteen
How will she fare—how will she re-assess her role in the procession of life,
as her great uncle so aptly puts it?

We know that Lev did not want us to grieve
He bluntly asked how long he had to live—
there close to the end.
And when the answer came back frankly,

He said—don't worry—
I want my ashes scattered
at the beach just like Big Mom.

Little Lev would have
had a great laugh at the
poignant touch in the obituary—

"Survived by his mother and father,
Mr. & Mrs. F. L. Neely, sister Emily, and
brother Roscoe."

That Lev really did love that dog.

F. L. Neely, Senior

Little Lev, age 15, died of lymphoma on May 13, 1985 after an illness of six months.